NEUTRON AND X-RAY SCATTERING IN ADVANCING MATERIALS RESEARCH

To learn more about AIP Conference Proceedings, including the
Conference Proceedings Series, please visit the webpage
http://proceedings.aip.org/proceedings

NEUTRON AND X-RAY SCATTERING IN ADVANCING MATERIALS RESEARCH

Proceedings of the International Conference on
Neutron and X-Ray Scattering - 2009

Kuala Lumpur, Malaysia   29 June – 1 July 2009

EDITORS

Ahmad Saat
Hassan Abu Kassim
Mohammad Hafizuddin Hj Jumali
Junita Mohamed Saleh
Mohd Roslee Othman
Azmi Ibrahim
Faridah Mohd Idris
Megat Harun Al-Rashid Megat Ahmad

TECHNICAL EDITORS

Hafizal Yazid
Azraf Azman
Rafhayudi Jamro
Anwar Abdul Rahman
Shalina Sheik Muhamad
Mohd Rizal Mamat @ Ibrahim
Zalina Laili
Julia Abdul Karim
Hishammudin Hussain

All papers have been peer-reviewed

SPONSORING ORGANIZATIONS

Malaysian Nuclear Agency (Nuclear Malaysia)
Malaysian Nuclear Society (MNS)
International Union of Crystallography (IUCr)

**American Institute
of Physics**

Melville, New York, 2009
AIP CONFERENCE PROCEEDINGS ■ VOLUME 1202

Editors:
Ahmad Saat
Hassan Abu Kassim
Mohammad Hafizuddin Hj Jumali
Junita Mohamed Saleh
Mohd Roslee Othman
Azmi Ibrahim
Faridah Mohd Idris
Megat Harun Al-Rashid Megat Ahmad

Technical Editors:
Hafizal Yazid
Azraf Azman
Rafhayudi Jamro
Anwar Abdul Rahman
Shalina Sheik Muhamad
Mohd Rizal Mamat @ Ibrahim
Zalina Laili
Julia Abdul Karim
Hishammudin Hussain

L.C. Catalog Card No. 2009912004
ISBN 978-0-7354-0739-8
ISSN 0094-243X
Printed in the United States of America

CONTENTS

vi

PREFACE

The International Conference on Neutron and X-ray Scattering 2009 (ICNX2009) was held from 29, 30 June and 1 July 2009 in Kuala Lumpur. Malaysia. This event is a continuation of the International Conference on Neutron and X-ray Scattering 2009 (ICNX2007) which held in Bandung, Indonesia. It is also part of the Malaysian Reactor Interest group annual meeting; the meeting organises annually since 2002 to promote neutron scattering techniques and applications in Malaysia by utilizing neutron scattering and radiography facilities established by Malaysian Nuclear Agency (Nuclear Malaysia) since 1994 in Bangi, Selangor. ICNX2009 is the conference focused on neutron and X-ray scattering in materials science and engineering. The theme of the ICNX2009 was *Sustainable Materials Research Advancement through Research Reactor and Synchrotron.*

The conference comprises of three-day symposium in Kuala Lumpur. The three-day symposium activity took place at the Putra World Trade Centre, Kuala Lumpur. Prominent speakers from all corner of the world in the field of neutron and X-ray scattering were invited to present their papers. There were more than 150 participants coming from local, regional and international universities, research institutes and companies. There were more 110 abstracts for the poster sessions coming from Algeria, Australia, Swissland, Norway, German, Netherland, Taiwan, Czech, France, India, Kuwait, Japan. Myanmar,. Poland, Republic of Korea, Russia, Turkey, United Kingdom, United States of America, Argentina arid Indonesia. They were grouped into two parallel sessions. Session A was mainly for Neutron and Session B was mainly for X-ray.

Financial support for this conference was provided by the Malaysian Ministry of Science, Technology and Innovation (MOSTI)-dinner, Malaysian Nuclear Agency (Nuclear Malaysia)-participants, the International Atomic Energy Agency (IAEA) - IAEA speaker, the International Union of Crystallography (IUCr)-Young Scientist grant, and Malaysian Nuclear Society-local student support. Support from international and local companies was very much appreciated. We would also like to acknowledge here the session Chairs and people who gave introductory remarks :, Robert Knott, Edy Giri Rachman Putra , Zaini Hamzah, Sung Mun Choi, .

We are very thankful to all members of the Organizing Committees - from Malaysian Nuclear Agency, universities and componies.. The Executive Committee promoted and selected Prof Chih-Hao Lee of TaiwAn as the Protem Chair for the next International Conference on Neutron and X-ray Scattering in 2011 (ICNX2011). Finally, all participation's support and contribution are gratefully appreciated, and we look forward to meeting you all in Taiwan.

Abdul Aziz Bin Mohamed (Chair)

October 2009

Bangi, Kajang, Malaysia.

COMMITTEES

INTERNATIONAL ADVISORY

ADVISOR

Dr. Robert Knott,
ANSTO-Bragg Institute, Australian Nuclear Science & Technology Organization, Australia

MEMBERS

Prof. A.R West, *Dept of Engineering Materials, University of Sheffield, UK*

Prof. Michihiro Furusaka, *Dept of Mechanical Intelligence Engineering, Grad School of Engineering, Hokkaido University*

Prof. Dr. David Baxter, *Indiana University Cyclotron Facility, USA*

Prof. Masatoshi Arai, *J-PARC Center, Japan Atomic Energy Agency, Japan*

Prof. Peter Timmins, *Large Scale Structures Group, Institute Laue-Langevin, France*

Prof. Bohari M Yamin, *Chemistry Dept, Science and Technology Faculty, UKM, Malaysia*

Prof. Rolando Granada, *Centro Atomico Bariloche e Instito Balseiro, Commision Nacional de Energia Atomica, Bariloche, ARGENTINA*

Prof. RNDR Pavol Mikula, *Nuclear Physics Institute AS CR, Czech Republic*

Prof. Hirokazu Hasegawa, *Dept of Polymer Chemistry, Grad School of Engineering, Kyoto University, Japan*

Prof. Peter Holden, *Neutron Deuteration Facility, Australian Nuclear Science & Technology Organization, Australia*

Assoc. Prof. Sung-Min Choi, *Dept of Nuclear and Quantum Engineering, Korea Advanced Institute of Science and Technology, Republic of Korea*

Dr. Edi Giri Rachman, *Neutron Scattering Laboratory, National Nuclear Energy Agency of Indonesia (BATAN), Indonesia*

Dr. Saibal Basu, *Solid State Physics Division, Bhabha Atomic research Center, Prof Homi Bhabha National Institute, India*

Dr. Sergy Kulikov, *Joint Institute for Nuclear Research (JINR), Dubna, Russian Federation*

Assoc. Prof. Mikihito Takenaka, *Dept of Polymer Chemistry, School of Engineering, Kyoto University, Japan*

Prof. Mohd Zaid Abdullah, *School of Electrical and Electronic Engineering, USM, Malaysia*

Dr. Abdul Fatah Awang Mat, *TM Research & Development, TMR&D Innovation Centre, Malaysia*

LOCAL ORGANIZING COMMITTEE

Advisor
Dr. Muhd Noor Muhd Yunus, *Deputy Director General (Technical), Malaysian Nuclear Agency*

Conference Chair/Director
Dr. Abdul Aziz Mohamed, *MNS (Secretary) / Malaysian Nuclear Agency,*

Conference Co-Chairman
Assoc. Prof. Dr. Ahmad Saat, *Institut Sains UiTM (IOS), UiTM,*

Members
Assoc. Prof. Dr. Hassan Abu Kassim, *Dept. of Physics, University Malaya*
Assoc. Prof. Dr. Mohammad Hafizuddin Hj Jumali, *Applied Physics, Sci. & Tech. Faculty, UKM*
Dr. Junita Mohamad Saleh, *School of Electrical & Electronic Engineering, USM*
Dr. Mohd Roslee Othman, *School of Chemical Engineering, USM*
Azmi Ibrahim, Electronic Materials Lab., TMRND
Sarimah Mahat, *MNS (Conference Treasurer) / Malaysian Nuclear Agency*
Faridah Mohd Idris, *MNS / Malaysian Nuclear Agency*
Megat Harun Al-Rashid Megat Ahmad, *Malaysian Nuclear Agency*
Azraf Azman, *MNS / Malaysian Nuclear Agency*
Rafhayudi Jamro, *MNS / Malaysian Nuclear Agency*
Anwar Abdul Rahman, *MNS, Malaysian Nuclear Agency*
Shalina Sheik Muhamad, *MNS, Malaysian Nuclear Agency*
Mohd Rizal Mamat @ Ibrahim, *MNS / Malaysian Nuclear Agency*
Zalina Laili, *MNS / Malaysian Nuclear Agency*
Julia Abdul Karim, *MNS / Malaysian Nuclear Agency*
Hafizal Yazid, *MNS / Malaysian Nuclear Agency*
Hishammudin Hussain, *MNS / Malaysian Nuclear Agency*
Julie Andrianny Murshidi, *Malaysian Nuclear Agency*
Cik Rohaida Che Hak, *Malaysian Nuclear Agency*
Nurhaslinda Ee Abdullah, *Malaysian Nuclear Agency*
Herlina Maskom, *Malaysian Nuclear Agency*
Zainida Zainal Abidin, *Malaysian Nuclear Agency*
Hazmimi Kassim, *Malaysian Nuclear Agency*

Message from the National Conference Advisor

Dear Readers,

Neutron science and engineering has proven to be a major contributor in knowledge and technical development. Similarly, the contribution of X-ray science has also brought a significant impact in science and technology advancement.

As Malaysia aspires to become a leading nation in nuclear technology in South East Asia region, she needs to upgrade its national capacity and capability in neutron and X-ray science and engineering, including the local academic and teaching contents.

By bringing experts from more than 20 countries aroud the globe whom engaged in neutron and x-ray scattering to ICNX2009 in Kuala Lumpur, it was indeed a major recognition for Malaysia; we realized that the convergence of great minds in neutron and x-ray scattering had spurred the quantum leap of neutron and x-ray science, academic excellence and technological advancement leading towards socio-economic benefit.

As neutron and x-ray complement each other, I have strong believed the impact through the conference would lead the the improvement of research and development in material research especially in advanced materials, chemical processes, catalyst, bio-materials and nano-materials, rubber/polymer materials, etc.

By bringing the ICNX2007 from Bandung, Indonesia, to Kuala Lumpur, I have seen the spirit of continuation and sustainability in R&D as well as friendships and networking that was born in Bandung, Indonesia, has been revived and further enhanced. In Malaysia, we have Reactor Interest Group (RIG) as a platform for local researchers interested in nuclear reactor development and utilization to converge and expand their minds, and have been successful in its endeavor since 2001, to propagate neutron science in the country. RIG benefitted so much from this event.

Since ICNX2009 conference was the continuation of the series of the international seminar on neutron and X-ray scattering, and the conference had achieved its objective which was aimed at building up strong user groups of neutron and X-ray scattering from the local and regional scientists and researchers,

Thank you for all your efforts in making this conference a success.

Sincerely yours,

Dr. Muhd Noor Muhd Yunus
For the Organising Committees ICNX2009

Conference Programme

	29 June 2009, Monday	
Time	**Program**	
08:00	Registration	
09:00	Opening Ceremony of INC'09 & ICNX2009 by Guest of Honor **Venue: PWTC (Main Hall)**	
	Venue A PWTC 1	**Venue B PWTC 2**
	Chairperson : Edi Giri Rachman Putra, Dr	Chairperson : Masatoshi Arai, Dr
11:00	**Keynote: Robert Knott, Dr.** Liposome structure using small angle X-ray and neutron scattering (SAXS and SANS) *by Robert Knott Dr.*	**Keynote: A.R.West, Prof.** Crystallography-Property Relations of BaTiO3-based Electroceramics *by Anthony R West, Prof.*
11:41	**Invited: Francoise Mulhauser, Dr.** An IAEA Coordinated Research Project on Improved production and utilization of short pulsed, cold neutrons at low-medium power spallation	**Invited: Jason S. Gardner, Dr.** Recent Neutron Scattering Results from Geometrically Frustrated Magnets *by Jason S. Gardner, Dr.*
	Chairperson : Michel Kenzelmann, Dr — Oral	Chairperson : Mikihito Takenaka, Dr — Oral
12:11	**Dobrin P. Bossev, Prof.** Capillary interactions in nano-particle suspensions *by Dobrin P. Bossev and Garfield Warren*	**Intikhab Ulfatt, Mr.** FIRST SUCCESSFUL RIXS MEASUREMENTS ON (GaMn)As
12:31	**Jim Low, Dr.** Dynamic Neutron Diffraction Study of Self-Recovery in Aluminium Titanate *by Jim Low and Z. Oo*	**Mohammad Abdul Barique, Dr.** Effect of Relative Humidity on the Morphology of Hydrocarbon Polymer Electrolyte Membrane for Fuel Cell *by M. A. Barique, A. Ohira, K. Kidena, N. Takimoto*
12:51	**It Meng LOW, Dr.** Effect of Grain Size and Controlled Atmospheres on the Thermal Stability of Al2TiO5 *by It Meng Low and Z. Oo*	**Sangappa Dr.** Microstructural parameters in 8 MeV Electron irradiated Bombyx mori silk fibers by wide-angle X-ray scattering studies (WAXS)
13:11	Lunch	
	Venue A PWTC 1	**Venue B PWTC 2**
	Chairperson : Robert Knott, Dr	Chairperson : A.R. West, Prof.
14:15	**Invited : Sung-Min Choi, Dr.** Small Angle Neutron and X-ray Scattering Studies of Self-Assembled Nanostructured Materials *by Sung-Min Choi, Dr.*	**Invited : Alexander Ioffe, Dr.** Larmor-precession based neutron scattering instrumentation *by Alexander Ioffe, Dr.*
	Chairperson : David V. Baxter, Prof. — Oral	Chairperson : J. Rolanda Granada, Prof. — Oral
14:46	**L. E. Bove** The making of salty ice *by L.E. Bove et al*	**Yohanes Edi Gunanto, Mr.** The influence of a Cu substitution on the structure and Magnetic Properties of La0.9Ca0.1Mn1-xCuxO3 (0 < x < 0.2) *by Yohanes Edi Gunanto*
15:06	**A.M. Shaikh** Neutron Radiographic Investigations on Hydride Formation in Zirconium based Pressure Tube Materials for Pressurised Heavy Water Reactors. *by A.M. Shaikh*	**Yehia Ahmed Lotfy Abdel-Hady, Mr.** De-excitation Decay Following 1s, and 2p Shell Ionization in Potassium and Calcium Atoms using Monte Carlo Simulation Method *by Yehia A. Lotfy1 , Adel M. Mohammedein2, Adel A.*
15:26	**Tommy Nylander, Dr.** Neutron Reflectometry Investigations of the Interaction of DNA-PAMAM Dendrimers with Model Biological Membranes *by Marie-Louise Ainalem, Adrian R. Rennie, Richard*	**Mitsuharu Yonemura Dr.** Two-dimensional time-resolved X-ray diffraction study of directional solidification in steels *by Mitsuharu Yonemura*

15:46	**Sergey Kuznetsov, Dr.** Determination of composites nano structure parameters by very cold neutrons scattering. *by I. L. Dubnikova, S. P. Kuznetsov, V. S. Litvin, I.V. Meshkov, A.V. Shelagin, A. I. Udovenko*	**Norio Ogata Dr.** Structural modeling and intermolecular correlation of liquid chlorine dioxide *by Norio Ogata, Hironori Shimakura, Yukinobu Kawakita, Koji Ohara, Shinji Kohara and Shinichi Takeda*
16:06	**Ryukhtin Vasyl, Dr.** In-situ high temperature SANS measurements of PYSZ and FYSZ turbine blade coatings *by V. Ryukhtin , D. Wallacher, B. Saruhan, R.*	**R. Brajpuriya** Photoemission and EDXRD study of annealed Fe/Al multilayers using synchrotron radiation *by R. Brajpuriya and T. Shripathi*
16:26	**Refaat Mahmoud Ali Maayouf, Dr.** On the use of Plexiglass Substrates for Neutron Mirrors *by R. M. A. Maayouf, Dr.*	**Shigeru OKAMOTO, Dr.** Small-Angle X-ray and Neutron Scattering Study on Microphase Separation Induced by Non-Solvent in a Semi-Dilute Solution of an Ultra-High-Molecular-Weight Block Copolymer
16:46	**Arum Patriati** Effect of a Long Chain Carboxylate Acid on Sodium Dodecyl Sulfate Micelle Structure: A Small-angle Neutron Scattering Study	**Wei Kong PANG, Mr.** Effects of vacuum annealing on the Ti3SiC2/TiC/TiSi2 ceramic composites *by W.K. Pang1, I.M. Low1, B.H. O'Connor1, A.J. Studer2*
17:06	Tea Break & Adjourn	

Time	Program	
	30 June 2009, Tuesday	
	Venue A **PWTC 1**	**Venue B** **PWTC 2**
	Chairperson : Abdul Aziz Mohamed, Dr	Chairperson : Francoise Mulhauser, Dr.
09:00	**Keynote : Masatoshi Arai, Dr.** J-PARC and the prospective Neutron Sciences *by Masatoshi Arai, Dr.*	**Keynote : Peter Timmins Prof** Structural Biology and Neutrons – The Shape of Things to Come
09:41	**Invited : Peter Holden, Dr.** Applications of deuteration for neutron scattering, at the Australian National Deuteration Facility *by Peter Holden, Dr.*	**Invited : J. Rolando Granada, Dr.** Development of Cold Neutron Scattering Kernels for Advanced Moderators *by J. Rolando Granada, Dr.*
10.11	**Tea break, Poster viewing & Booth visiting**	
	Venue A **PWTC 1**	**Venue B** **PWTC 2**
	Chairperson : Sung-Min Choi, Dr Oral	Chairperson : Alexander Ioffe, Dr. Oral
10:45	**Martin Rusňák, Mr.** Study of structural discontinuity in (Ce,Y)PdAl compounds at low and high temperatures *by Martin Rusňák et al*	**Weifeng Shang, Dr.** Automation and remote access of EMBL small angle X-ray scattering beamline X33 dedicated to biological macromolecules *by Weifeng Shang, Manfred Roessle, Clement Blanchet,*
11:05	**Chih-Hao Lee, Dr.** The commissioning of a neutron depolarization beamline for studying the magnetic correlation lengths of magnetic materials at Tsing Hua Open Pool Reactor *by Chih-Hao Lee, Hui-Chia Su, Hsin-Hao Chang, Yu-Han Wu , Yi-Ting Sie , Chih-Wei Hu, Lieh-Jeng*	**Chin Wei Wang, Mr.** Short Range Magnetic Order Induced by La-doping in $HoMn_2O_5$ *by C.-W. Wang, C.-M. Wu, S. K. Karna, C.-Y. Li, C.-K. Hsu, C.-H. Hung, S.-B. Liu, and W.-H. Li**
11:25	**Norlida Kamarulzaman, Dr** Neutron and X-Ray Diffraction Studies of LiMn2O4 Nano Material and its Electrochemical Performance *by N. Kamarulzaman1, R.Y. Yusoff1, N.A. Abdul Aziz1, M.A. Bustam2, N. Blagojevic3, M. Elcombe3, M.*	**Adel Aly Ahmed Ghoneim, Dr.** Computation of Ion Charge State Distributions After Inner-shell Ionization in Ne, Ar and Kr Atoms Using Monte Carlo Simulation *by Adel Aly Ahmed Ghoneim, Dr.*
11:45	**Sohrab Abbas** Sharpest angular collimation of monochromatic neutrons *by Sohrab Abbas1*, Apoorva G. Wagh1 and Wolfgang*	**Adel M. M. Mohammedein, Dr.** Calculation of Ion Charge State Distributions After Inner-Shell Ionization in Xe Atom *by Adel M. Mohammedein, Adel A. Ghoneim, Kandil M.*
12:05	**Ali Pazirandeh** The neutron radiography facility at Tehran Research Reactor (TRR) *by Ali Pazirandeh*	**Khine Nyunt, Dr.** Density Functional Approximation for Spin Dependent Quantum Transport in Magnetic Nanostructures *by Khine Nyunt, Dr.*
12:25	**Peter Willendrup, Dr.** McXtrace - An X-ray Monte Carlo Ray-tracing software package *by Erik Knudsen1, Peter Willendrup1, Søren*	**Abhijit Chatterjee, Dr.** Au-Nano-particle Deposition on alumina surfaces for environmental application – a density functional study. *by Abhijit Chatterjee, Dr.*
12:45	**Lunch**	
	Venue A **PWTC 1**	**Venue B** **PWTC 2**
	Chairperson : Peter Holden, Dr	Chairperson : Peter Timmins, Dr.
14:15	**Invited : Michel Kenzelmann, Dr.** Ferroelectricity from magnetic order - a perspective from neutron and resonant X-ray measurements *by Michel Kenzelmann, Dr.*	**Invited : Mikihito Takenaka, Dr.** Structure analyses of swollen rubber-filler systems by using contrast variation SANS *by Mikihito Takenaka, Shotaro Nishitsuji, Naoya Amino, Yasuhiro Ishikawa, DaisukeYamaguchi, and Satoshi Koizumi*
	Chairperson : Pavol Mikula, Dr. Oral	Chairperson : Michihiro Furusaka, Dr. Oral

Time		
14:46	**Ned Blagojevic, Mr.** Characterisation of Explosives by X-Ray Diffraction and Neutron Scattering Techniques: Phase Transformation Study by Synchrotron Radiation XRD of Forensically Sourced Ammonium Nitrate Prills *by Brian O'Connor1, Ned Blagojevic2*	**Mohammed Noori Ridha, Mr.** Study of wetting angle and Intermetallic compound for 63Sn–37Pb and Pb-free solder Sn-3Ag-0.5Cu on Cu substrate *by Mohammed Noori Ridha, Ervina Efzan Mhd Noor, Ahmad Badri Ismail, Azmi Rahmat*
15:06	**Ivan Bobrikov, Dr.** Magnetostructural Phase Separation and Giant Isotope Effect in R0.5Sr0.5MnO3 *by A. M. Balagurov1, I. A. Bobrikov1, V. Yu. Pomjakushin2, D. V. Sheptyakov2, N. A. Babushkina3, O. Yu. Gorbenko4. M. S. Kartavtseva4 and A. R. Kaul4*	**Mousavi Shirazi** The Measurements of the Absorbed Dose and Energy in the Liver Tissue and Other Component of Dosimeter System from the Monte Carlo Method (MCNP4C Code) and Analytical Method and Compare of Produced Results Together
15:26	**Refaat Mahmoud Ali Maayouf, Dr.** Neutron Diffraction Measurements Using the CFDF for Studying the Residual Stress. *by R.M.A.Maayouf*	**Sunil K. Karna, Mr.** Magnetic Ordering of Fe in Na-intercalated FeOCl *by Sunil K. Karna,C.M. Wu, C.W. Wang, C.Y. Li, C.K. Hsu, C.H. Hung, S. B. Liu, and W.-H. Li**
15:46	**Sistin Asri Ani** Particle Size Distribution Models of Small Angle Neutron Scattering Pattern on Ferrofluids *by Sistin Asri Ani a. Darminto ^{a}. Edy Giri Rachman*	**Mukesh Sharma, Dr.** Role of importance of X-ray fluorescence analysis of forensic samples *by Shailendra Jha and M. Sharma*
16:06	**ANANDARAJ V, Mr.** Enhanced Reliability of Inspection of Aero Space and Nuclear Components by Neutron Radiography *by V. Anandaraj, N.Raghu, T. Johny and K.V.Kasiviswanthan*	**Valerio Scagnoli, Dr.** Analysis of azimuthal-angle scans in resonant X-ray Bragg diffraction and parity even and odd atomic multipoles in the multiferroic modification of the terbium manganate TbMnO3
16:26	**Karami MOHSEN, Mr.** Decreasing of the detection limit for gamma-ray Spectrometry with the influence of sample treatment *by M. Karami2, A. Sadighzadeh1, F. Asgharizadeh1, D. Sardari2. A. Tavassoli1. A. Arbabi2. O.*	**Hafiz Zin, Mr.** Application of Active Pixel Sensors in Advanced Radiotherapy *by H Zin, P M Evans, E J Harris, J P F Osmond, J R N Symonds-Tayler. S E Bohndeik. A Konstantinidis. R D*
16:46	Tea Break & Adjourn	

Time	Program	
	1 July 2009, Wednesday	
	Venue A **PWTC 1**	**Venue B** **PWTC 2**
	Chairperson : Hirokazu Hasegawa, Dr.	**Chairperson : Jason S. Gardner, Dr.**
09:00	**Keynote : David V. Baxter, Prof.** Small neutron sources as centers for innovation and science	**Keynote : Michihiro Furusaka, Dr.** Development of mini-focusing small-angle neutron instruments (mfSANS)
09:41	**Invited : Sergey Kulikov, Dr.** Development of cryogenic moderators for neutron sources	**Invited : Saibal Basu, Dr.** Neutron Scattering At BARC Reactor *by Saibal Basu, Dr.*
10.11	**Tea break, Poster viewing & Booth visiting**	
	Venue A **PWTC 1**	**Venue B** **PWTC 2**
	Chairperson : Ahmat Saat, Dr **Oral**	**Chairperson : Mikihito Takenaka, Dr** **Oral**
10:45	**Azreena Binti Mastor, Ms.** In-Vitro Enzymatic Degradation of ã-irradiated Porous Chitosan Scaffold : Cristallinity and Degree of Deacetylation *by Ismail Zainola, Azreena Mastorb, Suhaida*	**Eleftheria Mavredaki** Using Synchrotron X-Ray Diffraction (SXRD) for Studying the BaSO4 Formation Kinetics and the Effect of Inhibitors on Barite Formation *by Eleftheria Mavredaki, Anne Neville, Ken S. Sorbie**
11:05	**Yee Mon Thu** Characterization of Titanium Silicide (TiSi2) for Complementary Metal Oxide Semiconductor.	**Suminar Pratapa, Dr.** X-ray Diffraction Microstructural Analysis of Bimodal-Size-Distribution MgO Nanopowder *by Suminar Pratapa, Budi Hartono*
11:25	**Saeed S. Jahromi, Mr.** Polarized neutron reflectometry at the presence of smooth interfacial potential *by Saeed S. Jahromi*, Seyed Farhad Masoudi*	**Hazizan Md Akil Dr** X-ray diffraction studies of cross linked chitosan with different cross linking agents for waste water treatment application. *by Nurhidayatullaili Muhd Julkapli, Zulkifli Ahmad and *Hazizan Md Akil*
11:45	**Andi Idhil Ismail, Mr.** X Ray Diffraction studies using synchrotron radiation of Mg-based bulk metallic glasses *by Andi idhil Ismail*	**Binoy Kumar Saikia, Dr.** X-ray structural analysis of some Indian coals *by 1. Binoy K Saikia 2.Rajani K Boruah*
12:05	**Young-Jin Kim, Dr** Neutron Beam Instruments for Neutron Science at HANARO *by Young-Jin Kim, Dr.*	**Adolf Asih Supriyanto, Mr.** Effect Of Milling Time On Microstructure And Lattice Structure Of Mechanically Alloyed Al-Ti Powders *by Adolf Asih Supriyanto1,2 and Abdul Razak Daud 2*
12:25	**Lunch**	
	Venue A **PWTC 1**	**Venue B** **PWTC 2**
	Chairperson : Sergey Kulikov, Dr.	**Chairperson : Saibal Basu, Dr.**
14:15	**Invited : Pavol Mikula Dr** Recent progress of neutron Bragg diffraction optics based on cylindrically bent perfect crystals and its unconventional use *by Pavol Mikula Prof.*	**Invited : Hirokazu Hasegawa, Dr.** SAXS and SANS Study on the Miscibility and Phase Behavior of Block Copolymers *by Hirokazu Hasegawa, Dr.*
14:45	**Invited : Abdul Aziz Mohamed, Dr.** Micro-focused Small Angle Neutron Scattering and Imaging for Science and Engineering Using RTP - - A Preliminary Study *by Abdul Aziz Mohamed, Faridah Md Idris, Azraf Azman, Rathavudi Jamro, Mohd Rizal*	**Invited : Edy Giri Rachman PUTRA, Dr.** A 36 m SANS BATAN spectrometer (SMARTer) for Materials Science and Biology *by Edy Giri Rachman PUTRA, Dr.*
	Chairperson : Abdul Aziz Mohamed, Dr. **Oral**	**Chairperson : Edy Giri Rachman Putra, Dr.** **Oral**

15:16	**M.M.Sinha, Dr.** Study of phonons in alkaline-earth hafnates BaHfO3 and SrHfO3	**Ulyanenkov, Dr.** Dynamical X-ray scattering from the relaxed structures *by A.Benediktovitch1, I.Feranchuk1 and A.Ulyanenkov2*
15:36	**Nur Shafiza A. Sharif** La-doped CaCu3Ti4O12 Prepared by Conventional and Microwave Processing *by N. Shafiza1, S.D. Hutagalung2,*	**Sabar D. Hutagalung** La-doped CaCu3Ti4O12 Prepared by Conventional and Microwave Processing *by Sabar D. Hutagalung, Nur Shafiza A. Sharif, Zainal A. Ahmad*
15:56	**S. Nair** Dynamical study of the ubiquitin protein *by S. Nair*	**Khairiah Yazid, Ms.** THREE-DIMENSIONAL IMAGING USING MICROCOMPUTED TOMOGRAPHY FOR STUDYING GAHARU MORPHOLOGY *by Khair'iah Yazid, Bert Masschaele , Mat Rasol Bin Awang ,*
16:16	**Lai San Kiong, Dr.** Magnetic Property of Gold Metallic Clusters *by Lai San Kiong, Dr.*	**Faridah Mohamad Idris, Ms.** Design and Simulation of FPGA-Based Readout Control A High-Energy Electron-Proton Collision Calorimeter *by Faridah Mohamad Idrisa,b, Wan Ahmad Tajuddin Wan Abdullahb, Zainol Abidin Ibrahimb, and Burhanuddin*
16:36	**Mohsen Karami, Mr.** Evaluation and comparison of digital methods of photopeak integration in gamma-ray spectrometry *by F. Asgharizadeh1, M. Karami2*, A.*	**Thant Zin Htwe, Mr.** Behaviour of copper in annealed Cu/Si systems for on -chip-interconnection *by Thant Zin Htwe*
16:46	**Tea Break & End of ICNX2009**	

INVITED PAPERS

Phase Separation in the Heisenberg Spin System, $Gd_2Ti_2O_7$

J S Gardner[1,2], J R Stewart[3], G Ehlers[4]

[1]*Department of Physics, Indiana University, Bloomington, IN 47408, USA*
[2]*NCNR, 100 Bureau Drive, Gaithersburg, MD 20899 -6102 USA*
[3]*ISIS Facility, Rutherford Appleton Laboratory, Chilton, Oxfordshire, OX11 0QX, U.K*
[4]*SNS, Oak Ridge National Laboratory, Oak Ridge, TN 37831-6475, USA*
e-mail: jsg@nist.gov

Abstract

$Gd_2Ti_2O_7$ is a geometrically frustrated antiferromagnetic system with two magnetic phase transitions at 1.1 K and 0.7 K. The determination of the magnetic structure in the ordered phases by a powder measurement is greatly complicated by the ambiguity between 1-**k** and 4-**k** structures resulting in identical structure factors. Here we will present data and new analyses showing that, as the system cools from the correlated, paramagnetic regime just above 1 K, (i) the magnetic system freezes into a partially ordered state, and (ii) the 4-**k** structure is maintained throughout down to a base temperature < 50 mK. This clears up the ambiguity in the magnetic structure and confirms the phase separation of the Gd-sites into two in equivalent sites with a 3:1 ratio.

Keywords: magnetic frustration, polarized neutron scattering, partial order

PACS: 75.10.-b, 75.25.+z, 75.50.Ee, 70.50.y

INTRODUCTION

As heat is removed from matter, it usually transforms from a gas to a liquid and then a solid. In other words, the atoms slow down, and as spatial correlations build up, the atoms freeze eventually into a regular array of a solid. At some non-zero temperature, all systems are expected to freeze, including magnetic spin systems. Recently, there has been significant interest in materials that do not follow this path through phase space. Geometrically frustrated magnets are hindered from minimizing their total classical ground state energy due to the underlying geometry of the lattice. These materials typically remain disordered to a freezing temperature, T^*, much lower than the magnitude of the Curie-Weiss (CW) constant, θ_{cw}. Indeed, the suppression of the transition temperature below that expected from the measured value of θ_{cw} is often used to gauge the degree of frustration in a magnetic system [1].

Among highly frustrated antiferromagnets, the three dimensional pyrochlore lattice of corner-sharing tetrahedra is a particularly interesting system. Theoretical studies [2,3] and Monte Carlo simulations [3,4] show that classical Heisenberg spins on the pyrochlore lattice, with nearest-neighbor antiferromagnetic exchange, will remain dynamic and disordered at non-zero temperatures. This is also true for mean-field calculations when dipolar couplings are considered. However, long-range order at various commensurate or incommensurate wave vectors is found when exchange interactions beyond nearest-neighbor exchange and dipolar coupling are included [5]. Palmer and Chalker [6] went further in their mean-field calculations and argued that, when quartic terms were included in the theory, a four sublattice Neel ordered state, at $\mathbf{k}_{ord} = (000)$ is selected.

Gd-based pyrochlores, where the magnetic Gd^{3+} ions have S = 7/2 and L = 0, resulting in a small single-ion anisotropy [7], are expected to be good experimental manifestations for Heisenberg spins on a pyrochlore lattice. The large moment associated with Gd^{3+} (H7 Γ_B) dictates that dipolar interactions will play a role in determining the magnetic ground state. Raju and co-workers determined that the dipole-dipole interaction is approximately 20% of the nearest-neighbor exchange in $Gd_2Ti_2O_7$ [8].

EXPERIMENTAL WORK & DISCUSSION

We have studied both $Gd_2Ti_2O_7$ and $Gd_2Sn_2O_7$ using neutron scattering despite the high neutron capture cross-section of the Gd ion [9-11]. Both pyrochlores have a CW constant of approximately -10 K. $Gd_2Sn_2O_7$ has been shown to magnetically order at approximately 1 K, into the state predicted by Palmer and Chalker [11,12]. Linear spin wave calculations [13] of this state have predicted a low lying excitation spectrum and the spin density of states which are consistent with that found by neutron scattering [11]. This is also consistent with new low temperature specific heat [14] and electron spin resonance [15] data. Although structurally very similar, $Gd_2Ti_2O_7$ has two phase transitions [16]

CP1202, *Neutron and X-Ray Scattering in Advancing Materials Research: International Conference – 2009*
edited by A. Saat, H. A. Kassim, M. H. H. Jumali, J. M. Saleh, M. R. Othman, A. Ibrahim, F. M. Idris, and M. H. A.-R. M. Ahmad
© 2009 American Institute of Physics 978-0-7354-0739-8/09/$25.00

and never enters the state predicted by Palmer and Chalker. Champion *et al.* [9] first published the magnetic diffraction pattern from $Gd_2Ti_2O_7$. A low temperature partially ordered magnetic structure was found to best describe the data. They concluded that the ground state was a noncollinear antiferromagnetic structure, with a propagation vector k_{ord} = (½ ½ ½). This model can be described as a set of ordered (frozen) kagomé planes separated by disordered interstitial moments that make up a triangular layer. Stewart *et al.* [10] later revisited this problem and confirmed Champion's magnetic diffraction pattern above 770 mK, but they also found a low Q magnetic Bragg peak indexed as the (½ ½ ½) below 770 mK and a significant amount of diffuse magnetic scattering down to 46 mK. This study highlighted a difficulty often encountered when trying to model a magnetic neutron scattering pattern. It turns out that the Bragg scattering form the static, correlated spins can be described equally well by the Champion, 1-**k** structure, as well as by several other multi **k** variants. Here, 2-**k** and 3-**k** models can be ruled out due to unrealistic moments being required on the Gd site. However, when the Bragg intensities alone are considered, the 1-**k** and 4-**k** structures proposed by Champion (1-**k**) and Stewart (4-**k**) respectively, are equally good descriptions of $Gd_2Ti_2O_7$ between 770 mK and 1 K. Both models require two inequivalent Gd sites: ¾ of all moments are large and well ordered, while ¼ are disordered in the high temperature phase above 770 mK and acquire only a small ordered moment in the low temperature phase. The disordered moments make up a ordered triangular lattice as described by Champion or form one disordered tetrahedron per unit cell as described by Stewart. However, when one considers the diffuse magnetic scattering in the diffraction pattern and attempts to model the short-range correlations, differences between the two models are clearly seen [10]. In our earlier paper we reported fits of the low temperature phase and showed that the 4-**k** model was more descriptive of the data. This result agreed with others [10,17] in that it required 2 different Gd moments, that the system has a local anisotropy, and put an ordered moment of 1.9 $\int_B$ at 250 mK (the lowest temperature refined) on the less ordered sublattice.

In this communication we will present data and new analyses showing that, as the system cools from the correlated, paramagnetic regime just above 1 K, (i) the magnetic system freezes into a partially ordered state, but (ii) the 4-**k** structure is maintained throughout. In the intermediate phase, three fully ordered spin tetrahedra (with moment 7.0 μ_B/spin at 800 mK) and one completely disordered spin tetrahedron (with no net spin on the Gd ions) coexist. On cooling further, the structure remains a 4-**k** structure but the disordered sublattice acquires a small ordered moment.

Isotopically enriched gadolinium oxide, $^{160}Gd_2O_3$ and titanium dioxide, TiO_2 were mixed and fired at 1350°C for 2 days like other titanate pyrochlores [18]. Room temperature X-ray powder diffraction confirmed that single phase $Gd_2Ti_2O_7$ was formed and specific heat measurements confirmed the presence of two transitions at 1.1 and 0.77 K.

Neutron diffraction measurements were performed on the D7 diffuse scattering spectrometer at the ILL, Grenoble, using three-directional *(XYZ)* neutron polarization analysis [19] in order to separate the magnetic diffuse scattering from nuclear coherent, nuclear spin-incoherent and background contributions. The data were collected below 2 K using a dilution refrigerator with an incident neutron wavelength of 3.1 Å.

Figure 1 shows the magnetic diffraction, determined from the polarisation analysis of the data at several temperatures below 1.4 K. The (3/2 1/2 1/2), (3/2 3/2 1/2) and (5/2 1/2 1/2) Bragg reflections are easily identified and consistent with earlier results measured on other instruments [9,10]. The broad diffuse scattering, peaked at <1.3 Å^{-1}, which is associated with the disordered spins, is clearly seen at 1.4 K where no long range order exists. This data was analysed in our earlier paper [10]. As the system cools, correlations develop between spins, Bragg scattering increases and the diffuse scattering is quenched. However, even in this plot one can see that the functional form of the diffuse scattering does not change over the entire temperature range studied. One should note that even at 46 mK, diffuse magnetic scattering is present implying that some fraction of the spins is not static. This is consistent with a neutron spin echo study [20], which probed the slow spin dynamics and indicated that <23% of the spin system remains dynamic at 110 mK.

In a centro-symmetric crystal, the diffuse magnetic neutron scattering intensity can be modeled using a Warren-Cowley type formalism in which the measured cross-section is described as a Fourier sum of spin-spin correlation functions over the first *n* near-neighbor atomic shells surrounding a central atom [21]. For a polycrystalline material this is written

$$\frac{d\sigma}{d\Omega} = \frac{2}{3}\left(\frac{\gamma r_0}{2}\right)^2 f^2(Q)g_J^2 S(S+1)\left[1 + \sum_n Z_n \frac{\langle \mathbf{S}_0 \cdot \mathbf{S}_n\rangle}{S(S+1)}\frac{\sin(QR_n)}{QR_n}\right]$$
Eq. (1)

where the Q is the wavevector transfer, γ = -1.913 is the neutron magnetic moment in nuclear magneton units, $r_0 = 2.82\times10^{-13}$ cm is the classical electron radius, $f^2(Q)$ is the squared neutron magnetic form factor of the magnetic species, $g_J^2 S(S+1)$ is the squared magnetic moment of the magnetic species, Z_n and R_n are the coordination number and radius of the *n*th near-neighbor shell, respectively, and $<S_0.S_n>$ is the average spin-correlation function between the zeroth and n^{th} near neighbor shell.

In fitting the diffuse magnetic scattering at 1.4 K to Eq. 1 we have used the "reverse Monte Carlo" (RMC) method previously successfully employed in Ref. [10,21].

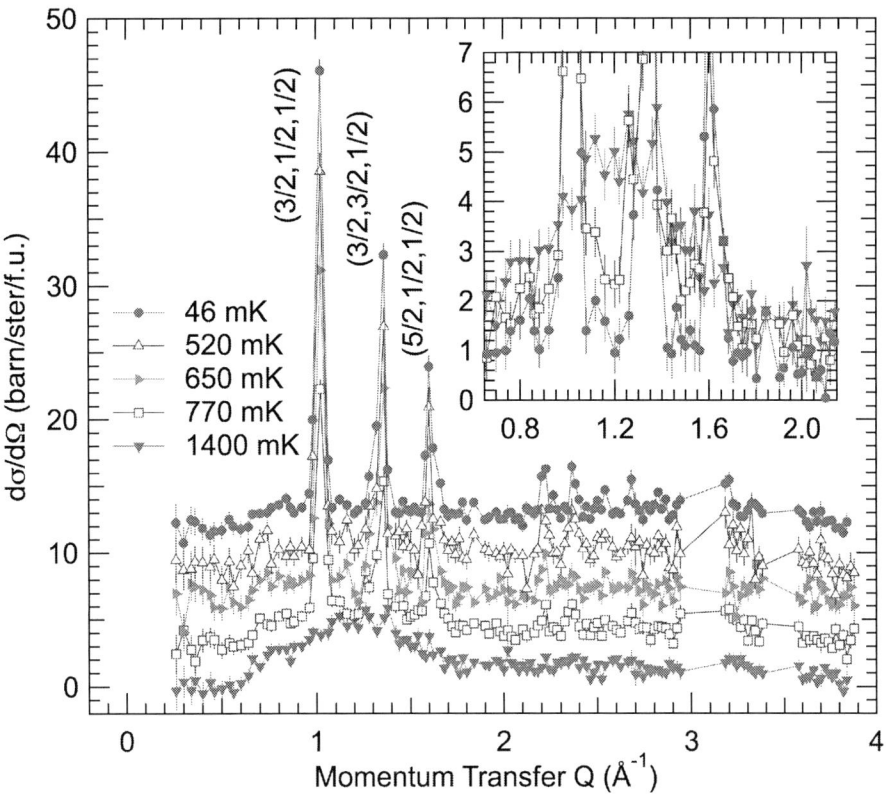

Figure 1: Magnetic diffraction from $Gd_2Ti_2O_7$ at 46, 520, 650, 770 and 1400 mK. Traces are vertically offset for visibility. The inset highlights the Q range where the diffuse scattering peaks. Here the traces are not offset to highlight the temperature dependence of the intensity.

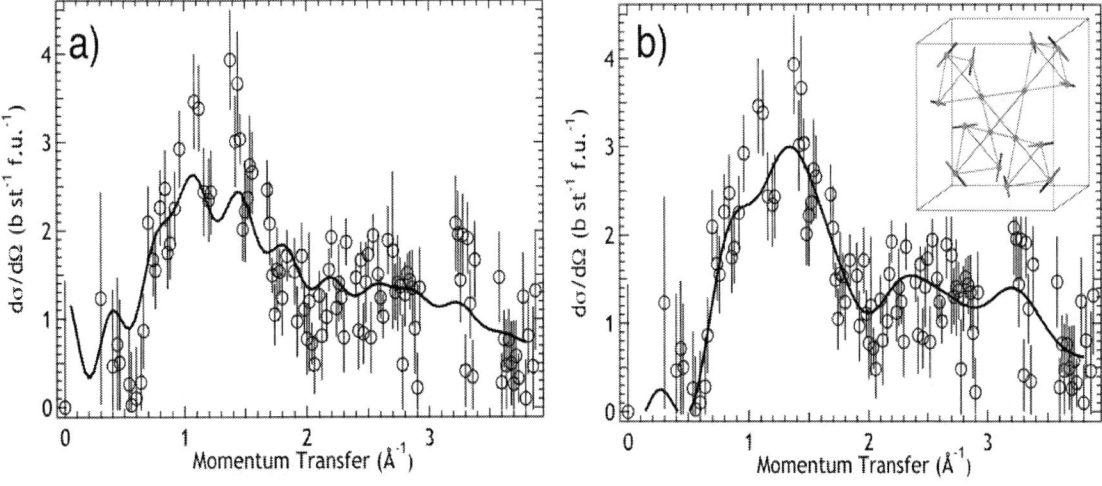

Figure 2: Diffuse magnetic scattering at 770 mK and the results from RMC fits to Eq. (1), (a) ignoring the ordered phase and (b) including a contribution from the ordered phase. The ordered (green) and disordered (orange) Gd spins in the primitive cell are depicted in the inset along with the directions of the ordered moments in the 4-k structure.

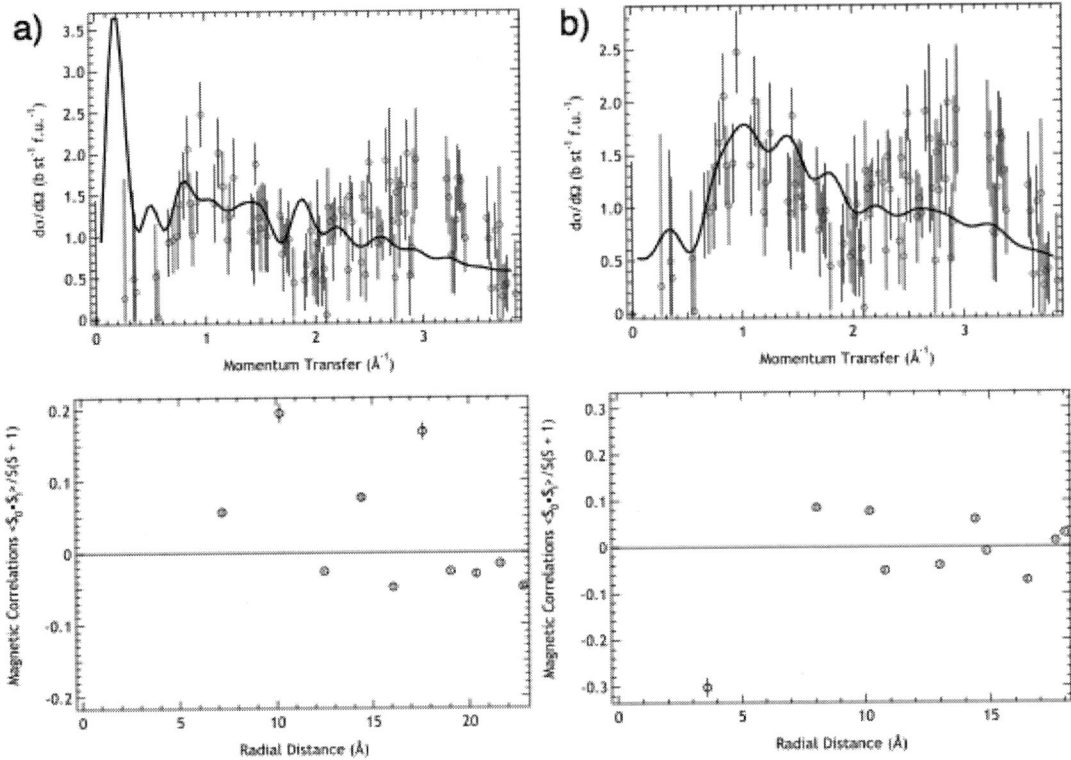

Figure 3. (top) RMC fits of Eq. (1) to the magnetic diffuse scattering in $Gd_2Ti_2O_7$ at 46 mK (the magnetic Bragg peaks arising from the long-range ordered Gd atoms have been removed from the data). (bottom) The first ten near-neighbor spin-correlation parameters extracted from the RMC models. The 1-**k** model (a) provides a poor description of the data, with a χ^2 of 2.0 and unphysical ferromagnetic correlations in the first two near-neighbor shells. The 4-**k** model (b) provides better description of the diffuse scattering ($\chi^2 = 1.55$), with a dominant antiferromagnetic spin-correlations and negligible further near-neighbour contributions

Both RMC models describe the data satisfactorily. The addition of a contribution to the disordered magnetic structure from the nominally ordered phase improved the model only slightly, reducing $\chi2$ from 1.64 to 1.24. However it is clear that a model close to that shown in the inset to Fig. 2b is appropriate: i.e. that of a central disordered Gd tetrahedron with dominant first near-neighbor antiferromagnetic interactions. By absolute calibration of the polarized neutron cross-section the total disordered Gd spin per formula unit was found to be $S(S+1)=11.7(5)$ implying a disordered moment per Gd atom of 4.8 (1) μB.

These simulations confirm that the short-range correlations at 770 mK and below are consistent only with the 4-**k** structure reported in Ref. [10], since the disordered spin structure factor S(Q) measured here is dominated by first near-neighbor interactions, and is hence incompatible with the previously proposed 1-**k** structure [9]. To confirm this, we have performed a similar fit at 46 mK. Due to the poor statistics, we have again performed the simpler model, ignoring the static correlations. Figure 3 shows these data with the 1-**k** and 4-**k** fits. Not only does the 4-**k** model visibly describe the data better, the 1-**k** data also enforces ferromagnetic

correlations between disordered spins, which are inconsistent with other data, including the lack of depolarization in this and the neutron spin echo data published elsewhere [19]. For the 46 mK data, we find $S(S+1) = 8.1(5)$ implying a disordered moment of 4.0 (1) ⌠B per Gd atom [22].

CONCLUSION

To summarize, all data presented here and in the literature indicate that $Gd_2Ti_2O_7$ has two magnetic phase transitions below 2 K. Since the discovery of these transitions by Ramirez *et al.* [16], both theoretical and experimental techniques have been employed to account for the two transitions and the two non-equivalent magnetic species in this cubic pyrochlore. Further studies (e.g., using a good single crystal sample that is enriched in a non absorbing Gd isotope) are required to elucidate the cause of the unusual properties of $Gd_2Ti_2O_7$. Another possibility would be to perform a low temperature structural study to search for a lattice distortion in a way similar to a study by Ruff *et al.* [23] on the spin liquid $Tb_2Ti_2O_7$. This might generate two different Gd sites. $Gd_2Ti_2O_7$ is a model magnet for the study of the competition between long-range order and short-range correlated disorder. At high temperatures (T > 2 K), the system is a cooperative paramagnet with dynamic, short-range correlations. Between 1.1 K and 770 mK long range spin correlations coexist with dynamic, disordered spins and below 770 mK the long range ordered spins interact with a well correlated, partially frozen set of spins. Such models are also relevant for the understanding of impurities in optics [24], relaxor behavior in ferroelectrics [25] and electronic transport through DNA [26].

ACKNOWLEDGMENT

The authors would like to thank our colleagues S T Bramwell, A S Wills and Y Qiu who helped in our earlier work on Gd pyrochlores, the technical help at the ILL during the experiment, and our collaborators B. D. Gaulin and M. J. P. Gingras for fruitful discussions on this topic. ORNL/SNS is managed by UT-Battelle, LLC, for the U.S. Department of Energy under contract DE-AC05-00OR22725.

REFERENCES

[1] A. P. Ramirez, B. S. Shastry, A. Hayashi, J. J. Krajewski, D. A. Huse and R. J. Cava, Phys. Rev. Lett. **89**, 067202 (2002).

[2] M. J. P. Gingras and B. C. den Hertog, Can. J. Phys., **79**, 1339 (2001).

[3] R. Moessner and J. T. Chalker, Phys. Rev. Lett. **80**, 2929 (1998), Phys. Rev. B **58**, 12 049 (1998).

[4] J. N. Reimers, Phys. Rev. B **45**, 7287 (1992).

[5] N. P. Raju, M. Dion, M. J. P. Gingras, T. E. Mason and J. E. Greedan, Phys. Rev. B **59** 14489 (1999).

[6] S. E. Palmer and J. T. Chalker, Phys. Rev. B, 62488 (2000).

[7] V. N. Glazkov, M. E. Zhitomirsky, A. I. Smirnov, H.-A. Krug von Nidda, A. Loidl, C. Marin and J.-P. Sanchez, Phys. Rev. B, **72** 020409(R) (2005).

[8] N. P. Raju, M. Dion, M. J. P. Gingras, T. E. Mason and J. E. Greedan, Phys. Rev. B, **59** 14489 (1999).

[9] J. D. M. Champion, A. S. Wills, T. Fennell, S. T. Bramwell, J. S. Gardner, and M. A. Green, Phys. Rev. B **64**, 140407 (2001).

[10] J. R. Stewart, G. Ehlers, A. S. Wills, S. T. Bramwell and J. S. Gardner, J. Phys.: Condens. Matter, **16** L321, (2004).

[11] J. R. Stewart, J. S. Gardner, Y. Qiu and G. Ehlers, Phys. Rev. B **78**, 132410 (2008).

[12] A. S. Wills, M. E. Zhitomirsky, B. Canals, J.P. Sanchez, P. Bonville, P. Dalmas de Reotier and A. Yaouanc, J. Phys.:Condens. Matter, **18**, L37 (2006).

[13] A. Del Maestro and M. J. P. Gingras, Phys. Rev. B, **76**, 064418 (2007).

[14] J. A. Quilliam, K. A. Ross, A. Del Maestro, M. J. P. Gingras, L. R. Corruccini and J. B. Kycia, Phys. Rev. Lett., **99**, 097201 (2007).

[15] S. S. Sosin, L. A. Prozorova, P. Bonville and M. E. Zhitomirsky, Phys. Rev. B, **79**, 014419 (2009).

[16] A. P. Ramirez, B. S. Shastry, A. Hayashi, J. J. Krajewski, D. A. Huse and R. J. Cava, Phys. Rev. Lett., **89**, 067202 (2002).

[17] P. Bonville, J. A. Hodges, M. Ocio, J. P. Sanchez, P. Vulliet, S. Sosin and D. Braithwaite, J. Phys.: Condens. Matter, **15** 7777, (2003) and S. S. Sosin, L. A. Prozorova, A. I. Smirnov, P. Bonville, G. Jasmin-Le Bras and O. A. Petrenko, Phys. Rev. B, **77**, 104424 (2008).

[18] J. S. Gardner, B. D. Gaulin and D. McK. Paul, J. Crystal Growth **191**, 740 (1998).

[19] O. Schärpf and H. Capellmann, Phys. Status Solidi a, **135**, 359 (1993).

[20] G. Ehlers, J. Phys.: Condens. Matter, **18**, R231 (2006)

[20] J. R. Stewart, K. H. Andersen and R. Cywinski, Phys. Rev. B **78** 014428 (2008)

[21] J. P. C. Ruff, B. D. Gaulin, J. P. Castellan, K. C. Rule, J. P. Clancy, J. Rodriguez, and H. A. Dabkowska, Phys. Rev. Lett., **99**, 237202 (2007).

[22] This value is lower than that presented in ref. [10]. Here we correctly account for 2 Gd ions per formula unit.

[23] P. Sheng, *Introduction to Wave Scattering, Localization and Mesoscopic Phenomena* (Academic Press, New York, 1995).

[24] K. Binder, Thin Solid Films, **20**, 36 (1976).

[25] L. E. Cross, Ferroelectrics, 76, 241 (1987).

[26] C.-K. Peng, S. V. Buldyrev, A. L. Goldberger, S. Havlin, F. Sciortino, M. Simons and H. E. Stanley, Nature **356**, 168 (1992).

Development of Cold Neutron Scattering Kernels for Advanced Moderators

J.R. Granada and F. Cantargi

Centro Atómico Bariloche and Instituto Balseiro Comisión Nacional de Energía Atómica (CNEA) ARGENTINA

Abstract

The development of scattering kernels for a number of molecular systems was performed, including a set of hydrogeneous methylated aromatics such as toluene, mesitylene, and mixtures of those. In order to partially validate those new libraries, we compared predicted total cross sections with experimental data obtained in our laboratory. In addition, we have introduced a new model to describe the interaction of slow neutrons with solid methane in phase II (stable phase below T=20.4K, atmospheric pressure). Very recently, a new scattering kernel to describe the interaction of slow neutrons with solid Deuterium was also developed. The main dynamical characteristics of that system are contained in the formalism, the elastic processes involving coherent and incoherent contributions are fully described, as well as the spin-correlation effects.

Keywords: scattering kernel, advanced moderator, hydrogeneous methylated aromatics

INTRODUCTION

Cold neutrons are widely used in different fields of research such as the study of the structure and dynamics of solids and liquids, the investigation of magnetic materials, biological systems, polymer science, and a rapidly growing area of industrial applications.

The development and optimization of advanced cold neutron sources require neutronic calculations involving thermal and subthermal neutron energies, which in turn demand the knowledge of reliable cross section data relative to the materials which conform the system under consideration. The compromise solution adopted in standard Nuclear Data Libraries involves the inclusion of scattering cross sections for a few common moderators at some selected temperatures, and data for any different material or physical condition must be 'constructed' from pieces of information actually corresponding to those few cases found in the existing files.

Condensed molecular systems often display a complex behavior due to translational, rotational and vibrational degrees of freedom – and their couplings – that animate the intra-and inter-molecular motions. However, a full account of those is not necessarily required in some special circumstances, for example when only some integral properties of the neutron scattering interaction must be properly described.

In this paper we describe some of the activities currently ongoing at Neutron Physics Division of Centro Atómico Bariloche (Comisión Nacional de Energía Atómica, Argentina), which involved our recent theoretical, calculational and experimental efforts related to the study of hydrogeneous materials of interest as cold moderators. The main motivations for those studies are driven by the large and increasing impact of Neutron Scattering techniques on Science and Technology, the strong requirement of long wavelength neutrons, and the development of advanced cold neutron sources.

We present here our work on some selected molecular systems including solid phases hydrogeneous methylated aromatics such as toluene, mesitylene, xylene, and mixtures of those. In order to produce a partial validation of some of those new libraries, we performed transmission experiments using our pulsed neutron source based on an electron LINAC. Also, our most recent work on methane in phase II (T < 20.4K) and solid deuterium is discussed, and our model predictions are compared with the available experimental information.

BASIC EQUATIONS

The Van Hove scattering function $S(\mathbf{Q},\omega)$ is directly related to the double-differential cross section [1]:

$$\frac{d^2\sigma}{d\Omega d\omega} = \frac{k}{k_0} S(\mathbf{Q},\omega) \qquad (1)$$

where $\mathbf{k}$, $\mathbf{k0}$ are the scattered and initial neutron wave vectors, $\hbar\omega$ is the neutron energy loss, and $\hbar\mathbf{Q} = \hbar(\mathbf{k0} - \mathbf{k})$ is the momentum transferred to the system.

CP1202, *Neutron and X-Ray Scattering in Advancing Materials Research: International Conference – 2009*
edited by A. Saat, H. A. Kassim, M. H. H. Jumali, J. M. Saleh, M. R. Othman, A. Ibrahim, F. M. Idris, and M. H. A.-R. M. Ahmad
© 2009 American Institute of Physics 978-0-7354-0739-8/09/$25.00

The scattering law of a molecular system

$$S(\mathbf{Q},\omega) = \frac{1}{2\pi\hbar} \int_{-\infty}^{\infty} dt \, e^{-i\omega t} \left\langle \sum_{l,l'} \sum_{v,v'} \overline{a_{lv}^* a_{l'v'}} \exp\{-i\mathbf{Q}.\mathbf{R}_{lv}(0)\} \exp\{i\mathbf{Q}.\mathbf{R}_{l'v'}\} \right.$$

(2)

where $\mathbf{R}lv(t)$ denotes the position of the atom v within the molecule l, can be written as the sum of inter $(l \neq l')$-and intra $(l = l')$ molecular contributions (also referred to as the *outer* and *inner* terms, respectively). That means expressing its Fourier transform as

$$\chi(\mathbf{Q},t) = \left\langle \sum_{l \neq l'} \sum_{v,v'} \overline{a_{lv}^* a_{l'v'}} \exp\{-i\mathbf{Q}.\mathbf{R}_{lv}(0)\} \exp\{i\mathbf{Q}.\mathbf{R}_{l'v'}(t)\} \right\rangle +$$
$$\left\langle \sum_{l} \sum_{v,v'} \overline{a_{lv}^* a_{lv'}} \exp\{-i\mathbf{Q}.\mathbf{R}_{lv}(0)\} \exp\{i\mathbf{Q}.\mathbf{R}_{lv}(t)\} \right\rangle$$

(3)

Here the brackets denote the average of the time-dependent operators over an equilibrium-distribution function in the full phase space of the scattering system. The structural complexity of the system together with the motion of the molecules' centres of mass are the main elements determining the intermolecular component. On the other hand, the complete effects of the molecule's conformation and degrees of freedom are contained in the inner term.

The scattering lengths alv appearing in the above equations are spin dependent quantities, and therefore must be in principle included within the expectation value brackets of the intermediate scattering function $\chi(\mathbf{Q},\omega)$. In terms of the usual coherent, bc_v, and $vincoherent$, bi, scattering lengths for nuclei $_v$, one obtains:

$$a_v = b_c^v + 2b_i^v (\mathbf{S}_v.\mathbf{s})[S_v(S_v+1)]^{-1/2}$$

(4)

where Sv and $\mathbf{s}$ are the spin operators for nuclei v and the neutron, respectively.

The structural and dynamical properties of a given system determine the characteristics of its interaction with slow-neutrons. In other words, the probability for the occurrence of a scattering process with the exchange of certain energy and momentum between the neutron and the scatterer is controlled by those properties, in turn contained in the scattering law $S(\mathbf{Q},\omega)$ of the system. In the frame of the Gaussian approximation [1] the dynamics of the material is enclosed in its generalized frequency spectrum, and this is in fact the important piece of information we need to predict scattering probabilities in the case of hydrogenous materials, where interference effects are negligible.

Under the assumption of no coupling between the (translational, rotational, vibrational) molecular modes, the intermediate scattering function is the product of

those associated to each mode. In addition, each of the factors is assumed to satisfy the Gaussian approximation

$$\chi_i(Q,t) = \exp\{-\gamma_i(t)Q^2\} \, ; \, i=1,2$$

(5)

where the time-dependent mean-square displacement $\gamma(t)$ is related to the frequency spectrum

$$\gamma(t) = \int_0^{\infty} \frac{Z(\omega)}{\omega} \left[\{n(\omega)+1\}e^{i\omega t} + n(\omega)e^{-i\omega t} \right]$$

(6)

and $n(\omega)$ is the occupation number.

We constructed frequency spectra for the different materials considered in this work, by combining experimental [2,3] and synthetic [4] contributions, together with a fitting parameter related to the weight of the translational and librational modes, and a normalization condition for the weights of all the dynamical modes.

Scattering law data files were generated by the LEAPR module of the NJOY code [5] using the adopted frequency spectra, whereas the cross section data libraries were produced by the modules THERMR and ACER, for the appropriate format for MCNP calculations.

APPLICATIONS

Solid Aromatic Hydrocarbons

Although solid methane is the best moderator in terms of cold neutron production, it has very poor radiation resistance, causing spontaneous burping even at fairly low doses. Such effect is considerably reduced in the aromatic hydrocarbons. We reviewed the dynamics of a group of aromatic hydrogenous solids, and making use of existing experimental information on the density of states for translational and rotational (librational) motions, we developed synthetic frequency spectra for them. Cross section libraries in ACE format were then generated using the NJOY code. They were validated by comparing with experimental data obtained in our laboratory and used to predict neutron spectra emerging from a typical TMR configuration, by using the MNCP code for the evaluation of the energy distribution of neutrons coming out from the cold moderator.

Natkaniec and coworkers performed a comprehensive study on the inelastic neutron scattering properties of aromatic hydrocarbons at low temperatures [2,6], and they were able to derive preliminary densities of states for those materials. As an example, we show in Fig. 1 the low-energy part of the results they obtained for the three phases of mesitylene that could exist at 20K [2].

This material, 1-3-5 trimethyl-benzene, has three methyl groups placed symmetrically around the benzenic ring, presumably the configuration with richest density of rotational states at low energies. The phase II showing the typical behaviour of a disordered one, is the most interesting as an efficient cryogenic moderator material.

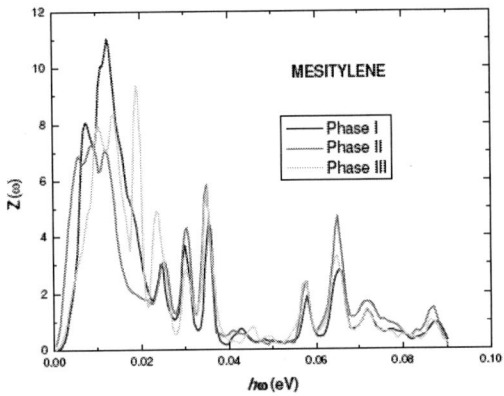

Fig. 1: The frequency spectrum of mesitylene used in our model for each of its three low temperature phases

Transmission experiments were performed at the 25 MeV Electron LINAC based pulsed neutron source at Centro Atomico Bariloche, operated in these experiments at a repetition rate of 50 Hz and 12 μA average current. The neutron energy was determined by the TOF technique, where the scale was corrected by the mean emission time of the moderator while the spectra were corrected by dead time effects from detectors and electronics.

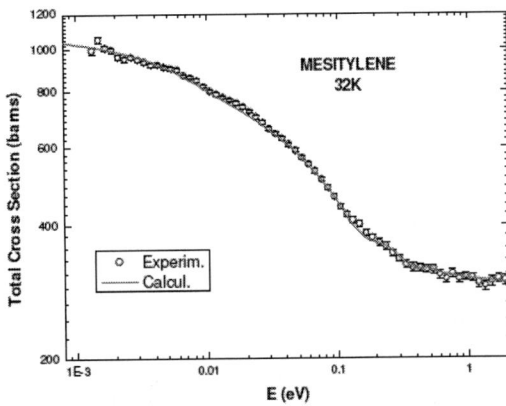

Fig 2: Total cross section of mesitylenee at 32 K measured at our laboratory, compared with the calculated curve from our new kernel.

Even though those aromatic materials are very easy to handle, the solid phases that produce an enhanced flux of cold neutrons correspond to amorphous structures rich in low-energy excitations, and they can be created through lengthy cooling processes requiring in many cases additional annealing stages. The 3:2 mesitylene-toluene mixture, that forms in a direct manner the appropriate disordered structure, constitutes an excellent cryogenic moderator material, as it is able to produce an intense flux of cold neutrons while presenting high resistance to radiation.

The comparison of experimental and calculated results for solid mesitylene and the mixture mesitylene 3:2 toluene at 32 K are shown in Figs. 2 and 3,

respectively, where a good agreement is obtained in each case with the optimized frequency spectrum. Once the new cross section libraries were produced from the validated kernels, spectra calculations were performed using the MCNP code [7], on a standard target-moderator-reflector configuration used at the Hokkaido LINAC.

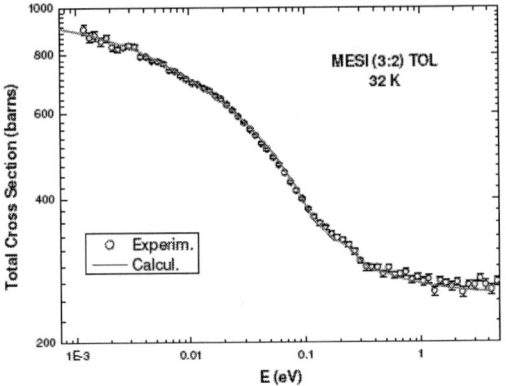

Fig 3: Total cross section of mesitylene (3:2) toluene at 32 K measured at our laboratory,

The normalized neutron spectra thus produced are shown in Fig. 4 for a temperature of 20 K, where it is evident that the methyl groups containing molecules produce a more intense and cooler spectrum than that due to benzene. In addition, the best of the group in that sense is mesitylene in phase II, closely followed by the mixture of mesitylene and toluene.

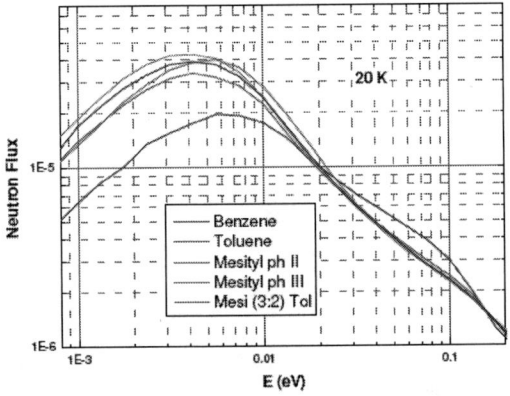

FIG. 4: Neutron spectra emerging from the aromatic hydrocarbons at 20 K studied here, using a standard TMR configuration.

Thus, by making use of existing experimental information on the density of states for translational and rotational (librational) motions, we developed synthetic frequency spectra for several aromatic hydrocarbons.

We presented the final forms of the frequency spectra for the aromatic materials considered, as well as their scattering kernels, cross section data and predicted neutron spectra emerging from a typical TMR configuration, by using the codes NJOY for data generation and MNCP for the Montecarlo evaluation of the energy distribution of neutrons coming out from the cold moderator.

Solid Methane in Phase II

We developed a simple model to describe the interaction of slow neutrons with solid methane in phase II, including the main dynamical features of the system and the effect of spin correlations. This effect occurs in molecules containing identical nuclei whenever spin and rotational states are coupled, thus imposing symmetry requirements on the molecular wave function. A central motivation for the development of this new scattering kernel has been the generation of cross section libraries appropriate for the calculation of neutron thermalization properties in CH4 II.

$$\chi^{inter}(\mathbf{Q},t) = \left\langle \sum_{\nu,\nu'} b_c^\nu b_c^{\nu'} f_{\nu\nu'} \right\rangle$$

$$\chi^{intra}(\mathbf{Q},t) = \left\langle \sum_{\nu,\nu'} \left\{ b_c^\nu b_c^{\nu'} + b_{inc}^\nu b_{inc}^{\nu'} \cdot \frac{(\mathbf{S}_\nu \mathbf{S}_{\nu'})}{[S_\nu(S_\nu+1)S_{\nu'}(S_{\nu'}+1)]^{1/2}} \right\} f_{\nu\nu'} \right\rangle \quad (7)$$

The scattering lengths b_ν are spin dependent quantities. After performing the spin averages appropriate to an unpolarized neutron beam, the scattering law can be written as the sum of inter ($l \neq l'$)- and intra ($l = l'$) molecular contributions (also referred to as the *outer* and *inner* terms, respectively). That means expressing its Fourier transform as

Detailed quantum calculations were performed to describe the low-level energy states in CH4 II (Ref. 8) and applied to the evaluation of neutron cross sections for that system [9,10]. Those formulations were then used to analyze neutron scattering and transmission experiments [11,12,13]. In our study we oriented our effort to the development of a simple model which should preserve the main dynamical features (as well as a proper description of spin correlation effects) of the system and therefore be able to make reliable predictions in terms of neutron fluxes emerging from such cold moderator material.

From the measurements performed by Harker and Brugger [14] at different temperatures in phase I and phase II of solid methane, we derived frequency spectra for free, hindered, and average (1:3) molecules in phase II, and they are displayed in Fig.5. It is observed that the spectrum corresponding to free molecules is richer at low energies, reflecting the availability of many well distributed levels over that region; on the contrary, the spectrum for hindered molecules shows a prominent peak around 0.0065 eV where the librational states coalesce. The third curve, the phase II spectrum, is the 1:3 weighted average of the other two.

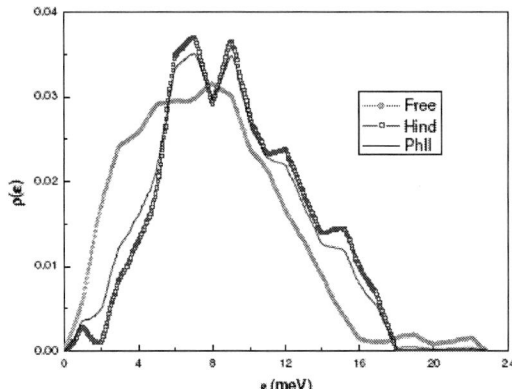

Fig 5: Frequency spectra (translational and rotational Motions) employed in our calculations, derived from the experimental data of Harker and Brugger [Ref. 14]

The calculation of the total cross section involves integration of the resulting scattering function over final neutron energies and scattering angles. There is a first term which is evaluated in a conventional way, starting from the phase II spectrum, normally through Fourier transforms or phonon expansion algorithms. The second, elastic term carrying the effects of spin correlations, only requires the integration of the (non rigid) molecular factor over the scattering angles.

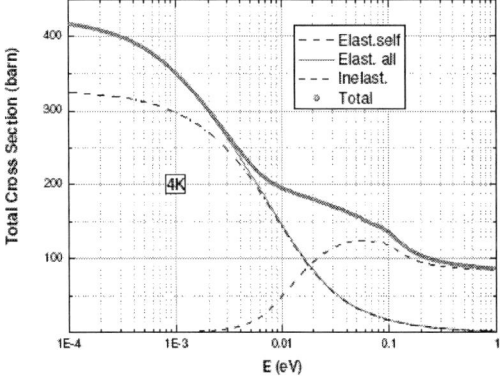

Fig 6: The calculated total cross section of CH4 at 4 K, indicating the elastic and inelastic contributions.

The prediction of the present model for CH4 II at 4K is displayed in Fig. 6, where the inelastic and elastic spincorrelated and uncorrelated contributions are also shown. In particular, our calculation indicates that, for example at 0.0001 eV the effect of spin correlation is to increase the total cross section at 4K from 324.5 b to 416.9 b.

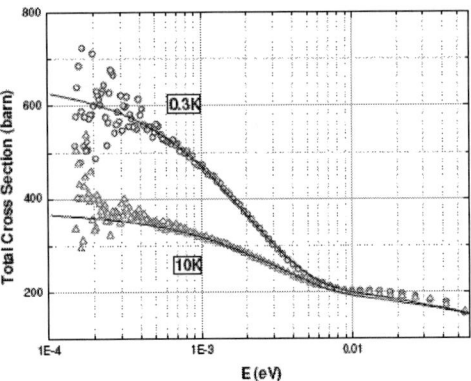

Fig 7: Comparison of the calculated total cross sections of methane at 0.3 K and 10 K, with the measured points of Grieger et al. [Ref. 11].

A comprehensive set of measurements on this system was performed by Grieger *et al*, [11]. at temperatures ranging from 19.5 K down to 0.3 K, and we compare in Fig. 7 our calculations with their experimental total cross sections at 0.3 K and 10 K, over the relevant energy range. The agreement is very good except at the lowest energies, where the scatter of data points does not allow to confirm such assessment.

The predictions of our model for methane in phase II are in good agreement with a full quantum mechanical calculation over the limited range where the latter was formulated, and with available experimental information over the complete thermal energy range. Bearing in mind the general nature of the approximations involved in the present prescription, and besides its predictive capacity demonstrated for solid methane in phase II, the model can be useful for the analysis of neutron scattering experiments designed to study spin species conversion of different rotational tunneling molecules at low temperatures, or in a wider context, to perform reliable multiple-scattering corrections in experiments oriented to precise determinations of density of states.

Solid Deuterium

Very recently, a new scattering kernel to describe the interaction of slow neutrons with solid Deuterium was developed. The main characteristics of that system are contained in the formalism, including the lattice's density of states, the Young-Koppel quantum treatment of the rotations, and the internal molecular vibrations. The elastic processes involving coherent and incoherent contributions are fully described, as well as the spin-correlation effects.

The deuterium molecule is formed by two bosons, and therefore its total wave-function must be symmetric under interchange of two identical nuclei [1]. Consequently, if the total nuclear spin **S** is even the spatial nuclear wave function must be symmetric, and antisymmetric if **S** is odd, which leads, respectively, to the existence of the *ortho* states, with S = 0, 2 coupled to J = 0, 2, 4,..., and *para* states with S = 1 coupled to J = 1, 3, 5, ..., where J denotes the molecule's total angular momentum.

The neutron scattering laws S($\mathbf{Q}$,ω), energy-transfer kernels σ(E,E′), and cross sections σ(E) for inelastic scattering in solid ortho-and para-deuterium were calculated using the code NJOY [5], which is based on a phonon expansion for the lattice motion, and the Young-Koppel [15] formalism for the quantum rotational description. As part of the code's input data, the density of states (DOS) for solid deuterium derived by Schmidt *et al.*[16] was used. The elastic incoherent component produced by NJOY was modified in order to include spin correlation effects. The lattice structure factor was calculated using our code CRIPO [17] and then affected by the Debye-Waller factor χ($\mathbf{Q}$,0) and the molecular structure factor to obtain the elastic coherent component. In our calculations we used the values σc = 5.59 b, σi = 2.05 b, for the bound atom coherent and incoherent cross sections respectively, and σa= 0.0005 b for the thermal neutron absorption cross section [18], as well as hωr = 0.0074 eV and hωv = 0.371 eV for the rotational and vibrational molecular energies.

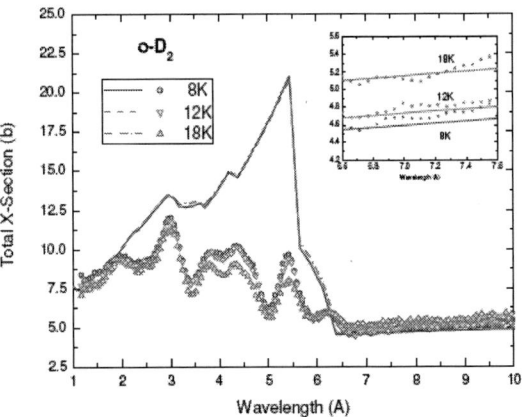

Fig.8: Calculated total cross section of solid o-D2, compared with experimental data from [19,20]. The inset shows that comparison beyond the first Bragg peak.

The total cross section of ortho-deuterium is shown in Fig.8 as calculated with the present model, compared with experimental data from Refs.[19] and [20] for a few temperatures over the thermal neutron wavelength range. The large elastic coherent contribution due to the *hcp* structure factor dominates the cross section at those energies, and the disagreement with the measured points is a clear indication of the lack of perfect polycrystallinity in the samples. However, and in spite of the significant difficulties to achieve a high precision normalization for the measured curves in those experiments, a very satisfactory agreement between both sets is observed beyond the first Bragg peak, a region dominated by total inelastic and incoherent (including spin correlation effects) elastic components of the total cross section (see inset in Fig.8).

The comparison between the present model and existing cross section data for solid deuterium is completed with the curves presented in Fig..9, where calculated values for solid ortho-and normal-deuterium are compared with the measurements from refs. [20] and [21] at around 18 K. At this temperature the inelastic cross section for ortho and para deuterium are very similar, and therefore the differences observed below the first Bragg peak for the cross sections of ortho-and normal-deuterium are purely due to spin correlation effects on the elastic incoherent contribution.

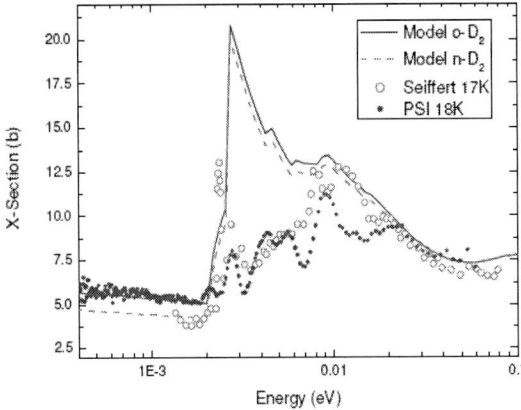

Fig.9: Calculated total cross sections of ortho-and normal-deuterium at 18 K, compared with the experimental data from ref. [20] and [21].

CONCLUSIONS

Cross section libraries of hydrogen bound in benzene, toluene, mesitylene and a solution 3:2 by volume of mesitylene/toluene were generated at different temperatures, in particular at 20K.The 3:2 mesitylene-toluene mixture, that forms in a simple and direct manner the appropriate disordered structure, constitutes an excellent cryogenic moderator material, as it is able to produce an intense flux of cold neutrons while presenting high resistance to radiation [22].

We developed a new scattering kernel for solid methane in phase II, including the main dynamical features of the system and the effect of spin correlations. A central motivation for this development has been the generation of cross section libraries for the calculation of neutron thermalization properties in CH4 II. Good agreement with a quantum mechanical calculation over the limited range where the latter was formulated, and with available experimental information over the complete thermal energy range [23].

A new scattering kernel to describe the interaction of slow neutrons with solid Deuterium has been developed. Scattering functions and cross sections for both *ortho*-and *para*-Deuterium have been evaluated for temperatures ranging from the freezing point (18.7 K) down to 5 K. The new model has been compared with the best available experimental data, showing a highly satisfactory agreement [24].

This work was partially supported by grant PICT 52963 from ANPCyT (Argentina) and IAEA RC N° 14161 funds.

REFERENCES

[1] S.W.Lovesey, Theory of Neutron Scattering from Condensed Matter, [Clarendon Press, Oxford, 1984].

[2] I. Natkaniec, K. Holderna-Natkaniec, 6th. Meeting of the International Collaboration on Advanced Cold Moderators, Juelich, Germany (2002).

[3] I. Natkaniec, Private communication (2002).

[4] J.R. Granada et al., 6th. Meeting of the International Collaboration on Advanced Cold Moderators, Juelich, Germany (2002).

[5] R.E. MacFarlane, D.W. Muir, "The NJOY Nuclear Data processing System", Los Alamos National Laboratory Report LA-12740-M (1994).

[6] I. Natkaniec, K. Holderna-Natkaniec, J.Kalus, Physica B 350, e651 (2004).

[7] MCNP, Monte Carlo Neutron and Photon Transport Code System, CCC200, RSIC (1992).

[8] T. Yamamoto, Y. Kataoka and K. Okada, J.Chem.Phys. 66, 2701 (1977).

[9] Y. Ozaki, Y. Kataoka and T. Yamamoto, J.Chem.Phys. 73, 3442 (1980).

[10] Y. Ozaki et al., Can.J.Phys. 59, 275 (1981).

[11] S. Grieger et al., J.Chem.Phys. 109, 3161 (1998).

[12] K.J. Lushington and J.A. Morrison, Can.J.Phys. 55, 1580 (1977).

[13] H. Friedrich et al., Physica B 226, 218 (1996).

[14] Y.D. Harker and R.M. Brugger, J.Chem.Phys. 46, 2201 (1966).

[15] Young J.A. and Koppel J.U., Phys.Rev. 135 (1964) A603; Koppel J.U. and Young J.A., Nukleonika 8(1966) 40.

[16] Schmidt J.W. et al., Phys.Rev.B 30 (1984) 6308.

[17] Kropff F. and Granada J.R., Unpublished Report CAB (1974).

[18] Neutron News 3 (1992) 29.

[19] Atchison F. et al., Phys.Rev.Lett. 95 (2005) 182502-1.

[20] Kasprzak M., PhD Thesis, University of Vienna (2008).

[21] Seiffert W.D., Report EUR 4455d (1970).

[22] F. Cantargi et al,. International Collaboration on Advanced Neutron Sources, ICANS.XVIII, Dongguan, Guangdong, China (April 2007).

[23] J.R. Granada, Nucl.Instr.Meth. B 266, 164 (2008).

[24] J.R. Granada, Eur.Phys.Lett. 86 (2009) 66007.

Small-Angle Neutron Scattering (SANS) Facility at BATAN for Nanostructure Studies in Materials Science and Biology

E. Giri Rachman Putra

Neutron Scattering Laboratory, Center for Technology of Nuclear Industrial Materials, National Nuclear Energy Agency of Indonesia (BATAN), Gedung 40 Kawasan Puspiptek Serpong, Tangerang 15314, Indonesia

ABSTRACT

A 36 meter small-angle neutron scattering (SANS) BATAN spectrometer (SMARTer) which is the second largest SANS spectrometer nowadays in the Asia-Oceania region was constructed at the neutron scattering laboratory (NSL) in Serpong, Indonesia. Lots of works on replacing, upgrading and improving the control system, experimental methods, data collection and reduction in the last three years have been carried out to revitalize and then optimize the performance of SMARTer. At first, some standard samples were measured for the inter-laboratory comparison and several kinds of substances such as liquid, gel, powder, and solid-state thin film have been investigated recently of proposed research interest. The morphological changes from ellipsoidal into cylindrical (worm-like) micelles of self-assembly amphiphilic molecules, sodium dodecyl sulfate (SDS) and transformation of disordered into ordered spherical micelle system from unimer Gaussian coils of PEO-PPO-PEO triblock copolymers (Pluronics) in solution by salt addition were also observed. Particle size and its distribution of spherical polystyrene latex and silica nanoparticles in dilute solution have been simply distinguished by applying a spherical calculation model. Bragg peaks which correspond to a lamellar structure was revealed from a powder sample of silver behenate $[CH_3(CH_2)_{20}COOAg]$ nanoparticle and a solid-state PS-PEP, polystyrene-*b*-poly(ethylene-*alt*-propylene), diblock copolymer film. The growth mechanism and fractal structures from aggregation of nanoparticles such as Fe_3O_4 ferrofluids or titanium-silica aerogels were investigated directly using a SANS technique through a power-law scattering of fractal structures approximation fitted at their scattering profiles. Meanwhile, magnetic structure from metal-alloys, CuNiFe showing anisotropic magnetic scattering structure properties up to 1 Tesla of external magnetic field was also accomplished confirming the nanocrystalline and magnetic domain sizes. The detail structure of *n*-dodecyl-β-D-maltoside (β-DMS) core-shell micelle has been revealed by applying a contrast variation, H_2O/D_2O mixture. Preliminary investigation of globular protein on folding-unfolding, protein denaturation and protein self-assembly studies is being performed. It can be concluded that SMARTer, a 36 m SANS BATAN spectrometer becomes a major tool for structural investigations in the effective length scale of 1 – 100 nm in materials science and biology.

Keywords: small-angle scattering, self-assembly, nanostructure, nanoparticle, fractals, protein, magnetic structure.

PACS: 61.05.fg; 61.46.-w; 64.70.km; 64.70.M-; 64.70.pv; 75.50.-y; 81.16.Dn

INTRODUCTION

Small Angle Neutron Scattering (SANS) is a powerful technique in characterizing the static and dynamic-structures of particles in the nanometer scale range of 1 – 100 nm. Information on the average size and its distribution, spatial correlation, as well as shape and internal structure of particles can be obtained from SANS scattering intensity profiles. The quantitative analyses on number or volume density from investigated structures in the surrounding medium can be determined from an absolute scale of scattering intensity. Thus,

SANS becomes a valuable technique for characterization in materials science and biology; e.g. for alloys, ceramics, polymers, colloids, vesicles, protein, viruses, etc.

A 36 m SANS BATAN spectrometer (SMARTer) has been completely installed in 1992[1] at the end of the 49 m long neutron guide and is located in the neutron guide hall of the neutron scattering laboratory (NSL) - BATAN in Serpong, Indonesia. SMARTer consists of an 18 m long collimator tube system which comprises four sections of movable guide-tubes and one section of fixed non-reflecting tube, see Fig. 1. Meanwhile, another 18 m long tube accommodates 128×128 channels of ^{3}He two-

CP1202, *Neutron and X-Ray Scattering in Advancing Materials Research: International Conference – 2009*
edited by A. Saat, H. A. Kassim, M. H. H. Jumali, J. M. Saleh, M. R. Othman, A. Ibrahim, F. M. Idris, and M. H. A.-R. M. Ahmad
© 2009 American Institute of Physics 978-0-7354-0739-8/09/$25.00

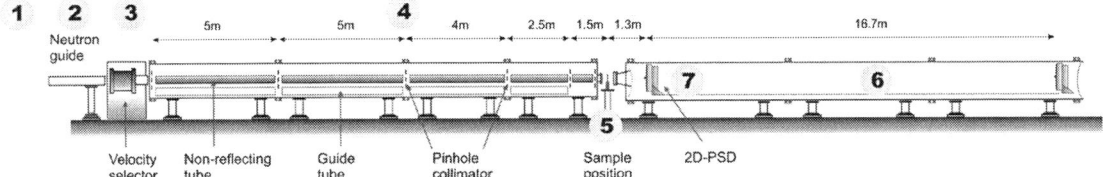

Figure 1. A schematic drawing of 36 m SANS facility at BATAN. (1) Neutron source from a multi-purpose reactor, G.A. Siwabessy (RSG-GAS); (2) neutron guide with a length of 49 m; (3) Mechanical velocity selector; (4) Collimation system; (5) Sample table and sample position; (6) Flight tube; (7) A two-dimensional position sensitive detector (2D-PSD)

dimensional position sensitive detector (2D-PSD). The detector can be moved continuously from 1.3 to 18 m from the sample position and can also be shifted in the lateral direction by 0.1 m covering a range of momentum transfer Q, $Q = (4\pi / \lambda) \sin(\theta/2)$ of $0.02 < Q$ (nm^{-1}) < 6 using a thermal neutron with the wavelength λ of $0.3 - 0.6$ nm. The effective size which can be observed by means of this spectrometer is $1 - 100$ nm.

Until 2004, the spectrometer was not well utilized due to a shortage in staff members, instrument failures and an undefined long-term research program. Then, a five years in-house work plan was proposed to replace, change and upgrade gradually the instrument, i.e. electronics, mechanics, computer software, etc.[2]. Therefore, in the last three years the control system, experimental methods, data collection and reduction were improved to optimize the performance as well as the data analyses[3-5]. Here, we highlight several experiments and research from micellar solution samples, PEO-PPO-PEO (Pluronics) triblock copolymers in solution, monodisperse polystyrene latex and silica nanoparticle, fine and porous ceramics, silver behenate nanoparticle, lamellar structure from PS-PEP, {polysterene-*b*-poly(ethylene-*alt*-propylene} diblock copolymers thin film, ferrofluid materials and hard magnetic alloys as well as biological materials. Those results show the capabilities and the improvement of the instrument on nanostructure investigation in materials and have been published elsewhere[6-10].

Micellar Solutions

Self-assembly which corresponds to a phenomenon of small molecules assembled spontaneously into a macromolecule or supra-macromolecule structure such as micelles, biological membranes, dendrimers, proteins, etc. is one of the most interesting issues in nanotechnology and bio-nanotechnology research. Hydrogen bonding, electrostatic force, van der Waals attraction, and hydrophobic chain interactions play an important role in structural changes of micelle such as a transformation of spherical-like into ellipsoidal and rod-like or cylindrical (worm-like) micelles or in the protein solutions. Phase transition and structural changes of amphiphilic molecules, i.e. sodium dodecyl sulfate (SDS), cetyltrimethyl ammonium bromide (CTAB) and PEO-PPO-PEO (Pluronics) triblock copolymers in solutions as a function of concentration, ionic strength (salt or counter ion), additive, and temperature have been

investigated intensively[7,9]. For the first time, a worm-like micelle structure with a length of 25 nm (250 Å) has been identified using SMARTer[11], Fig 2. The inter-correlation peak at momentum transfer $Q \sim 0.085$ Å^{-1} gradually disappears in addition of NaCl up to 0.3 M in which corresponds to disorder arrangement of micelles in solution.

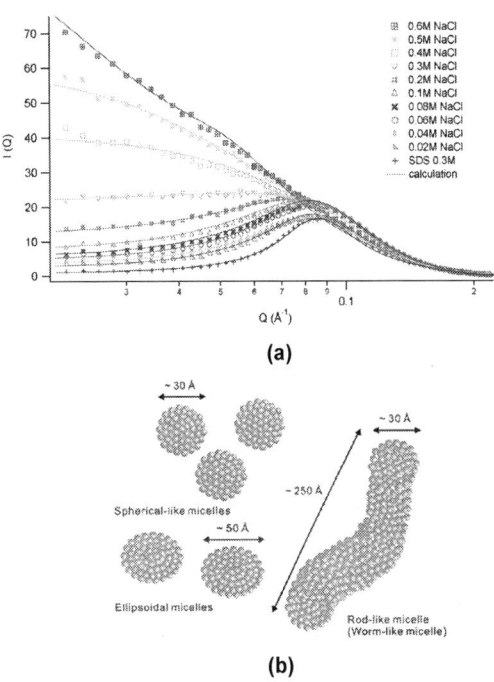

Figure 2. (a) Scattering profiles from 0.3 M SDS micellar solution with NaCl addition fitted by a calculation model from a simple geometrical shape such as sphere, ellipsoid and cylinder or rod. (b) A schematic model of micelle structural changes from spherical-like to rod-like micelle.

Unlikely 0.3 M SDS micellar solution, at a room temperature of 5 wt% Pluronics F88 [(EO)$_{103}$-(PO)$_{39}$-(EO)$_{103}$] micellar solution is unimers with a Gaussian coil structure that has a radius of gyration of 2.2 nm (22 Å), see Fig. 3. Adding salt such as KCl above the concentration of 0.6 M into that Pluronics micellar solution will induce entropically a micellization or aggregation to form a micelle with a core-shell structure where PPO and PEO chain blocks respectively become core and corona or shell of a micelle. These micelles have an average size of $4.3 - 5.1$ nm ($43 - 51$ Å) and the density number of the micelles in solution increases with increasing KCl concentration that correspond to increasing intensity of the scattering profile at low momentum transfer Q. In the end, at high KCl addition, i.e. 1.5 M it is noticeably clear that the inter-correlation Bragg peak appears at $Q \sim 0.03$ Å⁻¹. It can be described that the interactions of incident neutron beam with each micelle were distributed orderly in solution with an average distance of 20 nm (200 Å) giving a constructive interference of coherent scattering. This system shows the transition from disordered to ordered structures in nanoscale range of micellar solution.

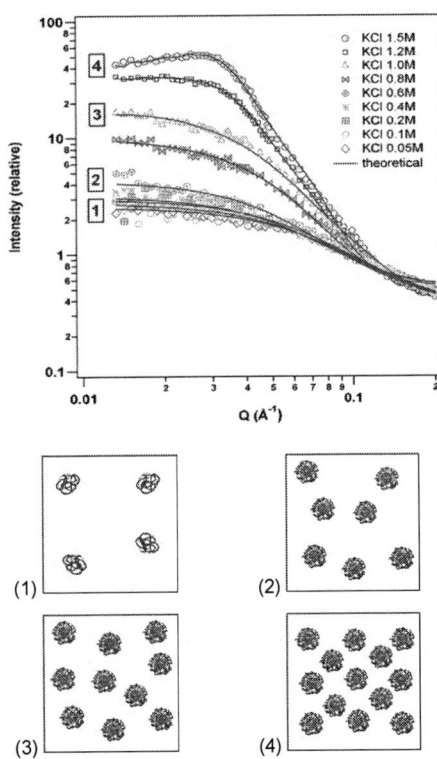

(1) (2) (3) (4)

Figure 3. Disorder to ordered system. A 5 wt% Pluronics F88 [(EO)₁₀₃-(PO)₃₉-(EO)₁₀₃] in solution at room temperature has a Gaussian coil structure (1). Micellization was taking place as KCl 0.6 M has been added to form a core-shell micelle structure as the block chain of PPO and PEO respectively become core and shell (2). A number of micelles increase by increasing the concentration of KCl (3). At KCl concentration of 1.5 M, the micelles form an ordered system in solution with an average distance between micelles is about 20 nm (200 Å), indicated by the appearence a Bragg peak at Q = 0.03 Å⁻¹ (4).

Polystyrene latex and silica nanoparticle samples in D₂O have been investigated up to the lowest momentum transfer range of SMARTer, $Q = 0.02$ nm⁻¹ to determine the size and its distribution, Fig. 4. The scattering patterns from those samples show the maxima and minima at the momentum transfer range that indicated the system is dilute where there is no inter-correlation amongst the particles and nearly monodispersed. A theoretical calculation of smeared monodisperse spherical model has been applied to fit the experimental data and in general showed a good agreement. From the best fitting on the monodisperse polystyrene latex

nanoparticle data it is found out that the particle radius r and its distribution σ are 61 nm (610 Å) and 0.1. While, for silica nanoparticle those number are 57 nm (570 Å) and 0.1. The volume fraction which corresponds to the density of particle can also be obtained from the calculation model. They are respectively 0.0026 and 0.0015 for polystyrene and silica nanoparticles.

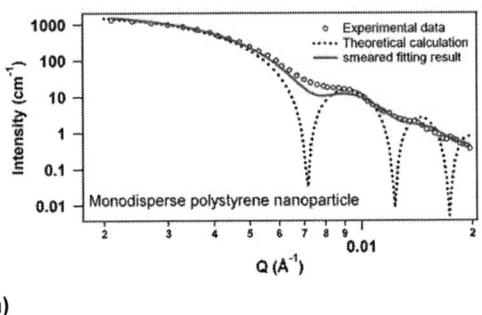

(a)

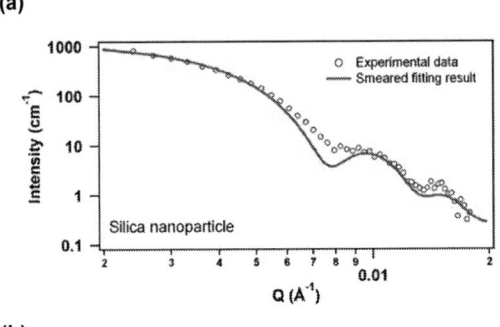

(b)

Figure 4. Scattering profile from a nearly-monodispersed polystyrene latex (a) and silica (b) nanoparticles in dilute solution. From the experimental data that has been analysed the radii r of spherical particles are correspondingly 61 and 57 nm for polystyrene latex and silica.

The SANS experiment on a plate-like silver behenate nanoparticle powder sample, silver behenate CH₃(CH₂)₂₀COOAg with a length of 0.2 – 2 μm and thickness of 0.1 μm has been completed. This sample gives several Bragg peaks in low angle scattering at a momentum transfer Q of 1.08, 2.17 and 3.25 nm⁻¹, Fig. 5. Those appeared Bragg peaks confirmed the emerging of third-order (00*l*) plane with the inter planes distance d = 5.84 nm (58.4 Å) in that silver behenate nanoparticle powder sample.

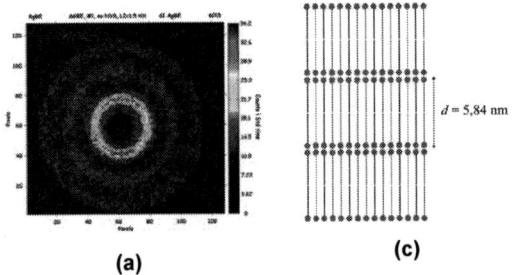

(a) (c)

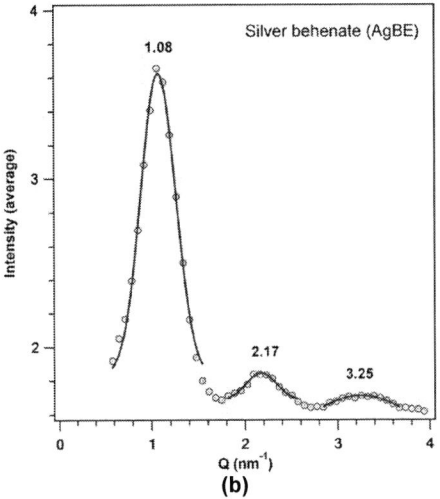

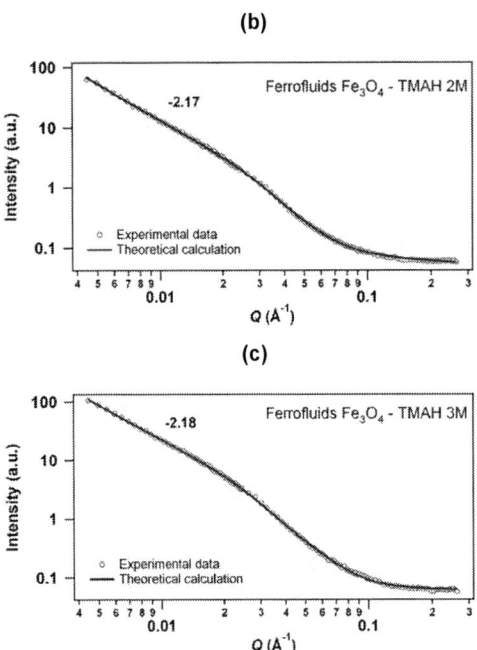

Figure 5. (a) A two-dimensional scattering pattern from a silver behenate nanoparticle powder sample with a neutron wavelength of 0.322 nm showing Bragg diffraction rings. (b) A full radial averaging from silver behenate nanoparticle powder sample in 1-dimension showing Bragg peaks at momentum transfer $Q = 1.08$, 2.17, 3.25 nm^{-1}. (c) A schematic lamellar structure from silver behenate nanoparticle powder sample with a (00l) diffraction plane and the inter diffraction plane distance $d = 5.84$ nm.

Magnetic Materials (Ferrofluid and Alloy)

Magnetic fluid (ferrofluid) which is a liquid dispersion of magnetic nanoparticle has a small size ~ 10 nm (100 Å) in diameter with a single magnetic domain in water (water-based ferrofluid) or oil (oil-based ferrofluid). This nanoparticle is coated or stabilised by surfactant molecules to prevent them to be aggregated. A SANS technique is a very suitable technique to determine the size of particle and its distribution as well as to reveal the thickness of surfactant layer by employing a contrast variation method.

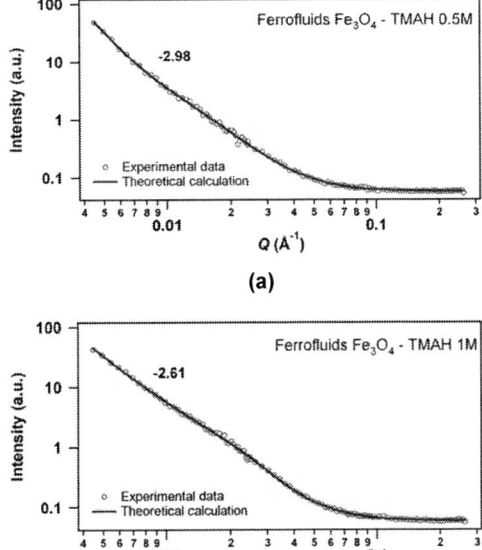

Figure 6. SANS distribution profiles from Fe_3O_4 ferrofluid sample which dispersed by tetramethyl hydroxide (TMAH) as a function of ferrofluid concentration (a) 0.5 M (b) 1 M (c) 2 M (d) 3 M. By increasing the ferrofluid concentration, the aggregation occurred which indicated by the slope of "power-law" scattering or fractal structure at a specific momentum transfer range. From power-law scattering approximation a fractal dimension changes from 3 to 2 with an increasing ferrofluid concentration. This indicates that the growth of aggregate likely spread in the space with a building block diameter of 20 – 30 nm.

However, due to the molecular force such as van der Waals interaction that is similar to a classical mechanism of colloidal coagulation, magnetite particles also has a possibility to aggregate and then form a cluster, though they were stabilized by surfactant molecules. The mechanism or phenomenon of aggregation and their internal structure of disordered system in nanometer scale termed as fractal structures can be obtained directly using a SANS technique. The fractal structure is often obeys a "power-law" scattering in the magnitude D of the momentum transfer Q in the range of $1/\xi \ll Q \ll 1/a$, where a is a typical chemical or bond distance related to local structure and ξ is the correlation length or average diameter of a scatterer. The fractal structures formed from Fe_3O_4 ferrofluid coated by tetramethyl ammonium hydroxide[12] as well as titanium-silica aerogles[13] have been determined by means of SMARTer, Fig. 6. The increasing of magnetite Fe_3O_4 concentration in ferrofluid and pH in synthesizing titanium-silica aerogels had an affect on the mass fractal dimension of a formed aggregate or cluster. This result confirmed the aggregation mechanism on ferrofluid or titanium-silica nanoparticles in which how they interact one to another, assembled or organised as a building block and then aggregated to form a large cluster.

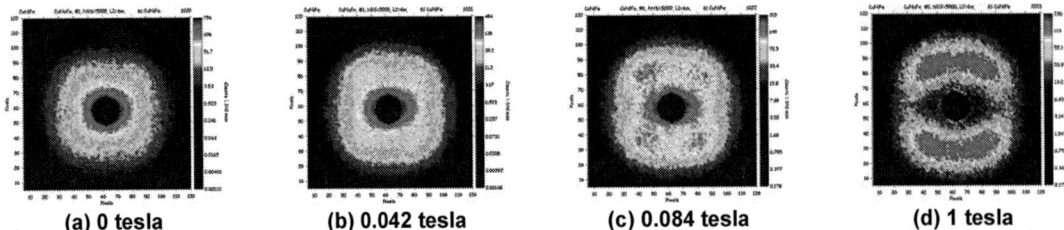

(a) 0 tesla **(b) 0.042 tesla** **(c) 0.084 tesla** **(d) 1 tesla**

Figure 7. Sequence of two-dimensional corrected SANS patterns from magnetic sample of Cu(NiFe) metal-alloy in external magnetic field (a) 0 (b) 0.042 (c) 0.084 and (d) 1 tesla. The external magnetic field direction (→) to the sample is in a plane perpendicular to the incident of unpolarised neutron beam. The data were taken using SMARTer at room temperature and at the momentum-transfer Q range of 0.1 < Q (nm^{-1}) < 1.

For magnetic sample, besides nuclear interaction, magnetic interaction is also occurred due to magnetic moment of neutron which is approximately equally strong as the nuclear interaction. With nuclear and magnetic interactions of neutrons they offer the opportunity to study both, compositional and magnetic structures and correlations. Thus, the magnetism in solid state physics and condensed matter research is a dominant subject on neutron application. Since SANS technique probing the structures on the nanometer scale, then it is applicable in finding applications in micromagnetism, magnetic clusters embedded in a solid nonmagnetic matrix or nanocrystalline, magnetic clusters suspended in fluids (e.g. ferrofluids), magnetism in nanostructured materials, vortex lattices in superconductors. The magnetic structure of hard magnetic metal-alloys such as Cu(NiFe), CuCo, and FeSiBNbCu (finemet) have been investigated by SMARTer in the external electromagnetic field of 1 tesla[8,14]. Two-dimensional scattering profiles showing nearly isotropic and anisotropic scattering patterns for Cu(NiFe) hard-magnetic sample under the external magnetic field up to 1 tesla are presented in Fig. 7. The isotropic and anisotropic scattering contributions are attributed to nuclear and magnetic scattering, respectively and they can be separated by fitting the intensity pattern. The significant changes from nearly isotropic pattern at zero magnetic field $B = 0$ tesla to anisotropic one at high field $B = 1$ tesla where the magnetic moment fully saturated parallel to applied external field direction are clearly seen in Fig. 7. From a full radial averaging scattering curve it has been analysed that the size of the magnetic domain or nanocrystalline is 10 nm and the average distance amongst domains is 15 nm. This result indicated that in the Cu(NiFe) sample each nanocrystalline mostly has a single magnetic domain.

Block Copolymers

A thin film of PS-PEP diblock copolymers sample has been also measured on SMARTer in the momentum transfer Q range of $0.03 < Q$ (nm^{-1}) < 0.3. An anisotropic scattered neutron which corresponds to the lamellar structure orientation with a (00l) diffraction planes is clearly shown in 2-dimension pattern, Fig. 8. The Bragg peaks related to inter-lamellar distance appeared at low $Q \sim 0.077$, 0.15 and 0.23 nm^{-1}. The second-order Bragg peak clearly appeared after applying a sectional radial

averaging from anisotropic scattering data with a 20° double fan mode. Those peaks correspond to a distance of 82 nm and associate to the separation between two polymer chain blocks: polystyrene and poly(ethylene-*alt*-propylene).

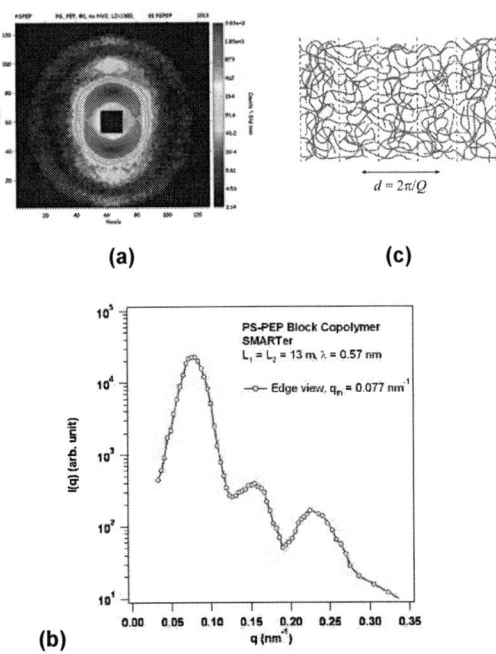

(a) **(c)**

$d = 2\pi/Q$

(b)

Figure 8. (a) Two-dimensional anisotropic scattering pattern from block copolymer PS-PEP with a neutron wavelength of 0.57 nm and sample-to-detector distance of 13 m showing the orientation of lamellar structure in the sample (b) One-dimensional I versus Q profile after appropriate corrections, sectorial radial averaging with a 20° double fan mode, showing three Bragg peaks at momentum transfer Q = 0.077, 0.15, 0.23 nm^{-1}. (c) A schematic lamellar structure of PS-PEP diblock copolymers with (00l) planes diffraction and the inter-lamellar plane distance of 82 nm.

Biomacromolecules and Protein

The *n*-dodecyl-β-D-maltoside (β-DMS) amphi-philic molecule consists of a polar head group (disaccharides) and a hydrophobic chain. It exhibits unique properties during spontaneously forming a micellar structure in aqueous solution and has played significant roles in isolation and purification of protein. Detail inner structures of the β-DMS micelle can only be explored by applying a contrast variation method on the H_2O/D_2O ratio as an aqueous solution. The method is surely useful to separate each part in the micelle structures to identify the part of micelle that interacts since a protein molecule where a study on protein structures and interactions which is fundamental to many aspects of biology such as metabolism is a major challenge in protein biochemistry.

(a)

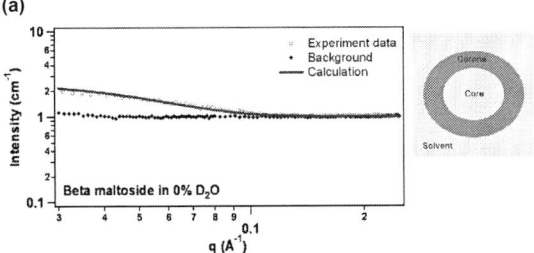

(b)

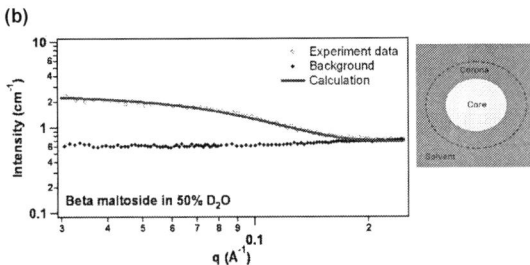

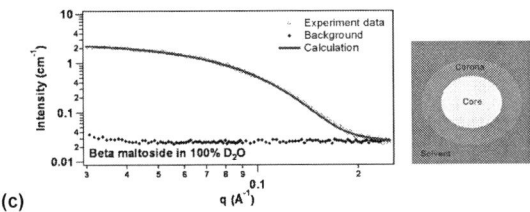

(c)

Figure 9. The radial averaging of corrected SANS scattering profiles from n-dodecyl-β-D-maltoside micellar (β-DMS) solutions in (a) 0%, (b) 50% and (c) 100% concentration of D_2O. The experimental data are fitted by the theoretical calculation model, an oblate core-shell structure model.

Three solutions of β-DMS were prepared in the 0% and 50% D_2O solution where their neutron scattering length density are respectively matched to hydrophobic tail and head group. While in 100% D_2O over all of the micelle structure, then over all shape and size can be distinguished. An oblate-core shell structure model was applied on the all scattering intensities data to reveal detail inner structure, Fig. 9. This structure model calculation is definitely appropriate with the micelle structure where the major-axis and minor-axis respectively are 3.4 nm and 1.8 nm. The thickness of the corona layer is 0.4 – 0.9 nm and the aggregation number is 121 – 130.

A contrast variation method, H_2O/D_2O mixture, is very important in SANS experiment to reveal detail structure on biological materials such as globular protein, protein coat and double-strain RNA of viruses or pore-forming toxin, protein-DNA junction, penetrated-

protein in bilayer membrane, etc. and their interactions in their native solution in the length scale of 1 – 100 nm. Therefore, the 3-dimensional structure and dynamics, e.g. the interaction of protein-surfactant, protein-protein, protein-nucleic acid and its self-assembly near their physiological environment of which the key to understanding the biomacromolecules mechanism can be certainly accomplished without crystallizing the sample and damage by neutron radiation. Seeing as the SANS technique emerges as a new and advanced technique in probing biological materials structure, a research proposal for studying folding-unfolding mechanism and their interactions of globular proteins in solution has been granted by TWAS[15].

CONCLUSION

We have described SMARTer, a 36 m SANS BATAN spectrometer for investigations in materials science and on biological materials in the nanometer length scale. All results above showed that there is no doubt, a nuclear technique i.e. SANS technique becomes one of the advanced techniques for solving the molecular morphology (structural properties-static) as well as magnetic structure in the relevant length scales, i.e. 1 – 100 nm of a wide range in solid state physics and chemistry, polymers, ceramics, alloys, as well as biological materials. It is a good sign for Indonesian as well as regional scientists and researchers to really utilize this instrument for their research in nanotechnology and bio-nanotechnology.

ACKNOWLEDGMENT

The author acknowledges all the staff members of neutron scattering laboratory (NSL) - BATAN in Serpong for helps and advices. The authors are delighted to thank IAEA for IAEA Expert Mission under INS/0/017 of the Human Resource Development and Nuclear Technology Support project in 2005, 2006, and 2007. This work was supported in part by National Nuclear Energy Agency (BATAN) for financial year of 2005 - 2007 on the Neutron Beam Utilization of G. A. Siwabessy Reactor for Materials Science Researches project. EGR Putra also acknowledges the support from IUCr and IAEA for attending the ICNX2009 in Kuala Lumpur, Malaysia, June 29 - July 1, 2009.

REFERENCES

1. Marsongkohadi, Ridwan, *Neutron News* **7**, 2 (1996)

2. E.G.R. Putra (2004). "*Pemberdayaan Berkas Neutron G. A. Siwabessy untuk Penelitian Bahan*", in-house research proposal, BATAN 2005 – 2009.

3. E.G.R. Putra, A. Ikram, E. Santoso, B. Bharoto, *J. Appl. Cryst.* **40**, s447 – s452 (2007)

4. E.G.R. Putra, Bharoto, E. Santoso, A. Ikram, *J. Nucl. Instrum. Method Phys. Res.* **A 600**, 198 – 202 (2009)

5. E.G.R. Putra, A. Ikram, Bharoto, E. Santoso, *J. Nucl. Related Technologies* **5 (2)**, 57 – 65 (2008)

6. E.G.R. Putra, Bharoto, E. Santoso, Y.A. Mulyana, *Neutron News* **18(1)**, 23 – 29 (2007)

7. E.G.R. Putra, A. Ikram, *Indonesian J. Chem.* **6 (2)**, 117 – 120 (2006)

8. E.G.R. Putra, A. Ikram, J. Kohlbrecher, *Pramana J. Phys.* **71(5)**, 1045 – 1050 (2008)

9. E.G.R. Putra, A. Ikram, *J. Nucl. Related Technologies* **5(1)**, 45 – 52 (2008).

10. M.H.A.R.M. Ahmad, A.A. Mohamed, A. Ibrahim, C.S. Mahmood, E.G.R. Putra, R. Jamro, R. Kasim, M.R.M. Zin, *J. Phys. Chem. Solids* **68**, 2349 – 2352 (2007)

11. A. Patriati, E.G.R. Putra, "*Ellipsoid to Worm-like Micelle Structures Transition Revealed by a Small-Angle Neutron Scattering Technique*", abstract ICMNS 2008, ITB, Bandung, 28 - 30 October 2008.

12. E.G.R. Putra, B.S. Seong, E. Shin, A. Ikram, S.A. Ani, Darminto,"*Fractal structures on Fe_3O_4 ferrofluids: A small-angle neutron scattering study*", abstract SAS2009, Oxford, UK, 13 – 18 September 2009; S.A. Ani, Darminto, E.G.R Putra, *Particle Size Distribution Models of Small Angle Neutron Scattering Pattern on Ferrofluids*, abstract ICNX2009, Kuala Lumpur, Malaysia, June 29 – July 1, 2009.

13. E.G.R. Putra, et al., *American Institute of Physics (AIP): Conference Proceeding 989*, April 2008, 130 – 133 (2008).

14. E.G.R. Putra, A . Ikram, *Neutron News* **19(4)**, 28 – 33 (2008)

15. "*Small-angle Neutron Scattering (SANS) Studies on Biological Macromolecules*", TWAS Research Grant, 30 April 2009. Principle investigator: Edy Giri Rachman Putra.

Micro-focused Small Angle Neutron Scattering and Imaging for Science and Engineering Using RTP – A Preliminary Study

Abdul Aziz Mohamed, Megat Harun Al Rashid Megat Ahmad, Faridah Md Idris, Azraf Azman, Rafhayudi Jamro, Mohd Rizal Mamat @ Ibrahim, Anwar Abdul Rahman

Agensi Nuklear Malaysia, Komplek PUSPATI, Bangi, 43000 Kajang, Selangor, Malaysia;
e-mail: aziz_mohd@nuclearmalaysia.gov.my

Abstract

Malaysian Nuclear Agency's (Nuclear Malaysia) Small Angle Neutron Scattering (SANS) facility - (MYSANS) - is utilizing low flux of thermal neutron at the agency's 1 MW TRIGA reactor. As the design nature of the 8m SANS facility can allow object resolution in the range between 5 and 80 nm to be obtained. It can be used to study alloys, ceramics and polymers in certain area of problems that relate to samples containing strong scatterers or contrast. The current SANS system at Malaysian Nuclear Agency is only capable to measure Q in limited range with a PSD (128x128) fixed at 4m from the sample. The existing reactor hall that incorporate this MYSANS facility has a layout that prohibits the rebuilding of MYSANS therefore the position between the wavelength selector (HOPG) and sample and the PSD cannot be increased for wider Q range. The flux of the neutron at current sample holder is very low which around 10^3 n /cm^2 / sec. Thus it is important to rebuild the MYSANS to maximize the utilization of neutron. Over the years, the facility has undergone maintenance and some changes have been made. Modification on secondary shutter and control has been carried out to improve the safety level of the instrument. A compact micro-focus SANS method can suit this objective together with an improve cryostat system. This paper will explain some design concept and approaches in achieving higher flux and the modification needs to establish the micro-focused SANS.

Keywords: SANS, TRIGA reactor, micro-focus SANS, nano-materials

INTRODUCTION

Since its introduction in the early 1970s, SANS has made significant contribution to understand the characteristics of many materials especially in the field of condensed and soft matters. It is very useful for probing nano- and mesoscale sample inhomogeneities .

At the Malaysian Nuclear Agency (Nuclear Malaysia), a SANS spectrometer (known as mySANS) is built at the agency's 1 MW TRIGA MARK II reactor (known as RTP). Construction started in 1995 and completed two years later. mySANS utilizes low flux of thermal neutron. Design nature of the instrument allows object resolution in the range between 5 and 80 nm to be obtained (Radiman *et al*, 1998). It can be used to study alloys, ceramics and polymers and to certain area of problems wherein the samples contain strong scatterers or contrast. The existing reactor hall layout does not allows the SANS instrument to use neutron beam from the tangential beamport as source because of space limitation. Instead a radial beamport and crystal monochromator were used and this permits only an 8-meter SANS instrument to be built. The Q range is thus narrow as the distance between the sample and the detector positions cannot be adjusted to obtain much wider Q range. The flux of the neutron at current sample position is very low which is about 10^3 n/cm^2/sec (Sufi *et al*., 1997). To obtain a better Q range and reasonable flux density, mySANS needs to be rebuilt to maximize the utilization of neutron. A neutron micro-focus method can be considered to achieve this objective. This paper will highlight the current mySANS design and the new concepts and approaches to realize higher flux and better Q range for our small research reactor, i.e. the RTP. This work is part of the IAEA Coordinated Research P project (IAEA No.14286/R0).

CURRENT STATUS OF mySANS

Figure 1 shows the cross-sectional diagram of TRIGA PUSPATI reactor beam ports. MySANS is constructed at radial beamport no 4.

CP1202, *Neutron and X-Ray Scattering in Advancing Materials Research: International Conference – 2009*
edited by A. Saat, H. A. Kassim, M. H. H. Jumali, J. M. Saleh, M. R. Othman, A. Ibrahim, F. M. Idris, and M. H. A.-R. M. Ahmad
© 2009 American Institute of Physics 978-0-7354-0739-8/09/$25.00

mySANS

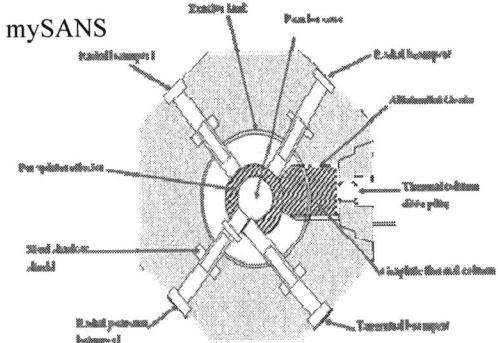

Figure 1 MySANS facility location at TRIGA PUSPATI reactor

The instrument consists in sequence starting from the neutron source: the course collimator, filter, monochromator system, collimation, sample, and detector system. More details of these instrumental parts and the whole SANS system reported by Sufi *et al.*, 1997. Description of the system that relates to the neutron flux and the minimum Q obtainable will be briefly discussed here.

The coarse collimator is placed in the beam port of 15.2 cm diameter and length of 200 cm. It is made from aluminium tube of diameter 11 cm and length of 150 cm. This serves to collimate the incoming neutron beam to within a divergence angle just sufficient to illuminate the monochromator. The distance from source to monochromator is 261.5 cm and this is shown in Figure 2.

The Be-filter is made of 16 polycrystalline Be bars with dimension of 3 cm x 3 cm x 15 cm (width x height x length) and assembled in a 4 x 4 rectangular array made from 1 mm thick Cd (Figure 3). This assembly is cryogenically cooled at 77K using liquid nitrogen. The distance between aluminum windows of the cryostat is 40 cm. The distance from the outlet cryostat window to the center of the monochromator is about 21.5 cm. The filter can effectively filter neutron with wavelength, $\lambda < 4\text{Å}$ (Sufi *et al.*, 1997; Wahba, 2002).

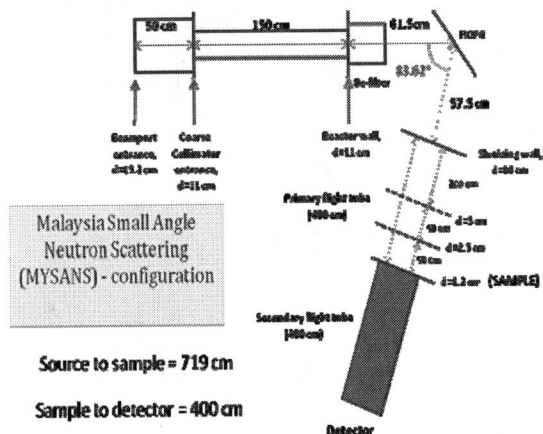

Figure 2 A schematic diagram of mySANS components and configuration

Monochromatic neutron centered at $\lambda = 5\text{Å}$ is obtained through (002) Bragg reflection of the neutron by highly oriented pyrolitic graphite (HOPG). The take-off angle, θ is 96.38°. The monochromator is composed of three layers of HOPG crystals (Fig. 4). Each layer is composed of six crystals (with area of 3 cm x 5.5 cm each) with a mosaic spread of 0.8° and misaligned at 0.4°. The total monochromator height is 9 cm and length is 11 cm. The thickness of each of HOPG crystal is 3 mm. The setup is mounted on a remote-controlled goniometer and a turntable.

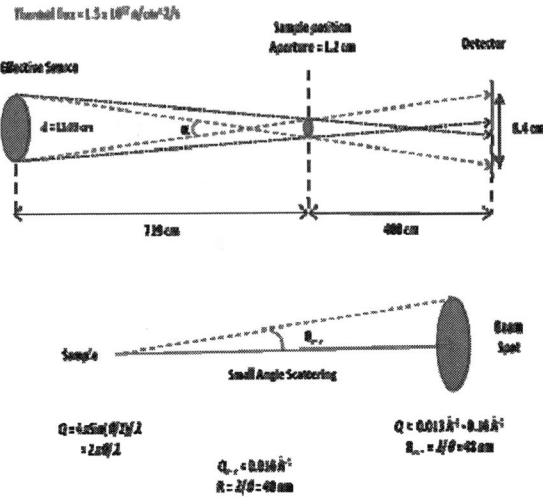

Figure 3 mySANS optical configuration

The collimation system or the primary flight tube is made of evacuated shielded steel tube, the total length is 400 cm and the diameter is 10 cm. The collimator is segmented by three tubes (one of 200 cm length and two of 100 cm length). At each joint between the tubes are the cadmium apertures with reducing sizes of 5 cm, 2.5 cm and 1.2 cm. A sample can be placed at the exit of the collimator. This primary flight tube collimation system allows a source divergence of 18.2 mrad (1.04°) to be transmitted (and scattered) at sample position and a calculated beamspot size of about 8.4 cm.

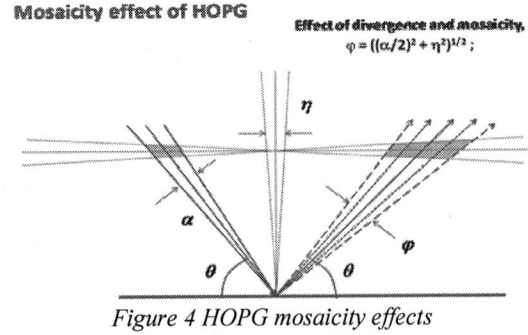

Figure 4 HOPG mosaicity effects

The secondary flight tube is made of evacuated shielded steel tube, of 100 cm in diameter and consists of two segments of 50 cm length, one of 100 cm and one of 200 cm to enable length adjustment. A neutron position-sensitive detector (PSD) is placed at the end section of the flight tube to measure the incoming scattered neutrons from samples.

The position-sensitive detector is a ^{3}He counter, manufactured by Risø Laboratories, Denmark which consists of 128 x 128 pixels with spatial resolution of about 0.5 cm size. The setup of this SANS system allows accessible Q range from about 0.013 Å^{-1} to 0.1 Å^{-1} at 400 cm sample to detector position. Higher Q range from about 0.4 Å^{-1} can be achieved by operating the SANS with sample to detector distance of 100 cm.

The use of HOPG allows the focusing of large divergence of the neutron source to the sample position. Nevertheless, because of the collimation system, only part of this focused neutron can arrives at the sample position

The $\Delta\lambda/\lambda$ of neutron beam reflected by the HOPG is much broader than from what is allowed by the collimation system to the sample position. At the same time, because of flux reduction in proportion of distance, the height of neutron intensity at sample position is about hundred times lower when compared with the neutron current at rightly after HOPG.

From here on, it is obvious that the collimation system used which is typical for conventional SANS system may not be suitable for SANS system at low flux reactor. If optimum condition is set for this collimation system, i.e. the size (diameter) of aperture at the beginning of the collimation system twice that of the size of aperture just before sample, the flux at sample is definitely becoming very low.

Thus, it is apparent that to increase the neutron current at sample position without forfeiting the Q minimum (and Q range) is to have a new focusing system. This new focusing system, apart from focusing the neutron, would also cause a gain of neutron current at sample position. Other than that, neutron current gain can also be achieved by increasing the $\Delta\lambda$ reflected by the monochromating system.

POSSIBLE CONFIGURATIONS FOR MYSANS WITH MICRO-FOCUSED NEUTRON BEAM

Focusing the neutron at a distance from the source can be achieved in a variety of ways as stated previously. It is our interest in this project to apply a system in which both focusing and increase of $\Delta\lambda$ can be achieved simultaneously, resulting both the increase of neutron current at sample position and the decrease Q minimum.

Beam focusing can be achieved in a variety of ways, each may suited to the type and structure of the source used (continuous source from research reactor or pulsed source from spallation method). Among the tested techniques include multi-hole converging collimators (Thiyagarajan et al., 1997), refractive lens (Daymond and Johnson, 2002), double focusing with bent perfect crystal (Strunz et al., 1997), ellipsoidal supermirror

(Furusaka et al., 2009), capillary optics (Kumakhov, 2004) and Kirkpatrick-Baez mirrors (Ice et al., 2005; Ice et al., 2006). Apart from that, there are several conceptual designs proposed, among them are lobster eye (Šaroun and Kulda, 2006) and torroidal mirror and their derivatives. There are also others that have not been considered but possible to be used. These include Schwarzschild objective, ellipsoidal ring mirror and zone plate (Bertolo et al., 2002).

For the construction of the new SANS instrument that utilizes focusing method, we see two possibilities. One would use horizontally and vertically bent perfect crystals to focus the beam to the detector. By using strongly and thick bent perfect crystal, significant $\Delta\lambda$ (that translate to higher integrated flux) can be reflected but this requires a crystal system that would not crack when bending force is applied to it. This can be tested by using layers of monochromating wafer crystals. The strong bending of the crystal system would results in much higher effective mosaicity, from which more neutron can be reflected to the sample. The other possibility is by using the Kirk-Patrickbaez mirror configuration. In this system, monochromating supermirrors are used to focus a neutron beam and to reflect $\Delta\lambda$ that can give higher integrated flux to the sample position. Figure 5 shows the schematic diagram of the possible setup.

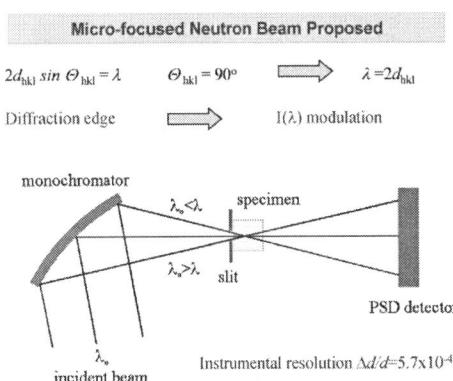

Figure 5 Possible focusing set-up for mySANS

SUMMARY

It is important to rebuild the MYSANS to maximize the utilization of neutron. Over the years, the facility has undergone maintenance and some changes have been made in order to improve the performance of the facility. Modification on secondary shutter and control has been carried out to improve the safety level of the instrument. A compact micro-focus SANS method can be implemented to increase the flux at the sample position together with an improve cryostat system. Some design concept and approaches in achieving higher flux and the modification have been studied in the needs to establish the micro-focused SANS.

ACKNOWLEDGMENT

The authors would like to thank to Malaysia Ministry of Science Technology and Innovation (MOSTI), Malaysian Nuclear Agency and International Atomic Energy Agency (IAEA) in supporting this work.

REFERENCES

Allen and Berk, Journal Applied Crystallography, 27, pp878, 1994. Barker, J.G., Glinka, C.J., Moyer, J.J., Kim, M.H., Drews, A.R. and Agamalian, M. (2005), Design and performance of a thermal-neutron double-crystal diffractometer for USANS at NIST, *J. Appl. Cryst.*, **38**, 1004-1011.

Bertolo, M., Gregoratti, L., Heun, S., Kaulich, B. and Kiskinova, M. (2002), *Science, technology and education of microscopy: An overview*, Vol. II, (pp. 776-686) A. Mendez-Vilas, (Ed.). Madrid, Spain: Formatex.

Brûlet, A., Thévenot, V., Lairez, D., Lecommandaux, S., Agut, W., Armes, S.P., Du and J., Désert, S. (2008), Toward a new lower limit for the minimum scattering vector on the very small angle neutron scattering spectrometer at Laboratoire Léon Brillouin, *J. Appl. Cryst.*, **41**, 161-166.

Crawford, R.K. and Carpenter, J.M. (1988), Tailoring beams for small-angle neutron diffractometers, *J. Appl. Cryst.*, **21**, 589-601.

Daymond, M.R. and Johnson, M.W. (2002), An experimental test of a neutron silicon lens, *Nucl. Inst. Meth. Phys. Res. A*, **485**, 606-614

Désert, S., Thévenot, V., Oberdisse, J. and Brûlet, A., (2007), The new very-small-angle neutron scattering spectrometer at Laboratoire Léon Brillouin, *J. Appl. Cryst.*, **40**, s471- s473

Furusaka, M., Fumiyuki, F., Homma, A., Kiyanagi, Y., Kamiyama, T., Hiraga, F., Kamada, K., Tanabe, K., Koyama, K., Hirota, K., Ikeda, K., Ikeda, S., Naito, S., Satoh, S., Shimizu, H., Sugiyama, M., Satoh, T., Mikula, P., Yoshizawa, H., Shibayama, M., Endoh, H., Kawamura, Y., Asami, T., Takahashi, H., Fujita, K., Kaneko, J., *Development of Mini-Focusing Small-Angle Neutron Scattering (mfSANS) Instruments*, Report IAEA Vienna, Proceedings of the meeting July 23-25, 2007, Hokkaido University, Japan

G.D. Wignell, Encyclopedia of Polymer Science and Engineering, John Wiley and sons, New York, 1987.

Hammouda, B. (2008), Probing nanosacle structures – the SANS toolbox

Hayter, J.B. and Mook, H.A. (1989), Discrete thin-film multilayer design for X-ray and neutron supermirrors, *J. Appl. Cryst.*, **22**, 35-41

IAEA Research Contract Report No.14286/R0 - Improved Production And Utilization Of Short-Pulsed, Cold Neutron Sat Low Medium Energy Spallation Neutron Sources: Development And Practical Small Angle Neutron Scattering Applications - (Development Of Micro-Focus Sans), 2007.

Ice, G.E., Hubbard, C.R., Larson, B.C., Pang, J.W.L., Budai, J.D., Spooner, S., Vogel, S.C., Bogge, R.B., Fox, J.H. and Donaberger, R.L. (2006), High-performance Kirpatrick-Baez supermirrors for neutron milli- and micro-beams, *Mat. Sci. Eng. A*, **437**, 120-125

Ice, G.E., Hubbard, C.R., Larson, B.C., Pang, J.W.L., Budai, J.D., Spooner, S. and Vogel, S.C. (2005), Kirpatrick-Baez microfocusing optics for thermal neutrons, *Nucl. Inst. Meth. Phys. Res. A*, **537**, 312-320

J.B. Hayfer, Physics of Amphiles, Micelles, Vesicles and Microemulsion, North Holland, Amsterdam, 1985.

Jach, T., Bakulin, A.S., Durbin, S.M., Pedulla, J. and Macrander, A. (2006), Variable magnification with Kirkpatrick-Baez optics for synchrotron x-ray microscopy, *J. Res. Natl. Inst. Stand. Technol.*, **111**, 219-225

Kumakhov, M.A. (2004), Neutron capillary optics: status and perspectives, *Nucl. Inst. Meth. Phys. Res. A*, 529, 69–72

Mazumder et al, Journal Applied Crystallography, 26 pp357, 1993.

Mazumder et al, Journal Physics Condensed Materials, 7, pp9737, 1993.

Moore, A.W., Popovici, M. and Stoica, A.D. (2000), Neutron reflectivity and lattice spacing spread of pyrolitic graphite, *Phys. B*, **276-278**, 858-859

Radiman S, Mahmood Z.U, and Sufi M A M, Malaysian Journal of Analytical Sciences, V4, 1, pp105-108, 1998.

Rehm, C. and Agamalian, M. (2001), Flux gain for next-generation neutron-scattering instruments resulting from improved supermirror performance, Neutron Optics, James L. Wood, Ian S. Anderson, Editors, Proceedings of SPIE, **4509**, 56-65

Safety Analysis Report RTP 2006, Agensi Nuklear Malaysia

Šaroun, J. and Kulda, J. (2006), MC ray-tracing optimization of lobster-eye focusing devies with RESTRAX, *Phys. B*, **385-386**, 1250-125

Strunz, P., Šaroun, J., Mikula, P., Lukáš, P. and Eichhorn, F. (1997), Double-Bent-Crystal Small-Angle Neutron Scattering Setting and its Applications, *J. Appl. Cryst.*, **30**, 844-848

Sufi M A M, Radiman S, Wiedenmann A and Mortensen K, Journal of Applied Crytallography, 30, pp884-888, 1997.

Sufi, M.A.M., Radiman, S., Wiedenmann, A. and Mortensen, K. (1997), Performance of a new small-angle neutron scattering instrument at the Malaysian TRIGA reactor, *J. Appl. Cryst.*, **30**, 884-888.

Thiyagarajan, P., Crawford, R.K. and Mildner, D.F.R. (1998), Neutron transmission of a single-crystal MgO filter, *J. Appl. Cryst.*, **31**, 841-844.

Thiyagarajan, P., Epperson, J.E., Crawford, R.K., Carpenter, J.M., Klippert, T.E. and Wozniak, D.G. (1997), The Time-of-Flight Small-Angle Neutron Diffractometer (SAD) at IPNS, Argonne National Laboratory, *J. Appl. Cryst.*, **30**, 280-293

Villa, M., Baron, M., Hainbuchner, M., Jericha, E., Leiner, V., Schwahn, D., Seidl, E., Stahn, J. and Rauch, H. (2003), Optimisation of a crystal design for a Bonse-Hart camera, *J. Appl. Cryst.*, **36**, 769-773.

Wahba, M. (2002), On the use of Beryllium as thermal neutron filter, *Egypt J. Sol.*, **25(2)**, 215-227.

Zsigmond, G., Mezei, F., Wechsler, D. and Streffer, F. (2001), Monte Carlo simulation of crystal monochromators/analysers – Application for the crystal-analyser neutron spectrometer IRIS, *Nucl. Inst. Meth. Phys. Res. A*, **457**, 299-308

ORAL PAPERS

EFFECT OF GRAIN SIZE AND CONTROLLED ATMOSPHERES ON THE THERMAL STABILITY OF ALUMINIUM TITANATE

I.M. Low and Z. Oo

Department of Imaging & Applied Physics, Curtin University of Technology, GPO Box U1987, Perth, WA 6845, Australia

ABSTRACT

Aluminium titanate (Al_2TiO_5) is an excellent refractory and thermal shock resistant material due to its relatively low thermal expansion coefficient and high melting point. However, Al_2TiO_5 is only thermodynamically stable above 1280°C and undergoes a eutectoid-like decomposition to α-Al_2O_3 and TiO_2 (rutile) at the temperature range of 900-1280°C. Hitherto, the effect of grain size and atmosphere on the kinetics of decomposition is poorly understood but experimental evidences suggest a nucleation and growth controlled process. In this paper, we describe the role of grain size and controlled atmospheres on the thermal stability of Al_2TiO_5. In particular, the effects of grain size and oxygen partial pressure on the rate of isothermal decomposition of Al_2TiO_5 at 1100°C have been investigated. Results show that the thermal stability of Al_2TiO_5 increases as the grain size and oxygen partial pressure increases. However, both the on-set temperature nor the temperature range of Al_2TiO_5 thermal decomposition are not affected by the variation of oxygen partial pressure present in the furnace atmosphere.

KEYWORDS: Thermal stability, Al_2TiO_5, grain size, atmosphere, neutron diffraction, decomposition.

INTRODUCTION

Aluminium titanate (Al_2TiO_5) is an excellent refractory and thermal shock resistant material due to its relatively low thermal expansion coefficient ($\sim 1 \times 10^{-6}$ °C^{-1}) and high melting point (1860°C). It is one of several materials which is isomorphous with the mineral pseudobrookite (Fe_2TiO_5) [1,2]. In this structure, each Al^{3+} or Ti^{4+} cation is surrounded by six oxygen ions forming distorted oxygen octahedra. These AlO_6 or TiO_6 octahedra form (001) oriented double chains weakly bonded by shared edges. This structural feature is responsible for the strong thermal expansion anisotropy which generates localised internal stresses to cause severe microcracking. Although this microcracking weakens the material, it imparts a desirable low thermal expansion coefficient and an excellent thermal shock resistance.

In addition, Al_2TiO_5 is only thermodynamically stable above 1280°C and undergoes a eutectoid-like decomposition to α-Al_2O_3 and TiO_2 (rutile) within the temperature range 900-1280°C [3-7]. This undesirable decomposition has limited its wider application. Hitherto, the mechanisms of decomposition are poorly understood but experimental evidences suggest a nucleation and growth controlled process. It is generally agreed that the decomposition rate peaks at 1100°C and that residual alumina particles might act as preferred nucleation sites for the decomposition [3]. The impact of this thermal instability can be improved through the use of various stabilisers such as MgO, Fe_2O_3 and SiO_2.

In recent studies by Low and co-workers [8-12], both grain size and the oxygen partial pressure of the furnace atmosphere has been observed to have a profound influence on the thermal stability of Al_2TiO_5. For instance, the decomposition rate of Al_2TiO_5 at 1100°C is significantly enhanced in vacuum (10^{-4} torr) or argon where >90% of Al_2TiO_5 decomposed after only 4 h soaking when compared to less than 10% in atmospheric air [8,12]. This suggests that the process of decomposition of Al_2TiO_5 is susceptible to environmental attack or sensitive to the variations in the oxygen partial pressure during ageing. The stark contrast in the mechanism of phase decomposition is believed to

CP1202, *Neutron and X-Ray Scattering in Advancing Materials Research: International Conference – 2009*
edited by A. Saat, H. A. Kassim, M. H. H. Jumali, J. M. Saleh, M. R. Othman, A. Ibrahim, F. M. Idris, and M. H. A.-R. M. Ahmad
© 2009 American Institute of Physics 978-0-7354-0739-8/09/$25.00

arise from the vast differences in the oxygen partial pressure that exists between air and vacuum.. In addition, it is still unclear whether the variation of oxygen partial pressure has any influence on the range and on-set of decomposition temperature of Al_2TiO_5.

A similar phenomenon, although less profound, has been observed for Al_2TiO_5 with a distinct difference in grain size. However, it is unclear whether there is a critical grain size associated with this phenomenon. The reason for this grain-size effect is unclear at this stage although it may be closely related to its greater tendency for microcracking as the grain size increases. The microcracking phenomenon is closely related to the material microstructure and thermal expansion anisotropy [13-15]. Below a critical grain size, the elastic energy of the system is insufficient to nucleate microcracks during cooling and thus causing no degradation to the mechanical strength. The density of microcracks increases drastically with grain size once the critical value is exceeded.

In this paper, we present results on the effect of grain size and controlled atmospheres on the isothermal stability of Al_2TiO_5 at 1100°C as well as its decomposition behaviour in the temperature range 20-1400°C. The temperature-dependent thermal stability and isothermal decomposition of Al_2TiO_5 have been dynamically monitored and characterized using neutron diffraction to study the structural changes occurring during phase decomposition in real time.

EXPERIMENTAL METHODS

Sample preparation

The starting powders used for the synthesis of Al_2TiO_5 (AT) consisted of high purity commercial alumina (99.9% Al_2O_3) and rutile (99.5% TiO_2). One mole of alumina powder and one mole of rutile powder were initially mixed using a mortar and pestle. The powder mixture was then wet mixed in ethanol using a Turbula mixer for 2.0 h. The slurry was then dried in a ventilated oven at 100°C for 24 h. The dried powder was uniaxially-pressed in a steel die at 150 MPa to form cylindrical bars of length 20 mm and diameter 15 mm, followed by sintering in a air-ventilated furnace at (a) 1400°C in air for 1 h to achieve a fine-grained microstructure (~1-3μm); (b) 1500°C in air for 2 h to achieve a medium-grained microstructure (~5-10μm), and (c) 1600°C in air for 4 h to achieve coarse-grained (~30-50μm) Al_2TiO_5.

Neutron diffraction (ND)

A medium resolution powder diffractometer (MRPD) located at the Australian Nuclear Science and Technology Organisation (ANSTO) in Lucas Heights, NSW was used for neutron diffraction study of the thermal stability of Al_2TiO_5. The effect of grain size on the isothermal stability of Al_2TiO_5 was dynamically monitored at 1100°C in air atmosphere for up to 12. Medium-grained Al_2TiO_5 samples were used for the study of isothermal stability in different atmospheres, namely air, argon (99.99% purity) and 50% argon - 50% oxygen. In addition, the influence of atmosphere on the temperature range and the onset of thermal decomposition of Al_2TiO_5 in the temperature range 20 – 1400°C was investigated. The operation conditions of the MRPD were λ = 1.667 Å, 2θ range = 4-138°, step size = 0.1°, counting time ~40-50 s/step, monochromator of 8 Ge crystals (115 reflection), and 32 ^{3}He detectors 4° apart. The relative abundance of phases present was computed using the Rietveld method. The models used to calculate the phase abundance for MRPD were Maslen et al. [16] for alumina, Epicier et al. [17] for Al_2TiO_5, and Howard et al. [18] for rutile. The software used to analyse the data was Rietica 1.7.7.

RESULTS AND DISCUSSION

Effect of Grain Size

Figure 1 shows the effect of grain size on the isothermal stability of in air at 1100°C. Coarse-grained Al_2TiO_5 exhibits a slowest rate of thermal decomposition when compared to its medium-grained and fine-grained counterparts. To the best of our knowledge, this is the first time that grain size has been shown to affect the propensity of thermal degradation in Al_2TiO_5. However, it is unclear whether there is a critical grain size associated with this phenomenon. The reason for this grain-size effect is unclear at this stage although it may be closely related to its greater tendency for microcracking as the grain size increases. The microcracking phenomenon is closely related to the material microstructure and thermal expansion anisotropy [13-15]. Below a critical grain size, the elastic energy of the system is insufficient to nucleate microcracks during cooling and thus causing no degradation to the mechanical strength. The density of microcracks increases drastically with grain size once the critical value is exceeded.

Fig. 2(a) shows the typical microstructure of as-sintered coarse-grained AT prior to isothermal ageing where the presence of fine microcracks within certain grains is clearly evident. The formation of these microcracks can be attributed to the pronounced thermal expansion anisotropy of AT during cooling from an elevated temperature. The presence of these microcracks is believed to impart a low fracture strength but high thermal shock resistance to AT. Following isothermal-ageing in air at 1000°C for 14 h, both needle-like and angular particles could be seen to form on the surface of Al_2TiO_5 grains [19]. Based on the energy dispersive spectrocopy (EDS) results [19], these nano-sized particles were identified as surface by-products (ie.

Al_2O_3 and TiO_2) of thermally decomposed AT. This may indicate that the initial nucleation process of thermal decomposition of AT is surface-initiated and the growth kinetics are both temperature and time dependent.

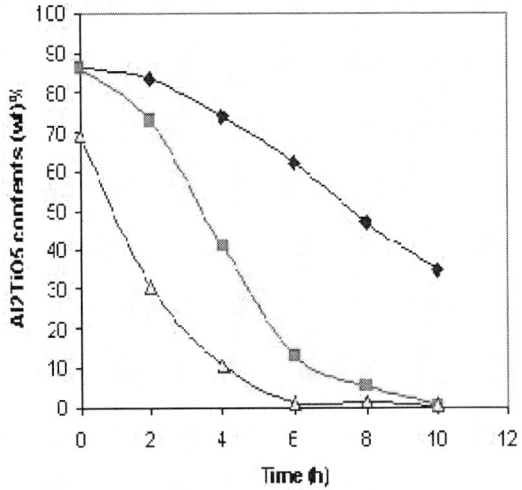

Fig. 1: Effect of Al_2TiO_5 grain size on the propensity of isothermal decomposition at 1100°C in air. [Legend: Coarse (♦); Medium (■); Fine (▲)]

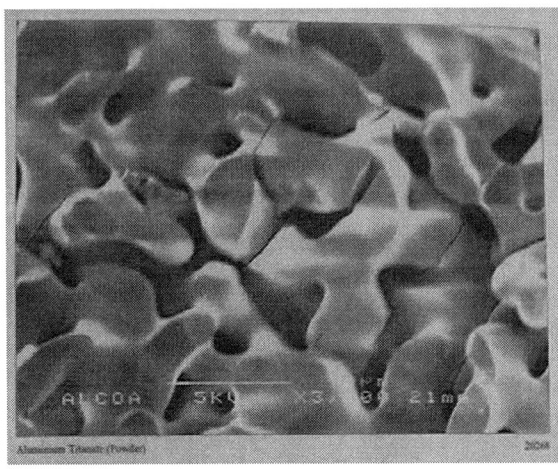

Fig.2: Scanning electron micrograph of as-sintered medium-grained AT. Note the presence of microcracks within certain grains.

Effect of Controlled Atmospheres

Figure 3 shows the isothermal stability of Al_2TiO_5 at 1100°C in air for 12 h duration. The sample remained fairly stable with no apparent phase decomposition for up to 5 h. Further ageing caused only ~5% decomposition. In contrast, substantial phase decomposition was observed when Al_2TiO_5 was aged in an argon atmosphere (Fig. 4) where more than 98% of the sample decomposed to form corundum (Al_2O_3) and

rutile (TiO_2) after only 5 h of ageing. When the ageing atmosphere was changed to 50% argon and 50% oxygen (Fig. 5), the decomposition rate was considerably reduced compared with the rate for 100% argon but more substantial than the rate for air.

A closer look at the results in Figs. 3-5 suggests that the propensity of phase decomposition of Al_2TiO_5 is dependent on the atmosphere or oxygen partial pressure during isothermal ageing. This implies that the oxygen partial pressure in the atmosphere plays a key role in triggering the thermal instability via oxygen nonstoichiometry changes and/or disordering of cations in Al_2TiO_5 [6]. Indeed, nitrogen atmosphere has also been observed to cause enhanced thermal instability in Al_2TiO_5 [20]. Similar observations have also been observed for the enhanced dissociation of Ti_3SiC_2 in vacuum and argon which can also be attributed to the role of oxygen partial-pressures [21,22].

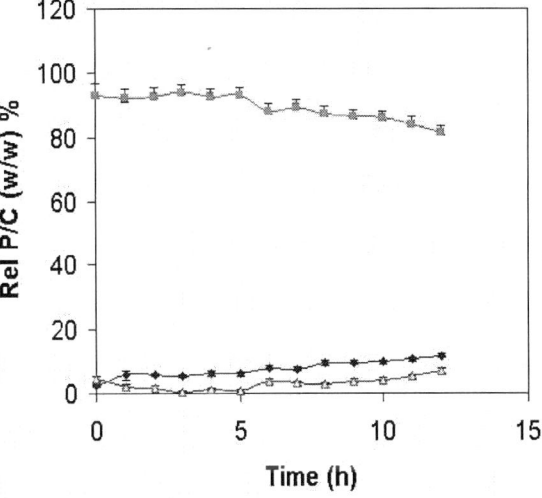

Fig. 3: Isothermal stability of Al_2TiO_5 at 1100°C in air. Errors bars indicate two estimated standard deviations ±2σ. [Legend: ■ = Al_2TiO_5; ♦ = Al_2O_3; Δ = TiO_2]

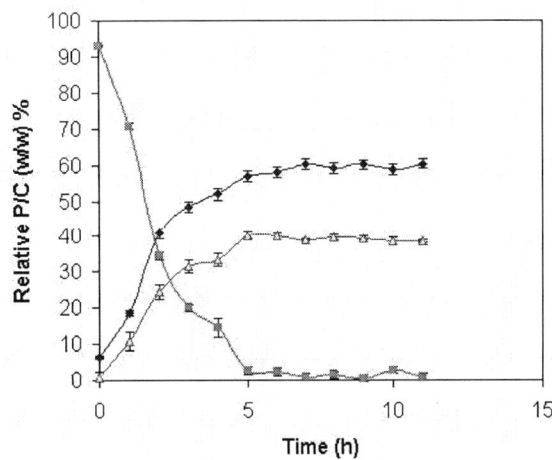

29

Fig. 4: Isothermal stability of Al_2TiO_5 at 1100°C in controlled argon atmosphere. Errors bars indicate two estimated standard deviations ±2σ. [Legend: ■ = Al_2TiO_5; ♦ = Al_2O_3; Δ = TiO_2]

It is postulated that in the presence of very low oxygen partial pressure, the titanium ions in TiO_2 are very susceptible to non-stoichiometry, thus triggering the release of oxygen atoms and the concomitant decomposition process. However, the exact mechanism of enhanced phase decomposition in argon or inert atmosphere remains unclear, especially in relation to the role of oxygen partial pressures in reducing the free energy change for thermal decomposition of Al_2TiO_5. If the oxygen partial pressure is the cause, then the decomposition rate of Al_2TiO_5 should depend on the variation of the oxygen partial pressure as indicated in Figs. 3 - 5. It follows that an increase in the oxygen partial pressure should reduce the rate of decomposition and vice-versa. This further implies that the thermal stability of Al_2TiO_5 will be improved in an atmosphere of 100% oxygen when compared to ageing in air.

The thermal stability of Al_2TiO_5 in the temperature range 20 – 1400°C in argon is shown in Fig. 6. Clearly, Al_2TiO_5 was stable up to ~1100°C and became unstable at between ~1150 - 1300°C. Beyond 1300°C, the thermal decomposition was arrested and the phase stability was restored. This implies that the process of thermal decomposition is reversible or recoverable provided the restricted temperature range of between ~1150 - 1300°C is not transgressed. The implication of this phenomenon is far-reaching whereby it may be possible to restore the decomposed Al_2TiO_5 to its original condition by thermal annealing at >1400°C. Interestingly, the use of either air, argon or oxygen atmosphere did not appear to alter the on-set temperature and the temperature range of thermal decomposition, although the propensity of phase decomposition was dramatically affected [8]. The reason for this phenomenon is unclear at this stage. To the best of our knowledge, this is the first time that the in-situ display of the temperature range for thermal decomposition of Al_2TiO_5 during ageing in controlled atmospheres as shown in Fig. 5 has been reported in the literature.

CONCLUSION

The effects of grain size and controlled furnace atmospheres on the thermal stability of Al_2TiO_5 at 1100°C and in the temperature range 20-1400°C have been dynamically examined by neutron diffraction. The thermal stability of Al_2TiO_5 increases as the grain size and oxygen partial pressure increases. The susceptibility of Al_2TiO_5 to thermal decomposition increases as the oxygen partial pressure of the furnace atmosphere decreases. However, neither the on-set temperature nor the temperature range of thermal decomposition is affected by the variation of oxygen partial pressure present in controlled furnace atmospheres.

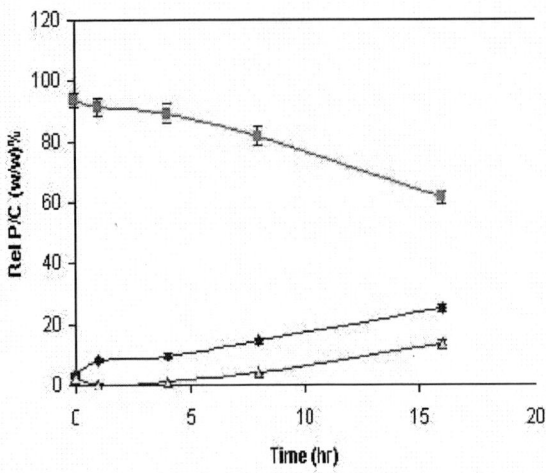

Fig. 5: Isothermal stability of Al_2TiO_5 at 1100°C in the controlled atmosphere of 50% oxygen and 50% argon. Errors bars indicate two estimated standard deviations ±2σ. [Legend: ■ = Al_2TiO_5; ♦ = Al_2O_3; Δ = TiO_2]

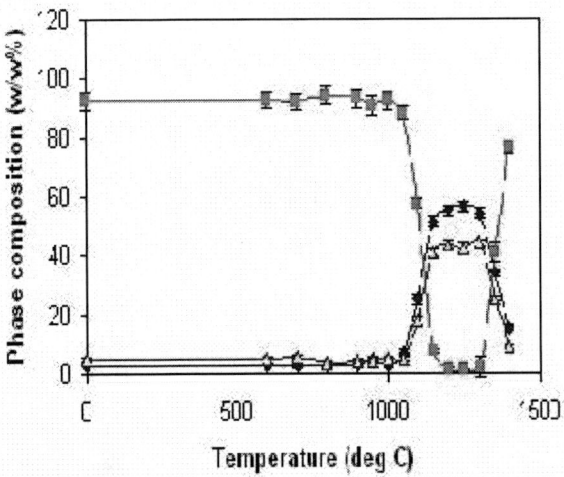

Fig. 6: Thermal stability of Al_2TiO_5 in controlled argon atmosphere over the temperature range 20 – 1400°C. Note the display of pronounced thermal decomposition at ~1150 - 1300°C. Errors bars indicate two estimated standard deviations ±2σ. [Legend: ■ = Al_2TiO_5; ♦ = Al_2O_3; Δ = TiO_2]

ACKNOWLEDGMENTS

This work was supported by funding from the Australian Institute of Nuclear Science and Engineering (AINSE Awards 04/207 & 05/206). We are grateful to our colleague, E/Prof. B. O'Connor, for advice on Rietveld analysis of XRD data. We thank Mr. M. Prior of the Bragg Institute of ANSTO for experimental assistance in the collection of MRPD data. We also thank Mr. A. Jones of Alcoa for assistance in SEM work.

REFERENCES

1. A.E. Austin and C.M. Schwartz, *Acta Cryst.* **6**, 812 (1953).

2. B. Morosin and R.W. Lynch, *Acta Cryst.* B. **28**, 1040 (1972).

3. H.A.J. Thomas and R. Stevens, *Br. Ceram Trans. J.* **88**, 144 (1989).

4. G. Tilloca, *J. Mater. Sci.* **26**, 2809 (1991).

5. E., Kato, K. Daimon and Y. Kobayashi, *J. Am. Ceram. Soc.* **63**, 355 (1980).

6. R.W. Grimes and J. Pilling, *J. Mater. Sci.* **29**, 2245 (1994).

7. M. Ishitsuka, et al., *J. Am. Ceram. Soc.* **70**, 69 (1987).

8. I.M. Low, D. Lawrence and R.I. Smith, *J. Am. Ceram. Soc.* **88**, 2957 (2005).

9. I.M. Low, P. Manurung, R.I. Smith and D. Lawrence, *Key Eng. Mater.* **224-226**, 465 (2002).

10. I.M. Low, Z. Oo & B. O'Connor, *Physica B.* **385-386**, 502 (2006).

11. I.M. Low and R.I. Smith, pp.175-176 in *Proc. of AUSTCERAM 2002* (Eds. I.M. Low & D.N. Phillips), 30 Sept - 4 Oct. 2002, Perth, WA.

12. I.M. Low, D. Lawrence, A. Jones and R.I. Smith, 29th *Int. Cocoa Beach Conference on Advanced Ceramics & Composites:* (D. Zhu & W.M. Kriven, Eds.) CESP. Vol. 26, Issue 3 & 4, pp. 303-310 (2005).

13. K. Hamano, Y. Ohya and Z. Nakagawa, pp. 129-137 in *Int. Journal of High Tech. Ceram.* Elsevier Science Publishers Ltd., UK. (1985).

14. Y. Ohya, Z. Nakagawa and K. Hamano, *J. Am. Ceram. Soc.* **71**, C23 (1988).

15. Y. Ohya and Z. Nakagawa, *J. Am. Ceram. Soc.* **70**, C184 (1987).

16. E.N. Maslen, V.A. Streltsov, N.R. Streltsova, N. Ishizawa and Y. Satow, *Acta Cryst. B.* **49**, 937 (1993).

17. T. Epicier, G. Thomas, H. Wohlfromm and J.S. Moya, *J. Mater. Res.* **6**, 138 (1991).

18. C.J. Howard, T.M. Sabine and F. Dickson, *Acta Cryst. B.* **47**, 462 (1991).

19. A. Jones and I.M. Low, pp.185-186 in *Proc. of AUSTCERAM 2002* (Eds. I.M. Low & D.N. Phillips), 30 Sept – 4 Oct. 2002, Perth, WA.

20. D.S. Perera and M.E. Bowden, *J. Mater. Sci.* **26**, 1585 (1991).

21. Z. Oo, I.M. Low and B.H. O'Connor, *Physica B.* **385-386**, (2006) 499-501.

22. I.M Low, *Mater. Lett.* **58**, 927 (2004).

MICROSTRUCTURAL PARAMETERS IN 8 MeV ELECTRON-IRRADIATED *BOMBYX MORI* SILK FIBERS BY Wide-ANGLE X-RAY SCATTERING STUDIES (WAXS)

Sangappa [a] *, S Asha [a], Ganesh Sanjeev [b], G Subramanya [c], P Parameswara [d] and R Somashekar [d]

[a]*Department of Studies in Physics, Mangalore University, Mangalagangotri – 574 199, India*
[b]*Microtron Center, Mangalore University, Mangalagangotri - 574 199, India*
[c]*Department of Studies in Sericulture, University of Mysore, Manasagangotri, Mysore – 570 006, India*
[d]*Department of Studies in Physics, University of Mysore, Manasagangotri, Mysore – 570 006, India*
*Corresponding Author:sangappa@mangaloreuniversity.ac.in

ABSTRACT

The present work looks into the microstructural modification in electron irradiated *Bombyx mori* P31 silk fibers. The irradiation process was performed in air at room temperature using 8 MeV electron accelerator at different doses: 0, 25, 50 and 100 kGy. Irradiation of polymer is used to cross-link or degrade the desired component or to fix the polymer morphology. The changes in microstructural parameters in these natural polymer fibers have been computed using wide angle X-ray scattering (WAXS) data and employing line profile analysis (LPA) using Fourier transform technique of Warren. Exponential, Lognormal and Reinhold functions for the column length distributions have been used for the determination of crystal size, lattice strain and enthalpy parameters.

Keywords: Irradiation, Microstructural parameters, WAXS, fiber

INTRODUCTION

Silk proteins are of practical interest because of their excellent intrinsic properties utilized in biotechnological and biomedical fields. There is a continued interest in silkworms and their genetic modifications which result in superior quality textile fibers (Lance et al., 1999). These genetic or other kinds of modification can be brought about using energetic electron irradiation. It is quite important to know microstructural changes in silk fibers due to electron irradiation as these parameters determine the property and strength of the fibers. Such studies have not been carried out except for the chemical effects on these fibers (Mohanthy et al., 1995, Kawahara et al., 1996, Freddi et al., 1996). Somashekarappa et al (1998) have reported the effect of degumming and dye processing on the microstructural parameters in pure Mysore silk, nistari, NB7 and NB18 silk fibers. Somashekar et al (2002) have studied the structure-property relation in varieties of acid dye processed silk fibers. Sangappa et al (2004, 2005a, 2005b) have reported microstructural parameters in Hosa Mysore (HM), pure Mysore (PMS), nistari and C.nichi silk fibers. Takeshita et al (2000) have studied the effect of electron beam irradiation on silk fibers. Effects of gamma irradiation on biodegradation of *Bombyx mori* silk fibers have been carried out by Tsukada et al and Kojthung's group (1994, 2008).

When a polymer is subjected to irradiation by ionizing radiation such as gamma rays, X-rays or accelerated electrons , various effects like modification and degradation are expected because of modification in polymer network. At the microscopic level, the polymer degradation is characterized by macromolecular chain splitting, creation of low mass fragments, production of free radicals, oxidation and cross-linking. These affects the macroscopic properties like mechanical strength, color, electrical conductivity and other physical properties (Wang et al., 1987). Such modified polymer will have extended range of applications in industry. In radiation chemistry, polymers are classified into two types: scission polymers and cross-linking polymers. Most biopolymers are classified as scission polymers (Sangappa et al., 2008). Recent developments in this field have proved however that a variety of biopolymers could be cross-linked by irradiation of high energy radiation, and P31 silk tends to exhibit such radiation cross-linking. The interaction of electron beam with matter results in changes in crystallinity as well as microstructure. Such changes are related to a particular type of treatment of the materials. Here we have irradiated silk fiber samples with 8 MeV electron beams for various radiation doses. X-ray

CP1202, *Neutron and X-Ray Scattering in Advancing Materials Research: International Conference – 2009*
edited by A. Saat, H. A. Kassim, M. H. H. Jumali, J. M. Saleh, M. R. Othman, A. Ibrahim, F. M. Idris, and M. H. A.-R. M. Ahmad
© 2009 American Institute of Physics 978-0-7354-0739-8/09/$25.00

recording of such irradiated samples have been used for Line profile analysis (LPA) of Bragg's reflections. Impact of such research studies are normally felt in development and quality control. Also it is usefull in understanding the physical, mechanical and chemical properties of materials which are strongly related to microstructural parameters. Hence in this paper we report profile analysis of X-ray Bragg's reflections observed in pristine and electron irradiated silk fibers.

MATERIALS AND METHODS

Sample Preparation

For our study we have used raw P31 silk fibers belonging to *Bombyx mori* family which comes under the classification Multivoltine on the basis of shape, color, denier and life cycle of the fibers/cocoons. Cocoons were collected from the germplasm stock of the Department of Sericulture, University of Mysore, India, which were then cooked in boiling water (100°C) for 2 min. to soften the sericin and transferred to water bath at 65°C for 2 min. Then the cocoons were reeled in warm water with the help of mono cocoon reeling equipment EPPROUVITE. The characteristic features of these fibers are that they are white in color with an average filament length of 350 meter and denier being in the range 1.8–2.0. These fibers were mounted on rectangular frame in *just taut* condition which does not involve any mechanical stretching of fibers. The whole process, starting from reeling to mounting of fibers, does not involve any type of mechanical deformation.

Electron Irradiation

Irradiation of fiber samples was carried out at Microtron Center; Mangalore University using the electron beam (by lanthanum hexa fluorite source). The monochromatic beam is made to fall on samples kept at a particular distance with the following beam features.

Table 1. Specifications of the electron beam accelerator and irradiation conditions

1. Beam energy	8 MeV
2. Beam current	20 mA
3. Pulse repetition rate	50 Hz
4. Pulse width	2.2 μs
5. Distance source to sample	30 cm
6. Dose range	0 -100 kGy
7. Atmosphere	air
8. Temperature	24 °C

The dose delivered to different samples is measured by keeping alanine dosimeter with sample during irradiation.

X-Ray diffraction measurements

The XRD diffractograms of the fiber samples were recorded using a Rigaku Miniflex-II X-ray diffractometer with Ni filtered, CuKα radiation of wavelength λ = 1.5406 Å, with a graphite monochromator. The specifications used for the recordings were 30kV, 15mA. The samples were scanned in the 2θ range 10-50° with a scanning speed and step size of 1° /min and 0.01° respectively.

THEORY

Micro structural parameters such as crystallite size ($<N>$) and lattice strain (g in %) are usually determined by employing Fourier method of Warren and Averbach (1950, 1955) and Warren (1969). The intensity of a profile in the direction joining the origin to the center of the reflection can be expanded in terms of Fourier cosine series;

$$I(s) = \sum_{n=-\infty}^{\infty} A(n)\cos\{2\pi nd(s-s_0)\} \qquad (1)$$

where the coefficients of the harmonics A(n) are

functions of the size of the crystallite and the disorder of the lattice. Here, s is $\sin(\theta)/(\lambda)$, s_0 being the value of s at the peak of a profile; n is the harmonic order of co-efficient and d is the lattice spacing. The Fourier coefficients can be expressed as;

$$A(n) = A_s(n).A_d(n) \qquad (2)$$

For a paracrystalline material, $A_d(n)$ can be obtained, with Gaussian strain distribution (Hall et al., 1991),

$$A_d(n) = exp\,(-2\pi^2\,m^2\,n\,g^2) \qquad (3)$$

Here, 'm' is the order of the reflection and g = $(\Delta d/d)$ is the lattice strain. Normally one can also defines mean square strain $<\varepsilon^2>$, which is given by g^2/n. This mean square strain is dependent on n, whereas not g (Popa et al., 1995 and Ribarik et al., 2001). For a probability distribution of column lengths P(i), we have;

$$A_s(n) = 1 - \frac{nd}{D} - \frac{d}{D}\left[\int_0^n iP(n)di - n\int_0^n P(i)di\right] \qquad (4)$$

$$A_s(n) = \frac{m^3 \exp[9/4)(2^{1/2}\sigma)^2}{3} erfc\left[\frac{\log(|n|/m)}{2^{1/2}\sigma} - \frac{3}{2}2^{1/2}\sigma\right] - \frac{m^2 \exp(2^{1/2}\sigma)^2}{2}|n| erfc\left[\frac{\log(|n|/m)}{2^{1/2}\sigma} - 2^{1/2}\sigma\right]$$
$$+ \frac{|n|^3}{6} erfc\left[\frac{\log(|n|/m)}{2^{1/2}\sigma}\right] \qquad (8)$$

where $D = <N>d_{hkl}$ is the crystallite size and 'i' is the number of unit cells in a column. In the presence of two orders of reflections from the same set of Bragg planes, Warren and Averbach (1950, 1955) have shown a method of obtaining the crystallite size ($<N>$) and lattice strain (g in %). But in polymer it is very rare to find multiple reflections. Hence, to determine the finer details of microstructure, we approximate the size profile by

simple analytical function for P (i) by considering only the asymmetric functions. Another advantage of this method is that the distribution function differs along different directions. Whereas, a single size distribution function that is used for the whole pattern fitting, which we feel, may be inadequate to describe polymer diffraction patterns (Popa et al., 1995 and Ribarik et al.,

2001). Here it is emphasized that the Fourier method of profile analysis (single order method used here) is quite reliable one as per the recent survey and results of Round Robin test conducted by IUCr (Balzar, 2002). In fact, for refinement, we have also considered the effect of background by introducing a parameter [see for details regarding the effect of background on the microcrystalline parameters (Somashekar et al., 1989)].

The Exponential distribution

It is assumed that there are no columns containing fewer than p unit cells and those with more decay exponentially. Thus, we have (Somashekar et al., 1997),

$$P(i) = \begin{cases} 0 & ;if\ p < i \\ \alpha\exp\{-\alpha(i-p)\} & ;if\ p \geq i \end{cases}$$
$$\dots\dots\dots(5)$$

where, $\alpha = 1/(N - p)$ Substituting this in equation

(4), we get;

$$A_s(n) = \begin{cases} A(0)(1-n/<N>) & ;if\ n \leq p \\ A(0)\{\exp[-\alpha(n-p)]\}/(\alpha N) & ;if\ n \geq p \end{cases}$$
$$\dots\dots\dots(6)$$

Here, α is the width of the distribution function, 'i' is the number of unit cells in a column, n is the harmonic number, p is the smallest number of unit cells in a column and $<N>$, the number of unit cells counted in a direction perpendicular to the (hkl) Bragg plane.

The Lognormal distribution

The Lognormal distribution function is given by;

$$P(i) = \frac{1}{(2\pi)^{1/2}\sigma i}\exp\left\{-\frac{[\log(i/m)]^2}{2\sigma^2}\right\} \qquad (7)$$

where, σ is the variance and m is the median of the distribution function.

Substituting for P(i) in equation (4) and simplifying [5], we get,

The above equation is the one used by Ribarik et al., (2001). The maximal value $A_s(0)$ is given by;

$$A_s(0) = \frac{2m^3\exp[(9/4)(2^{1/2}\sigma)^2]}{3}$$

$$(9)$$

The area-weighted number of unit cells in a column is given by

$$<N>_{surf} = \frac{2m\exp[(5/4)(2^{1/2}\sigma)^2]}{3} \qquad (10)$$

and the volume- weighted number of unit cell in a column is given by

$$<N>_{vol} = \frac{3m\exp[(7/4)(2^{1/2}\sigma)^2]}{4} \qquad (11)$$
$$(5)$$

The Reinhold Distribution

With the exponential distribution function, $P(i)$ rises discontinuously at p, from zero to its maximum value. In contrast, the Reinhold function allows a continuous change by putting,

$$p(i) = \begin{cases} 0 & ;if\ i \leq p \\ \beta^2(i-p)\exp\{-\beta(i-p)\} & ;if\ i > p \end{cases} \qquad (12)$$

where $\beta = \frac{2}{N-p}$ substituting these in eq. (4), we obtain

34

$$A_s(n) = \begin{cases} A(0)(1-n/<N>) & ;if\ n\leq p \\ [A(0)(n-p+2/\beta)/N]\{exp[-\beta(n-p)]\} & ;if\ n\geq p \end{cases}$$
$$\ldots\ldots\ldots\ldots(13)$$

where, β is the width of the distribution which has been varied to fit the experimental results. p is the smallest number of unit cells in a column, $<N>$ is the number of unit cells counted in a direction perpendicular to the (hkl) Bragg plane; d is the spacing of the (hkl) planes; λ is the wavelength of X-rays used; i is the number of unit cells in a column; n is the harmonic number and D_s is the surface weighted crystal size ($<N>d_{hkl}$).

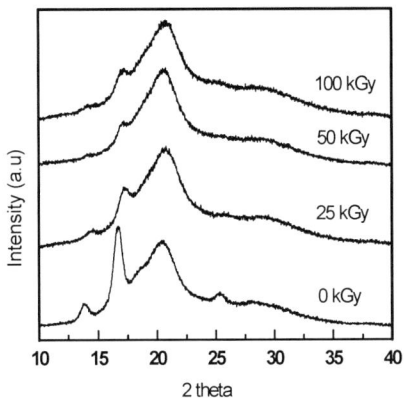

Figure. 1 XRD scans of pure and 8 MeV electron irradiated P31 Silk fiber samples.

All the distribution functions were put to test in order to find out the most suitable crystallite size distribution function for the profile analysis of the X-ray diffraction. The procedure adopted for the computation of the parameters is as follows. Initial values of g and N were obtained using the method of Nandi et al (1984). With these values in the equations give numbers earlier give the corresponding values for the width of distribution. These are only rough estimates, so the refinement procedure must be sufficiently robust to start with such values. Here we compute;

$$\Delta^2 = [I_{cal} - (I_{exp} + BG)]^2 / npt \quad (14)$$

where, BG represents the error in the background estimation, npt is number of data points in a profile, I_{cal} is intensity calculated using equations (1)-(13) and I_{exp} is the experimental intensity. The values of Δ were divided by half the maximum value of intensity so that it is expressed relative to the mean value of intensities, and then minimized.

X-Ray profile analysis

For the analysis, we have used X-ray diffraction (Figure 1) data in the above equations to simulate the intensity profile by varying the necessary parameters till one gets a good fit with the experimental profile. For this purpose, a multidimensional algorithm SIMPLEX is used for minimization (Press et al., 1986). We have used pure and 8 MeV electron beam irradiated silk fiber samples.

Figure 1

RESULTS AND DISCUSSION

Figures 2(a-d), 3(a-d) and 4(a-d) shows the comparison between simulated and experimental profiles for 8 MeV electron irradiated and pure fiber samples for Bragg's reflection. The simulated profile was obtained with the using appropriate model parameters. This procedure was followed for all the other treated samples at different radiation doses. The computed microcrystalline parameters such as crystallite size $<N>$(number of unit cells), lattice strain g in %, the width of the crystallite size distribution (α) and the standard deviation are given in Table 2. It is evident from Table 2. that all the asymmetric distributions used, give more or less similar results. By and large, Exponential distribution function gives a better fit than Reinhold/Lognormal distributions. Since Exponential distribution function gives a better fit than others, we used the corresponding results given in Table 2. to infer some important conclusions. They are

(i) The value of the surface weighted crystallite size Ds is increasing with increasing irradiation dose compared to pure sample; (ii) The value of the crystallite size is more for electron irradiated fibers.

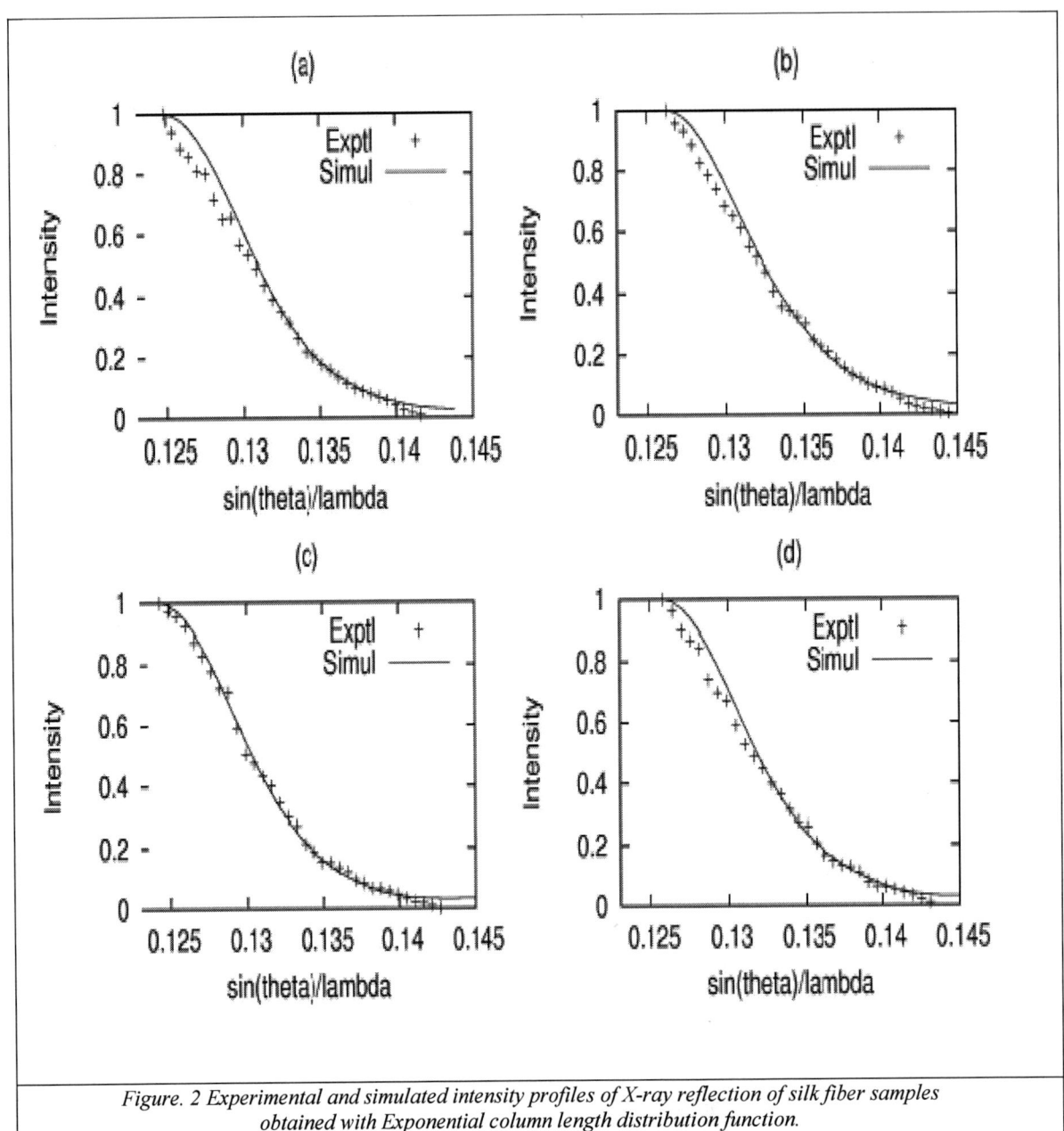

Figure. 2 Experimental and simulated intensity profiles of X-ray reflection of silk fiber samples obtained with Exponential column length distribution function.

Increasing crystallite size with increasing irradiation dose shows that there is an increase in sample tensile strength of the fiber (Uenoyama et al., 2002). This aspect suggests that for 100 kGy integral dose irradiated fiber, there is an increase in formation of cross-linking bonds between inter polymer chains than unirradiated one. Also, it essentially implies, the efficiency of cross-linking in these fibers is improved to a large extent by means of electron irradiation.

The variation of lattice strain (g) lies between 0.2 -0.5 % in the case of Exponential distribution for polymer samples. From the obtained micro crystalline parameters ($<N>$, g in %) one can estimate the minimum enthalpy (α^*), which defines the equilibrium state of microparacrystals in all the polymer samples, using the relation postulated by Hosemann (1988).

$$\alpha^* = (<N>^{1/2} g) \qquad (15)$$

Table 2 : Microstructural parameters of Electron irradiated polymer samples computed by various distribution functions

Exponential	Reinhold	Lognormal
Sample $<N>$ g in %$\alpha^*$$D_s$ (Å)delta	$<N>$$g$ in %α^* D_s (Å)delta	$<N>$ g in %α^* D_s (Å)delta
0 kGy7.01±0.34 0.5±0.02 0.013 30.35 0.049	6.99±0.35 0.5±0.03 0.013 30.26 0.051 6.30±0.19 0.5±0.02 0.012 26.96 0.029	7.53±0.45 0.5±0.03 0.014 32.59 0.060
25 kGy 6.34±0.27 0.5±0.02 0.012 27.13 0.042	6.86±0.34 0.5±0.02 0.013 29.38 0.053	6.84±0.41 0.5±0.03 0.013 29.23 0.060
50 kGy 6.90±0.34 0.5±0.02 0.013 29.56 0.049	7.15±0.43 0.5±0.03 0.014 30.45 0.065	7.55±0.53 0.5±0.04 0.014 32.34 0.070
100kGy 7.11±0.43 0.1±0.01 0.003 30.88 0.061		7.68±0.61 0.5±0.04 0.015 32.71 0.080

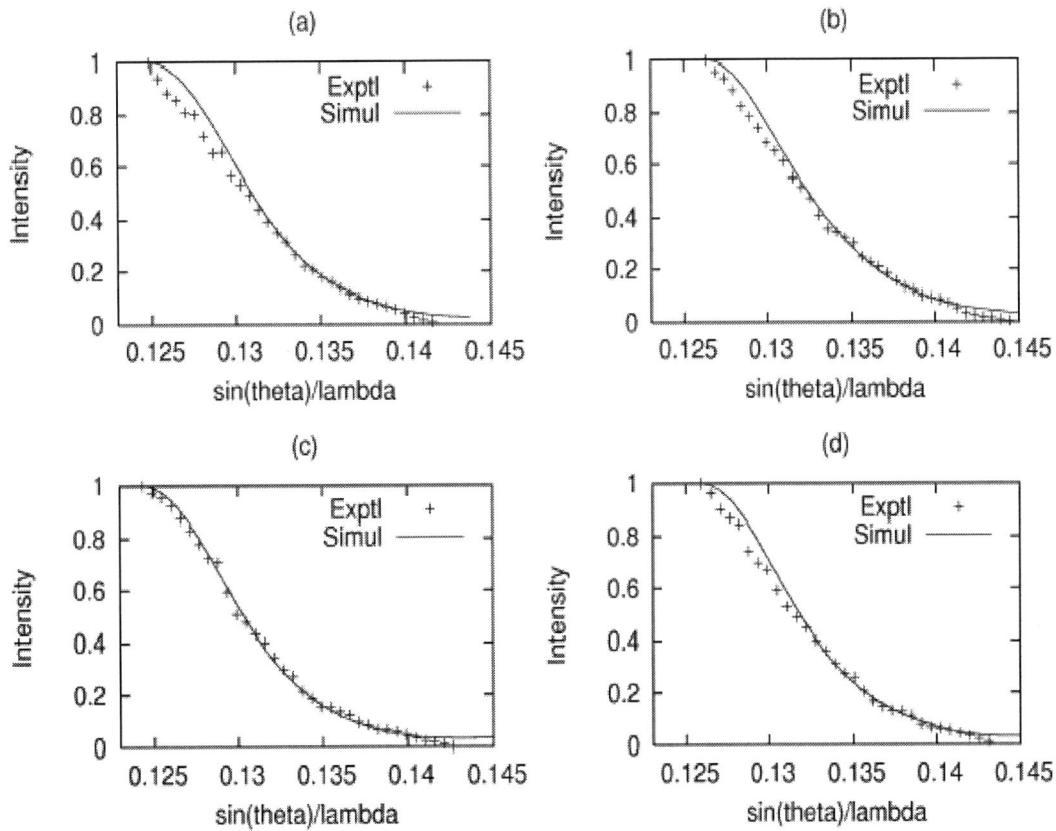

(a) (b)

(c) (d)

Figure. 3 Experimental and simulated intensity profiles of X-ray reflection of silk fiber samples obtained with Reinhold column length distribution function.

The estimated minimum enthalpy is given in Table 2. It is noted here that the value of α^* lies between 0.003 and 0.013 for these fiber samples. The value of enthalpy decreases with increasing dose rate which corresponds to the state with well ordered polymer network. We have observed that the lattice strain and its variation for

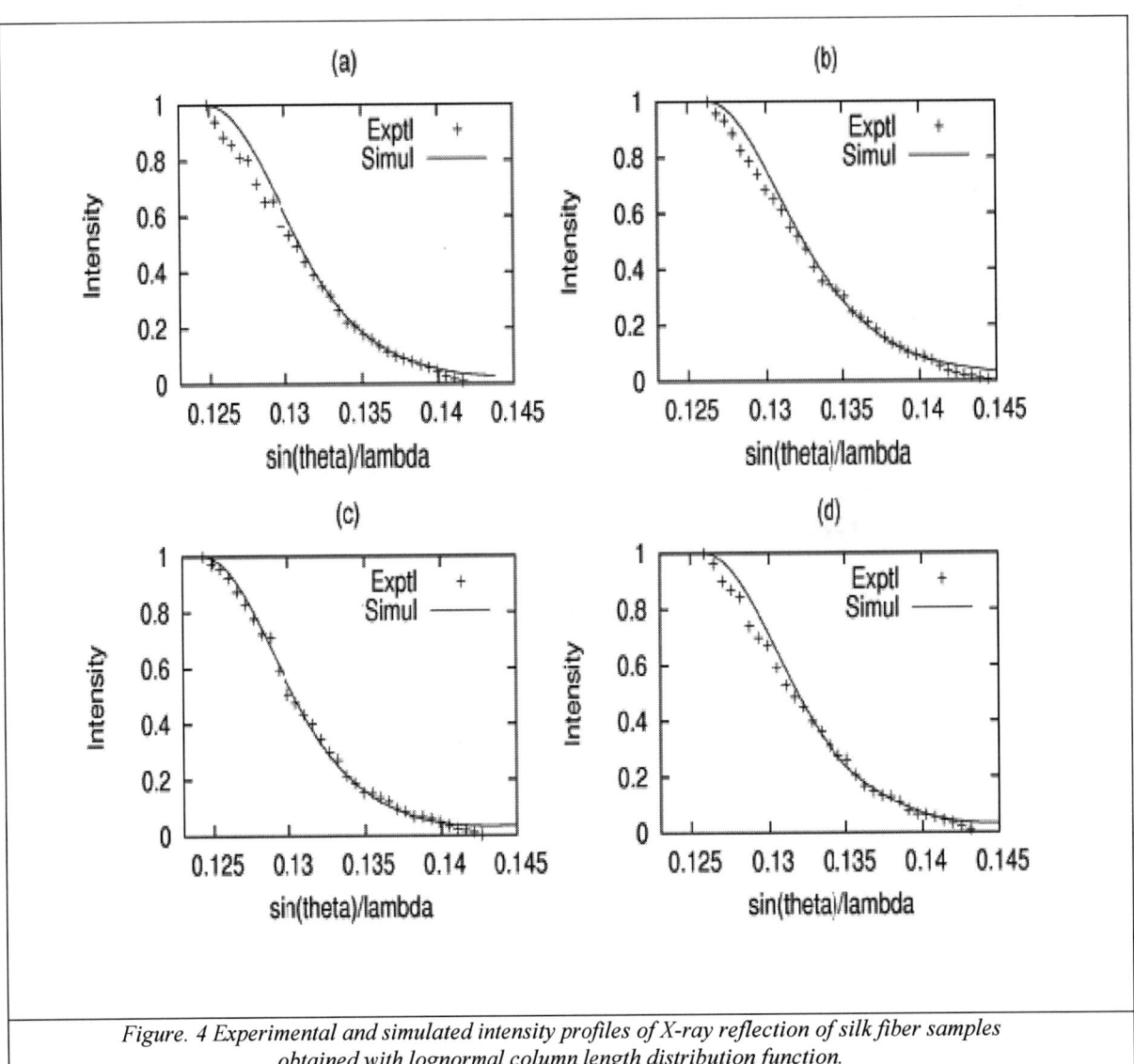

Figure. 4 Experimental and simulated intensity profiles of X-ray reflection of silk fiber samples obtained with lognormal column length distribution function.

various values of the radiation doses (kGy) in polymer samples are very small and this may be due to inherent model dependent factor.

CONCLUSION

From the wide angle X-ray scattering (WAXS) study of electron irradiated P31 silk fiber (*Bombyx mori*) samples, we have observed a significant change in the values of micro structural parameters occurs. This causes the cross-linking of small polymer units leading to the formation of a rigid 3-dimensional network. We have shown that among the three asymmetric crystallite size distributions, Exponential gives a better fit in polymer samples. The only justification for the good fit that we observed with Exponential distribution in these polymers can be interpreted on the basis of extensive usage of this function in condensed matter to explain various phenomenon's like dielectric relaxation, luminescence decay law and other physical properties. Single order

method that we have used here is capable of estimating both the size and the distortion parameters and could in general measure crystallite size, only onto a certain limit. The changes in natural polymer network with different dose rates are quantified here in terms of microstructural parameters. Surprisingly we observe that the intrinsic strains are very small. It is evident from this study that irradiation of silk fibers changes the polymer network and hence the physical properties, leading to a better quality polymer, depending on the nature of application.

ACKNOWLEDGMENTS

The authors are thankful to University Grant Commission, New Delhi, Govt of India, for providing financial assistance through a project F. No. 33-14/2007 (SR).

REFERENCES

Balzar D., (2002), Report on the Size-Strain Round Robin, *News Letter IUCr.* 228:14.

Freddi G, Massafra M R, Beretta S, Shibata S, Gotch Y, Yasui H and Sukuda M T., (1996), Structure and properties of Bombyx mori silk fibers grafted with methacrylamide (MAA) and 2-hydroxy ethylmethacrylate (HEMA), *J. Appl Poly Sci.* 60:1867-1876.

Hall I H and Somashekar R., (1991), The determination of crystal size and disorder from the X-ray diffraction photograph of polymer fiber.2. Modelling intensity profiles, *J. Appl Cryst.* 24:1051-1059.

Hosemann R., (1988), The α* - law in colloid and polymer science, *Prog. Colloid and Polymer Sci.* 77:15-25.

Kojthung A, Meesilpa P, Sudatis B, Treeratanapiboon L, Udomsangpetch R and Oonkhanond B., (2008), Effects of gamma irradiation on biodegradation of Bombyx mori silk fibroin, *Int.Biodeterioration and Biodegradation.* 62(4):487-490.

Kawahara Y, Shioya M and Takaku A., (1996), Influence of Swelling of Non-crystalline Regions in Silk Fibers on Modification with Methacrylamide, *J.Appl Poly Sci.* 59: 51-56

Lance D Miller, Putthanarat S, Eby R K and Adoms W W., (1999), Investigation of the nanofibrillar morphology in silk fiber by small angle X-ray scattering and atomic force microscopy, *Bio Macromolecules,* 24:159-165.

Mohanthy N, Das H K, Mohanthy P and Mohanthy E., (1995), Modification of Muga silk by Mythyl-Methacrylate-II, *J. Macromol. Sci.* A32:1103-1111.

Nandi R K, Kho H K, Scholsberg W, Wissler G, Cohen J B and Crist B Jr,   (1984), Single-peak methoda for Fourier analysis of peak shapes, *J. Appl Cryst.* 17:22-26.

Popa N C, Balzar D., (1995), An analytical approximation for a size-broadened profile given by the lognormal and gamma distributions, *J. Appl Cryst.* 35:338-346.

Press W, Flannery B P, Teukolsky S and Vetterling W T., (1986), Eds., *Numerical Recipes,* Cambridge University press.

Ribarik R, Ungar T, Gubicza J., (2001), MWP-fit: a program for multiple whole-profiles fitting of diffraction peak profiles by *ab initio* theoretical functions, *. J. Appl Cryst.* 34:669-676.

Sangappa, Okuyama K and Somashekar R., (2004), Stain – Tensor Components, Crystallite Shape and Their Effects on Crystalline Structure in Silk I, *J. Appl Poly Sci.* 91:3045-3058.

Sangappa, Mahesh S S and Somashekar R., (2005), Crystal structure of raw pure Mysore silk fiber based on (Ala-Gly)₂ –Ser –Gly peptide sequence using Linked- Atom-Least-Squaresmethod, *J Bioscience.* 30(2):259-268.

Sangappa, Mahesh S S, Somashekar R and Subramanya G., (2005), Analysis of Diffraction Line Profile from Silk Fibers Using Various Distribution Functions, *J. Polymer Research.* 12:465-472.

Somashekar R, Hall I H and Carr P D., (1989), The determination of crystal size and disorder from X-ray diffraction photographs of polymer fibers.1. The accuracy of determination of Fourier coefficients of the intensity profile of a reflection, *J. Appl Cryst.* 22:363-371.

Somashekar R, Somashekarrappa H., (1997), X-ray Diffraction-Line Broadening Analysis:Paracrystalline Method, *J. Appl Cryst.* 30:147-152.

Somashekarappa H, Nadgir G S, Somashekar T H, Prabhu J and Somashekar R., (1998), WAXS studies on silk fibers treated with acid (blue) and metal complex (brown) dyes, *Polymer.* 39:209-213.

Somashekarappa H, Annadurai V, Sangappa, Subramanya G and Somashekar R., (2002), Structure-property relation in varities of acid dye processed silk fibers, *Matrials Letters.* 53:415-420.

Takeshita H, Ishida K, Kamiishi Y, Yoshii F and Kume T., (2000), Production of fine powder from silk by radiation, *Macromolecular Materials and Engg.* 283(1):126-131.

Tsukada M, Freddi G and Minoura N., (1994), Changes in the Fine Structure of Silk Fibroin Fibers Following Gamma Irradiation, *J. Appl Polym. Sci.* 51:823-829.

Wang G, Pan G, Dow L, Jiang S, Yu R, Zhang T and Dai Q., (1987), Proton beam modification of isotactic polypropylene, *Nucl.Instr. and Meth B.* 27:410-416

Warren B E and Averbach BL.,(1950), Introduction to the Riveted method, *J. Appl Phys.* 21: 595-599.

Warren B E., (1955), A generalized treatment of cold work in powder patterns, *Acta Cryst.* 8:483-486.

Warren B E., (1969), *X-ray Diffraction,* Addison-Wesley, New York, Chap 13, pp 251-314.

Uenoyama M, Shukushima S, Hayami H and Nishimoto S., (2002), SEI TECHNICAL REVIEW, Number

Effect of A Long Chain Carboxylate Acid on Sodium Dodecyl Sulfate Micelle Structure: A SANS Study

Arum Patriati[a]*, Edy Giri Rachman Putra[a,b], Baek Seok Seong[b]

[a]Neutron Scattering Laboratory, National Nuclear Energy Agency of Indonesia (BATAN),
Gedung 40 Kawasan Puspiptek Serpong, Tangerang 15314, Indonesia
[b]Neutron Science Division, HANARO Center, Korea Atomic Energy Research Institute
(KAERI)
1045 Daedok-daero, Yuseong-gu, Daejeon 305-353, Republic of Korea

ABSTRACT

The effect of a different hydrocarbon chain length of carboxylate acid, i.e. dodecanoic acid, $CH_3(CH_2)_{10}COOH$ or lauric acid and hexadecanoic acid, $CH_3(CH_2)_{14}COOH$ or palmitic acid as a co-surfactant in the 0.3 M sodium dodecyl sulfate, SDS micellar solution has been studied using small angle neutron scattering (SANS). The present of lauric acid has induced the SDS structural micelles. The ellipsoid micelles structures changed significantly in length (major axis) from 22.6 Å to 37.1 Å at a fixed minor axis of 16.7 Å in the present of 0.005 M to 0.1 M lauric acid. Nevertheless, this effect did not occur in the present of palmitic acid with the same concentration range. The present of palmitic acid molecules performed insignificant effect on the SDS micelles growth where the major axis of the micelle was elongated from 22.9 Å to 25.3 Å only. It showed that the appropriate hydrocarbon chain length between surfactant and co-surfactant molecules emerged as one of the determining factors in forming a mixed micelles structure.

Keywords: micellar solution, micellar structure, mixed micelles, small angle neutron scattering

PACS: 64.75.Yz; 61.46.-w; 82.33.Nq; 61.05.fg

INTRODUCTION

In colloidal system, the interaction between surfactant molecules to form micelles depends on the balance of the hydrophobic attraction and the hydrophilic (electrostatic) repulsion. This leads in minimizing the degree of mixing between hydrophobic tails of surfactant and water, and then the hydrophobic head-groups are in contact with the water [1,2]. If the head groups are forced too closely they will repel each others by electrostatic repulsion. On the other hand, if the head groups are separated too far, it forces the hydrophobic tail to come into contact with the water.

The interaction between non-polar molecules like oil in the 0.3M sodium dodecyl sulfate, SDS micellar solution system has been studied previously using a small angle neutron scattering (SANS) spectrometer in BATAN (SMARTer). It was observed that the non polar organic molecules, for all oil types, i.e. hexane, octane, and decane, were solubilized deeply in the core of micelle [3]. They strongly interact with the non polar chain of the SDS surfactant. The ellipsoidal micelle has a propensity to be more spherical due to the existing of oil

pool in the core of micelle. However, in the increasing of the oil concentration, the micelle returns to be ellipsoid. It is known that the oil penetrates between the surfactant chains and reduce the packing parameter of the surfactant.

Meanwhile, the polar organic compounds, like alcohol, observably interact with both the hydrophilic and hydrophobic moieties of the surfactant molecules in the micellar solution [4,5]. For polar organic molecules with a hydrocarbon chain length is shorter than the surfactant, it is observed that the hydrocarbon chain length determined the way of how the co-surfactant induces the micellar system [6]. The alcohol with a short hydrocarbon chain under four of C-C bonding, tend to localize in the aqueous phase. Whereas, the longer hydrocarbon chain lies from four to eight of C-C bonding, tend to distribute between aqueous and micellar phases. As the hydrocarbon chain length increase, the alcohol molecules tend to favour in the micellar phase. This leads to the formation of mixed micelles and affect the growth of micelle. Furthermore, the presence of alcohol molecules with more than ten of C-C bonding of hydrophobic chain length, is observed to give effect on their physical properties changes [7,8]. That indicates a

CP1202, Neutron and X-Ray Scattering in Advancing Materials Research: International Conference – 2009
edited by A. Saat, H. A. Kassim, M. H. H. Jumali, J. M. Saleh, M. R. Othman, A. Ibrahim, F. M. Idris, and M. H. A.-R. M. Ahmad
© 2009 American Institute of Physics 978-0-7354-0739-8/09/$25.00

strong interaction between the long chain polar organic molecules and the surfactant exist.

In this work, the interaction of a long polar organic molecule such as carboxylate acid, which is relatively more polar than alcohols with approximate hydrocarbon chain length to its surfactant was studied. The aim of this work is to get a better understanding in the interaction of the surfactant with the co-surfactant, particularly the long chain polar organic molecules. A series of two carboxylic acid, dodecanoic acid or lauric acid and hexadecanoic acid or palmitic acid with a various concentration were added into 0.3 M SDS micellar solution, in which it will give a condensed micellar system and provide the inter-correlation between micelle. The effect of the interaction between carboxylic acid and SDS molecules on the structural micelles changes was then investigated by a small angle neutron scattering (SANS) technique due to its ability to observe the size and shape of micelle changes in the solution.

EXPERIMENTAL SETTING

All chemicals, i.e. sodium dodecyl sulfate (SDS), sodium chloride (NaCl), and 99.9% deuterium oxide (D_2O), which used in this experiment were purchased from Aldrich without further purification. D_2O was used as a solvent to enhance the contrast as certainly required for SANS experiment. A various amount of each lauric acid and palmitic acid in concentration of 0.01, 0.02, 0.04, 0.06, 0.08, 0.1 M was added into a fixed 0.3 M SDS/D_2O system.

The SANS experiments were performed at ambient temperature 22°C on SMARTer at neutron scattering laboratory (NSL) BATAN in Serpong, Indonesia [9]. Each sample was measured on two detector distances, 1.5 m and 3 m, with the neutron wavelength λ of 3.9 Å. This configuration gives a range of momentum transfer Q, from 0.03 to 0.3 Å^{-1}. Samples with 2 mm thickness were placed in the quartz cell. The raw scattering data of the samples were then corrected for incoherent scattering; i.e. background of detector, quartz cell and solvent scattering; and sample transmission by a data reduction program, GRAPS [10].

Data Analysis

The intensity $I(Q)$ of small angle scattering as a function of momentum transfer Q, for a monodisperse interacting micelle system can be expressed as

$$I(Q) = n(\rho_m - \rho_s)^2 V^2 P(Q) S(Q)$$

where n denotes the number density of micelles, ρ_m and ρ_s are the scattering length densities of the micelle and the solvent, respectively [11]. The term $(\rho_m-\rho_s)$ is called contrast factor. V is the volume of a micelle. The aggregation number N of the micelle related to the micellar volume V by the relation $V = Nv$, where v is the volume of a surfactant monomer. For ellipsoid and cylinder shape it is calculated by $N = 4\pi R_a^2 R_b / 3v$, where R_a and R_b are respectively minor and major axis.

$P(Q)$ is the intraparticle structure factor and depends on the shape and size of the particles. $S(Q)$ is the interpaticle structure factor and is decided by the interparticle distance and the particle interaction. For ellipsoid micelle with two equal semi-axis, Ra and principal axis $zRa = Rb$ where $Ra = Rc$, $P(Q)$ is formulated by

$$P_{el}(Q) = \int_0^{\pi/2} [j_1(x)]^2 \sin \beta \, d\beta$$

where $x = Qa[\cos^2 \beta + z^2 \sin^2 \beta]$ and β is the angle between the scattering vector and the direction of the symmetry axis of ellipsoid. It is noted that $z > 1$ or $z < 1$ depending on whether the ellipsoidal particle is prolate or oblate. Meanwhile, for isotropic system $S(Q)$ can be written as,

$$S(Q) = 1 + 4\pi n \int [g(r) - 1] \frac{\sin Qr}{Qr} r^2 dr$$

where $g(r)$ is the radial distribution function, a probability of finding another particle at a distance r from a reference particle centered at the origin.

The corrected data was then analyzed by a screen Coulomb model provided by NIST data analysis program [12]. Here, it is determined the fractional charge α and the major axis, since the minor axis is fixed at 16.7 Å, which is the same as the length of the extended hydrophobic chain of SDS molecule based on the Tanford's formulation [13].

RESULT AND DISCUSSION

The SANS distribution profile of 0.3 M SDS with an addition of lauric acid is shown in Figure 1. In the addition of lauric acid, the inter correlation peak Q_m appears to be shifted to a lower Q. This indicates that there is increasing in micellar distance in the solution due to the growth of micelle. The growth of micelle is also observed by increasing the intensities as the lauric acid concentration increases.

In the high Q region, the curves overlapped which indicates that the micelles grow in the one-dimension [14]. This is appropriate with the result of fitting data, Table 1 and the properties of SDS micelle itself. In the growth of micelle, the hydrocarbon chains will tend to get close each other inside the micelle to prevent from contact with the water. The minor axis R_a, is fixed at 16.7 Å due to the length of the fully extended conformation hydrophobic chain of SDS molecule based on the Tanford's formulation [13]. The major axis R_b increases in the increasing of lauric acid concentration. The increasing of R_b value leads on increasing of aggregation number N, which is the average number of surfactant molecules assembled in one micelle.

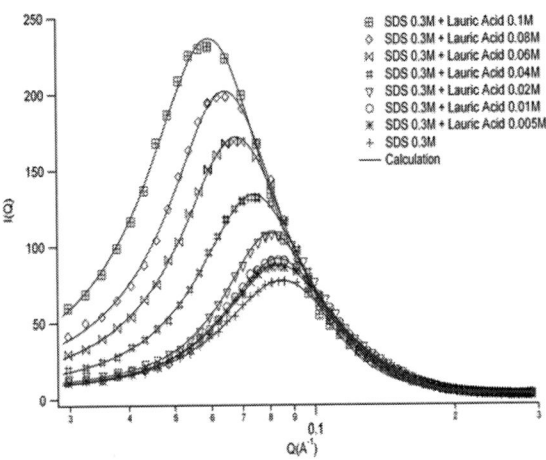

Figure 1. The pattern of SANS distribution profile from 0.3 M SDS solution in the present of lauric acid.

It is observed from the scattering distribution profile that the addition of lauric acid in the SDS micellar system effectively induces the micelle start from 0.02 M concentration. Under this concentration the inter correlation peak does not appear to be shifted to low Q. It appears that the intensities of the peak increase due to the increasing of the contrast, i.e. the scattering length density different, between the mixed micelle to their environment.

However, the growth of SDS micelle does not occur in the present of palmitic acid at the same concentration range. The SANS distribution profiles seem to overlap at whole concentration of palmitic acid which is added to the system. In Figure 2, the inter correlation peak shift insignificantly and so does the increasing of intensities as addition of palmitic acid.

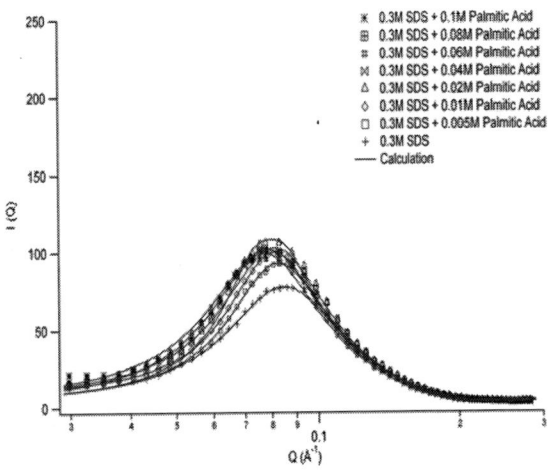

Figure 2. The overlapping SANS distribution profile of 0.3M SDS in addition of palmitic acid.

The growth of micelle slightly appears in the addition of palmitic acid up to 0.02 M. The major axis of micelle stay constant at a higher palmitic acid concentration added into the SDS solution. The insignificant effect of the palmitic acid in the growth of SDS micelle is clearly shown in the aggregation number where is relatively constant at above the 0.02 M of palmitic acid concentration. The result is given in Table 1.

Table 1. Data parameters of SDS in the addition of lauric acid and palmitic acid. The minor axis fixed at 16.7 Å

0.3M SDS + Carboxylic Acid (M)	Lauric Acid				
	R_b (Å)	R_b/R_a	N	α	Q_m (Å^{-1})
0	21.1	1.26	70	0.38	0.084
0.005	21.9	1.31	73	0.35	0.082
0.01	22.1	1.32	75	0.34	0.082
0.02	23.3	1.39	77	0.31	0.079
0.04	25.5	1.53	85	0.26	0.072
0.06	29.1	1.74	97	0.22	0.067
0.08	31.6	1.89	105	0.19	0.064
0.1	34.7	2.08	115	0.15	0.058
0.3M SDS + Carboxylic Acid (M)	Palmitic Acid				
	R_b (Å)	R_b/R_a	N	α	Q_m (Å^{-1})
0	21.1	1.26	70	0.38	0.084
0.005	23.1	1.38	77	0.34	0.083
0.01	23.4	1.41	78	0.32	0.082
0.02	24.3	1.45	81	0.30	0.079
0.04	24.8	1.49	82	0.29	0.078
0.06	25.1	1.49	83	0.29	0.078
0.08	25.4	1.51	84	0.29	0.077
0.1	25.6	1.53	85	0.29	0.076

Since the different surfactant molecules in a system have a tendency to form a mixed micelle [15], the growth of micelle in the addition of lauric acid shows that they tend to lies in the micellar phase. The lauric acid induces the SDS micellar system (Figure 3a) and form a mixed micelle in the solution (Figure 3b). As the fundamental rule in the form of micelle is the balance of hydrophobic attraction and hydrophilic (or electrostatic) repulsion, the present of carboxylic acid like lauric acid enhances the propensity of surfactant and co-surfactant molecules to form a mixed micelle. The uncharged of carboxylic head group of lauric acid molecules in the mixed micelles shield the repulsion between sulfate head groups, indicated by decreasing of the fractional charge α. Furthermore, the small head group area of carboxylic head groups affects the average packing parameter ($p = v/(l_c.a_0)$), i.e. the optimum head groups area a_0, the critical chain length l_c, and the hydrocarbon volume v, of the molecules in the micelle (Jones, 2002). The increasing v/a_0 of average packing parameter of the molecules in the micelle, in which for lauric acid is higher than SDS, then the micelle favorably grow to form a longer ellipsoidal shape (Figure 3c).

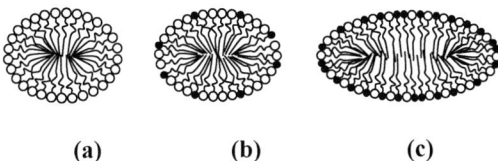

(a)　　　　(b)　　　　(c)

Figure 3. The schematic model of a mixed micelle {SDS (white), carboxylic acid (black)}. The addition of carboxylic acid in the SDS micellar phase leads to the micelle elongation.

On the other hand, the domination of a large hydrophobic attraction between carboxylic acid in aqueous solution decreases its solubility (Gelbart and Ben-Shaul, 1996). This inhibits the forming of a mixed micelle, like as in the present of palmitic acid in the micellar solution. At first, the addition of palmitic acid in the 0.3 M SDS micellar solution induces the SDS micelle and form a mixed micelle. The longer hydrophobic chain tail of palmitic acid than lauric acid causes the size of mixed micelle formed in the addition palmitic acid up to 0.02 M relatively higher than in an addition of lauric acid. However, as increasing the number of palmitic acid molecules in the system, they become saturated. The excess of palmitic acid molecules in the more of 0.02 M concentration cannot be solubilized in the aqueous system. The micelles emerge to stop growing in the addition on more than 0.02 M of palmitic acid.

From those results above clearly show the interaction of the long polar organic compounds in the micellar system. They induce the micelles and lead to the growth of micelles. Therefore, the effect of a series addition of the short – medium – long chain of polar organic molecules (alcohols) in the 0.3 M SDS micellar solution is an interesting subject. Subsequently, to complete the understanding of the interaction alcohols and surfactant in the micellar system was set up for further studies [16].

CONCLUSION

The addition of carboxylic acid in the SDS micellar system have lead to the growth of the micelles due to their interaction that tend to induce and lies in the micellar phase. The uncharged carboxylic head group of carboxylic acid molecules would shield the repulsion force between sulfate head groups and furthermore their tiny size make the micelle more packed, then tend to be elongated. However, the domination of large hydrophobic attraction between carboxylic acid would reduce their solubility in aqueous solution and then reduces their ability to induce the micelle. Those make the present of lauric acid gives a significant positive effect to the growth of SDS micelle, but not with the present of palmitic acid.

ACKNOWLEDGMENT

The work was supported by National Nuclear Energy Agency (BATAN) on financial year of 2006 under the Neutron Beam Utilization of G.A. Siwabessy Reactor program and in part by HANARO, Korea Atomic Energy Research Institute R&D Program of the Ministry of Science and Technology, Republic of Korea. The authors would thank to Dr. V.K. Aswal from Bhabba Atomic Research Centre, India for the beneficial discussions. A. Patriati acknowledges the support from ICNX2009 committees for attending the ICNX2009 in Kuala Lumpur, Malaysia, June 29 – July 1, 2009.

REFERENCES

1. Jones, R.A.L., (2002), *Soft Condensed Matter*, Oxford University Press.

2. Gelbart, W.M., Ben-Shaul, A., (1996), *J. Phys. Chem.* **100**, 13169-13189.

3. Putra, E.G.R., Ikram, A., Seong, B. S., (2009), *Nucl. Instr. and Meth.* **A 600**, 291-293.

4. Griffith, P.C., Whatton, M.L., Abbott, R.J., Kwan, W., Pitt, A.R., Howe, A.M., King, S.M., Heenan, R.K., (1999), *J. Colloid Interface Sci.* **215**, 114-123.

5. Almgren, M., Swarup, S., (1983), *J. Colloid Interface Sci.* **91**, 256-266.

6. Caponetti, E., Martino, D.C., Floriano, Triolo, M.A., R. (1995), *J. Molec.Struc.* **383**, 133-143.

7. Patist, A., Axelberd, T., Shah, D.O., (1998), *J. Colloid Interface Sci.* **208**, 259-265.

8. Djakovic, L., Milosevic, Marjanovic, S., V., (1996), *J. Colloid. Interface Sci.* **182**, 289-291.

9. Putra, E.G.R., Ikram, A., Santoso, E., Bharoto, (2007), *J. Appl. Cryst.* **40**, 447-452; Putra, E.G.R., Ikram, A., Santoso, E., Bharoto, (2009), *Nucl. Instr. and Meth.* **A 600**, 291-293.

10. Dewhurst C., (2001–2007), Institut Laue Langevin "GRASP: Graphical Reduction and Analysis SANS Program for Matlab", http://www.ill.eu/ fileadmin/users_files/ Other_Sites/lss-grasp/grasp_main. html,

11. Goyal, P.S., (1995), Small Angle Neutron Scattering, RCA/IAEA Workshop on Small Angle Neutron Scattering, BARC-Bombay, India, April.

12. Kline, S.R., (2006), *J. Appl. Cryst.* **39**, 895-400.

13. Tanford, C., (1980), The Hydrophobic Effect: Formation of Micelle and Biological Membranes, Willey, New York.

14. Flood, C., Dreiss, C.A., Croce, V., Crosgrove, T., (2005), *Langmuir* **21**, 7646-765.

15. Joshi, J.V., Aswal, V.K., Goyal, P.S., (2008), *PRAMANA J. Phys.* **71**, 1039-1043.

16. A.Patriati, in preparation.

DIFFRACTION STUDY ON THE THERMAL STABILITY OF Ti_3SiC_2/TiC/$TiSi_2$ COMPOSITES IN VACUUM

W.K. Pang[1], I.M. Low[1], B.H. O'Connor[1], A.J. Studer[2], V.K. Peterson[2] and J.-P. Palmquist[3]

[1]Centre for Materials Research, Department of Imaging and Applied Physics, Curtin University of Technology, GPO Box U 1987, Perth WA, Australia.
[2]The Bragg Institute, ANSTO, PMB 1, Menai, NSW 2234, Australia.
[3]Kanthal AB, Heating Systems R&D, P.O. Box 502, SE-734 27 Hallstahammar, Sweden.

ABSTRACT

Titanium silicon carbide (Ti_3SiC_2) possesses a unique combination of properties of both metals and ceramics, for it is thermally shock resistant, thermally and electrically conductive, damage tolerant, lightweight, highly oxidation resistant, elastically stiff, and mechanically machinable. In this paper, the effect of high vacuum annealing on the phase stability and phase transitions of Ti_3SiC_2/TiC/$TiSi_2$ composites at up to 1550°C was studied using *in-situ* neutron diffraction. The role of TiC and $TiSi_2$ on the thermal stability of Ti_3SiC_2 during vacuum annealing is discussed. TiC reacts with $TiSi_2$ between 1400-1450°C to form Ti_3SiC_2. Above 1400°C, decomposition of Ti_3SiC_2 into TiC commenced and the rate increased with increased temperature and dwell time. Furthermore, the activation energy for the formation and decomposition of Ti_3SiC_2 was determined.

KEYWORDS: Ti_3SiC_2, thermal stability, *in-situ* neutron diffraction.

INTRODUCTION

Many attempts have been made to produce new materials with a unique combination of the ductility, conductivity, and machinability of metals, and with the high strength, high modulus, high thermal stability, and superior high-temperature-oxidation resistance of ceramics. Ternary carbides, such as Ti_3SiC_2 and Ti_3AlC_2, are hexagonal layered compounds belonging to a family with the general formula: $M_{n+1}AX_n$, where n is 1, 2 or 3, M is an early transition metal, A is an A-group (mainly group III-A and IV-A) element, and X is either carbon or nitrogen [1-12]. Ti_3SiC_2 has high toughness, high Young's modulus, low hardness, and moderate flexural strength. Furthermore, it exhibits plasticity at high temperature, good electrical conductivity, high thermal shock resistance, and good machinability [3, 5, 7, 9, 11, 13]. The salient combination of properties makes ternary carbides ideal candidate materials for high-temperature applications.

The thermal stability of Ti_3SiC_2 in vacuum has attracted little attention [10, 14-16]. Emmerlich *et al.* [14] investigated the thermal stability of Ti_3SiC_2 thin films and reported that the rapid decomposition of Ti_3SiC_2, associated with Si out-diffusion and de-twining of as-relaxed Ti_3C_2 slabs into oriented $TiC_{0.67}$ layers, is observed when annealing at 1100-1200°C. Radhakrishnan *et al.* [16] reported similar results for the decomposition of Ti_3SiC_2 in vacuum and stated that Ti_3SiC_2 did not dissociate up to 1800°C but was found susceptible to carburization and oxidation. Oo *et al.* [15] studied the thermal stability of Ti_3SiC_2 in argon with low oxygen partial pressure and reported that Ti_3SiC_2 decomposed into TiC and $Ti_5Si_3C_x$. Sun *et al.* [10] reported the occurrence of transformation between α-and β- Ti_3SiC_2, and that α-Ti_3SiC_2 is more stable than β- Ti_3SiC_2.

It is difficult to synthesize Ti_3SiC_2 with 100 % purity due to its high propensity to dissociate into TiC. As indicated in the Ti-Si-C phase diagram (Fig. 1), the equilibrium state of a single phase of Ti_3SiC_2 only occupies a small area which is intersected by the boundaries of TiC, SiC, and $TiSi_2$. In this study, two Maxthal Ti_3SiC_2 samples with different amounts of TiC and $TiSi_2$ were used. The role of TiC and $TiSi_2$ on the thermal stability and phase transition of Ti_3SiC_2 was investigated using *in-situ* high-temperature neutron diffraction. Furthermore, synchrotron radiation diffraction (SRD) was used to characterize the surface composition in as-received and vacuum-annealed samples.

EXPERIMENTAL SAMPLE PREPARATION

Maxthal Ti_3SiC_2 samples (15 mm in diameter and 50 mm in length) were fabricated using a proprietary method developed by Kanthal AB, Sweden. The density of these samples was ~4.47 g/cm^3 with ~1 % porosity. Two Maxthal Ti_3SiC_2 samples (A and B) with different amounts of TiC and $TiSi_2$ were used. The contents of TiC and $TiSi_2$ present in Samples A and B were 34.8 and 7.0 mol %, and 50.8 and 9.9 mol %, respectively.

In-situ neutron diffraction

The collection of high-temperature *in-situ* neutron diffraction data was conducted using Wombat (the high-intensity neutron powder diffractometer) at the OPAL source in Australia. Data were collected using neutrons with incident wavelength of λ = ~1.660 Å from 15 to 135° 2θ at a separation of 0.125°, with the use of the oscillating tertiary collimator. Rietica 1.7.7 was used for phase identification and Rietveld refinement. The optimized parameters during refinement were background coefficients, zero-shift error,

CP1202, *Neutron and X-Ray Scattering in Advancing Materials Research: International Conference – 2009*
edited by A. Saat, H. A. Kassim, M. H. H. Jumali, J. M. Saleh, M. R. Othman, A. Ibrahim, F. M. Idris, and M. H. A.-R. M. Ahmad

peak shape parameters, cell parameters, and anisotropic thermal factors. The residual values of the refinement, statistical reliability factor of Bragg (R_B), R-weighted pattern (R_{wp}), R-expected (R_{exp}), and the goodness-of-fit (χ^2), were evaluated. In Rietica, χ^2 is defined as the square of the ratio of R_{wp} to R_{exp}. Solid cylindrical bars with dimensions 15 mm (Diameter) x 20 mm (Height) cut from as-received samples were used in this study. The temperature of the sample environment was controlled by a closed cylindrical niobium vacuum furnace (10^{-6}-10^{-8} torr). The sample was held by vanadium wire and heated to 1550°C according to the heating protocol shown in Fig. 2. Diffraction patterns were collected every minute.

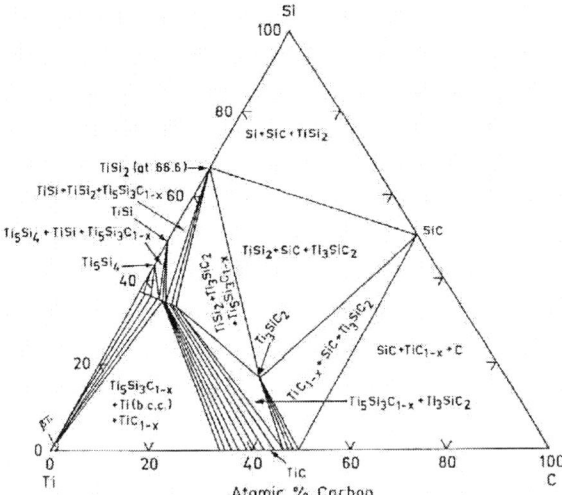

Fig. 1: Ti-Si-C ternary phase diagram at isothermal section at 1250 °C [17].

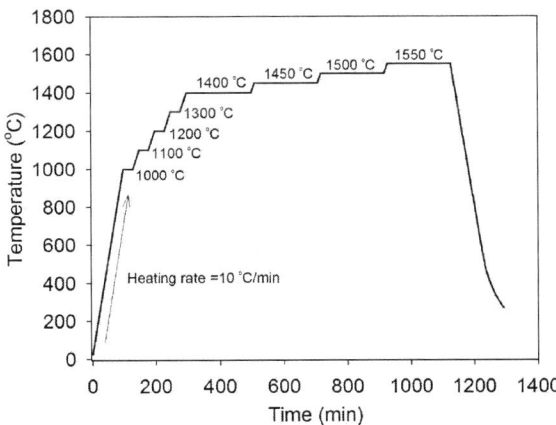

Fig. 2: Heating protocol for in-situ high-temperature neutron diffraction experiment.

Synchrotron radiation diffraction (GISRD)

The diffraction patterns were collected using BIGDIFF (the synchrotron powder diffractometer) at the Australian Synchrotron Research Program Facility (ANBF), beam-line BL-20B, at the Photon Factory, KEK, Tsukuba, Japan. The diffractograms were recorded from 3° to 145° 2θ with a step size of 0.01°. The computer program "*DIFFRAplus* EVA"

was used to identify the crystalline phases present. A diamond blade was used to cut thin slices (~1 mm thick) from the as-received and as-annealed samples. The slices were cleaned ultra-sonically prior to SRD experiment. The compositional information was measured using flat-plate SRD and image plates were used to record the diffraction patterns at 3.0° with a fixed wavelength of 0.7Å.

RESULTS AND DISCUSSION

Phase Transition during Vacuum Annealing

Results in Figures 3a and 3b show the phase transition of Sample A and Sample B annealed in vacuum at temperature up to 1550°C. For Sample A, $TiSi_2$ reacted with TiC during vacuum annealing to form Ti_3SiC_2 at temperatures above 1300°C as follows:

$$TiSi_{2\,(s)} + 2TiC_{(s)} \rightarrow Ti_3SiC_{2\,(s)} + Si_{(g)} \qquad (1)$$

This reaction resulted in the increased amounts of Ti_3SiC_2. By 1500°C, $TiSi_2$ was fully consumed to form Ti_3SiC_2, and at the same time, a pronounced decrease in Ti_3SiC_2 content from 65.4 ±1.22 to 58.3 ±1.22 mol % was observed. The reduction of Ti_3SiC_2 with a complementary increase in TiC suggests that the decomposition of Ti_3SiC_2 into TiC occurs (*via* the sublimation of Ti and Si gaseous) as follows:

$$Ti_3SiC_{2\,(s)} \rightarrow 2TiC_{(s)} + Ti_{(g)} + Si_{(g)} \qquad (2)$$

On the other hand, Sample B, consisting of more TiC than Sample A, showed better stability in vacuum. Here, Ti_3SiC_2 also formed from the reaction of $TiSi_2$ and TiC at 1400-1500°C. Once $TiSi_2$ was depleted, there was only a small decrease in Ti_3SiC_2 content from 1500 to 1500°C. However, within experimental and/or calculation errors, the Ti_3SiC_2 decomposition was not clearly observed, which implies better thermal stability for Sample B in vacuum at up to 1550°C.

A comparison of phase abundances between Samples A and B before and after vacuum annealing is summarized in Table 1. It shows that the consummation of $TiSi_2$ did not contribute to the increase of Ti_3SiC_2 in Sample A, but did increase the amount of Ti_3SiC_2 in Sample B. In other words, Sample A underwent more decomposition, even though the relative molar fraction of Ti_3SiC_2 content remained 58.3±1.19 mol%.

Figures 4 and 5 show the isothermal phase transition at 1400, 1450, 1500, and 1550 °C for Sample A and Sample B, respectively. The results show that phase abundance varies as a linear function of dwell time (< 200 minutes). The slope of Ti_3SiC_2 regression lines can be treated as the rate for either decomposition (negative) or reformation (positive). It is noted that the decomposition rate increases with temperature.

Table 1: Phase abundances before and after vacuum treatment for samples A and B.

Phase abundance (rel. mol%)				
	Sample A		Sample B	
	Before	After	Before	After
Ti_3SiC_2	58.2±1.48	58.3±1.19	39.3±1.46	48.2±1.22
TiC	34.8±1.40	41.8±1.41	50.8±1.34	51.8±1.36
$TiSi_2$	7.0±1.28	0	9.9±1.30	0

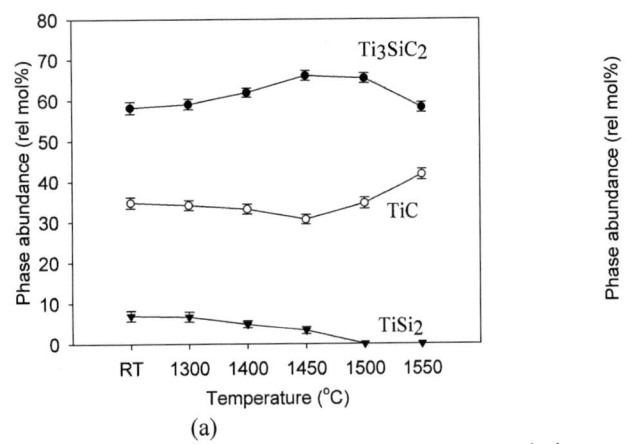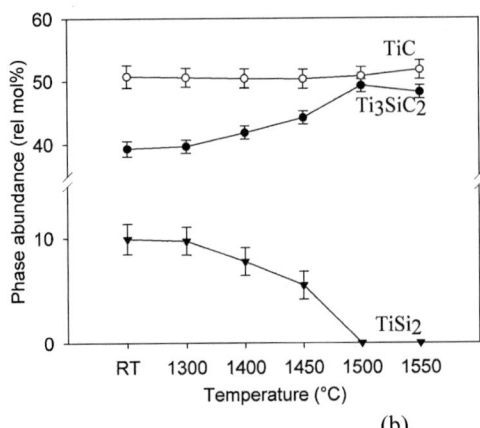

(a) (b)

Fig. 3: Phase abundance as a function of temperature during vacuum annealing for (a) Sample A and (b) Sample B. Errors bars indicate two estimated standard deviations ±2σ.

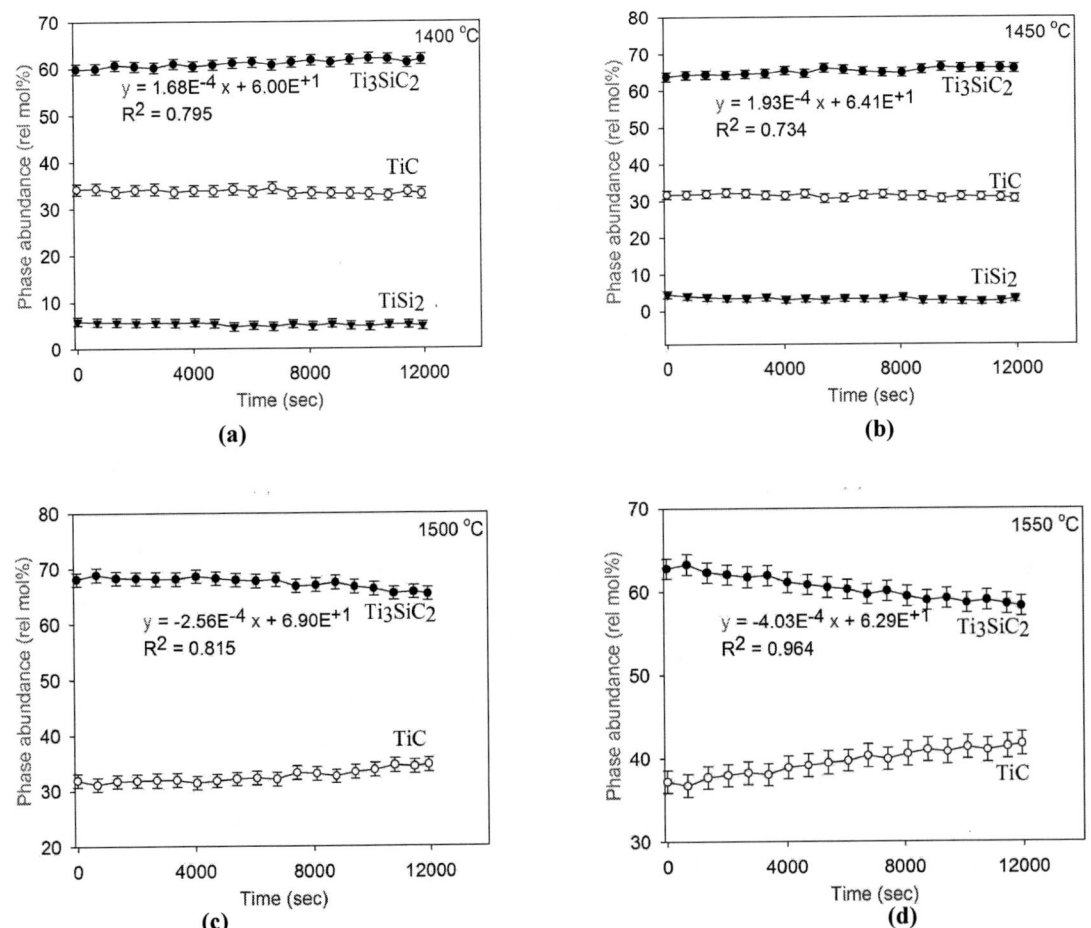

Fig. 4: Phase abundance as a function of dwell time at (a) 1400, (b) 1450, (c) 1500, and (d) 1550°C for sample A. Errors bars indicate two estimated standard deviations ±2σ.

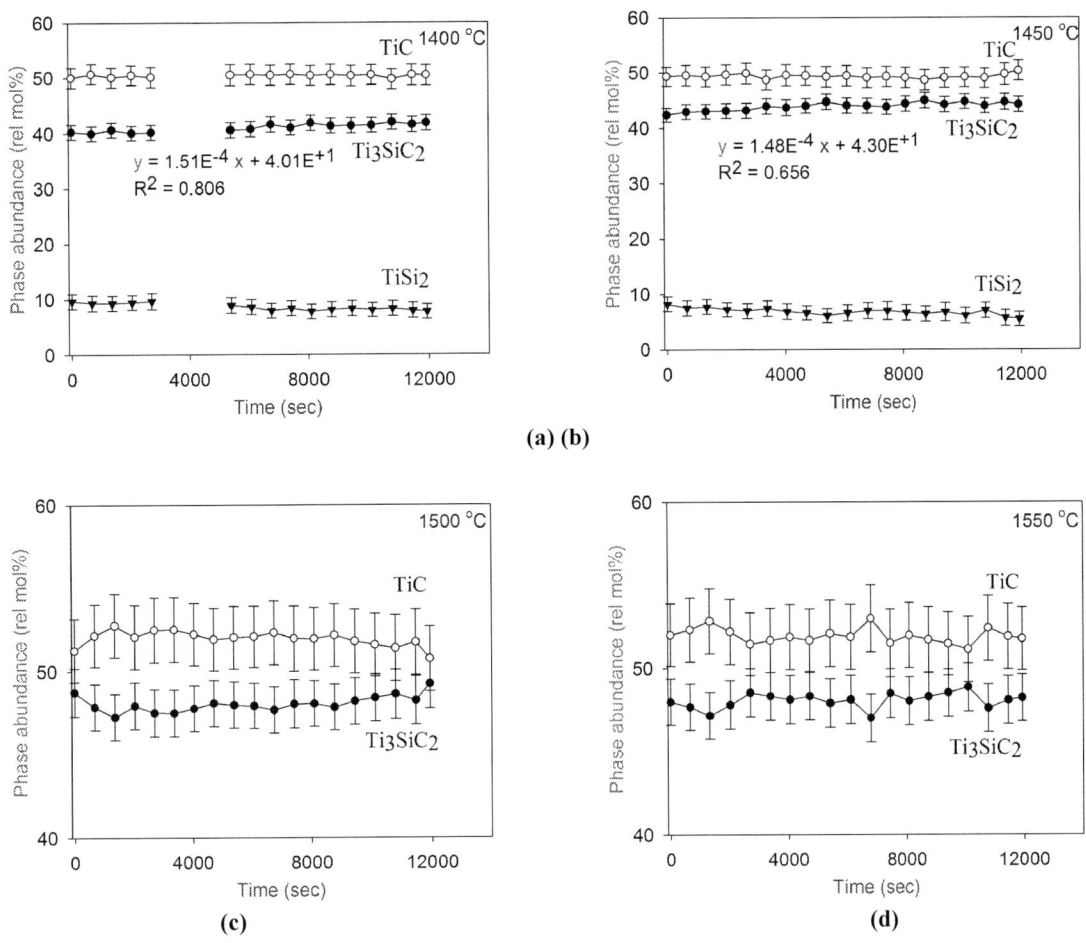

Fig. 5: Phase abundance as a function of dwell time at (a) 1400, (b) 1450, (c) 1500, and (d) 1550°C for sample B. Note that sections of data in (a) were not obtained due to equipment malfunction. Errors bars indicate two estimated standard deviations ±2σ.

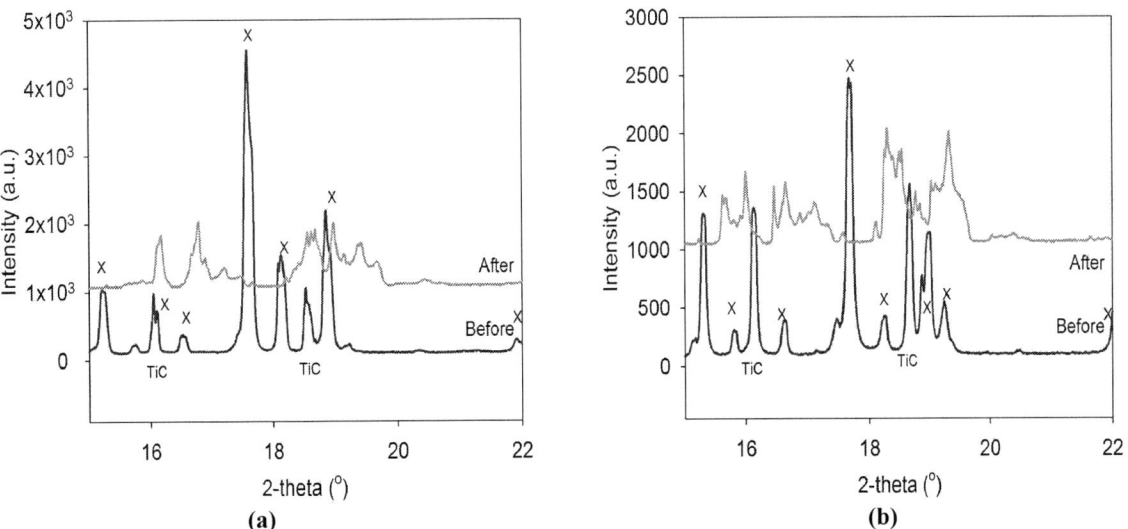

Fig. 6: Comparison of surface composition in (a) Sample A and (b) Sample B before and after vacuum annealing. (Legends: X: Ti₃SiC₂)

For Sample A, the plots show slopes with positive values at 1400 and 1450°C and negative at 1500 and 1550°C, consistent with reformation and decomposition of Ti_3SiC_2 occurring at these temperatures, respectively. Assuming that the decomposition of Ti_3SiC_2 commenced at ≤1400°C, the increased Ti_3SiC_2 content for Sample B implies that the reformation and decomposition of Ti_3SiC_2 appeared rapidly and that the reaction rate for the decomposition was smaller.

For Sample B, only the formation of Ti_3SiC_2 was clearly observed. At 1500 and 1550°C, TiC and Ti_3SiC_2 were in an equilibrium state, no decomposition or reformation was observed. The ratio of TiC: Ti_3SiC_2 = ~1:1 may infer that the concentration of TiC affects the rate of Ti_3SiC_2 decomposition and a compound with 50 mol% Ti_3SiC_2 and 50 mol % TiC may be stable in vacuum up to 1550°C.

Comparison of surface composition in as-received and as-annealed Samples A and B

Fig. 6 shows the comparison of near-surface composition (diffraction pattern) in Samples A and B before and after vacuum annealing up to 1550°C. Certain unknown peaks were observed in the annealed samples. These unknown peaks cannot be identified by the "*DIFFRA^plus EVA*" program, which may be not reported in the current ICSD or PDF database system. This suggests that the thermal dissociation process is more complex than anticipated. More future works will be required to dig out the truth of the thermal dissociation of Ti_3SiC_2 composites.

CONCLUSIONS

A comparative study on the thermal stability of Ti_3SiC_2 with various contents of TiC and $TiSi_2$ was conducted using *in-situ* neutron diffraction. When annealed in vacuum at an elevated temperature, TiC and $TiSi_2$ reacted to form Ti_3SiC_2, but at the same time Ti_3SiC_2 decomposed into TiC *via* the sublimation of Ti and Si. We find here that the reformation of Ti_3SiC_2 is more rapid than its decomposition and that Ti_3SiC_2 with a higher content of TiC and $TiSi_2$ is less susceptible to thermal decomposition in vacuum at up to 1550°C. SRD results show that the thermal dissociation process of Ti_3SiC_2 is much more complicated than expected. More work is required to understand the mechanism of phase dissociation in Ti_3SiC_2.

ACKNOWLEDGEMENTS

This work formed a part of a much broader project on the thermal stability of ternary carbides which is funded by an ARC Discovery-Project grant (DP0664586) and an ARC Linkage-International grant (LX0774743) for one of the authors (IML). The collection of diffraction data was conducted at the OPAL research reactor in ANSTO with financial support from AINSE (08/329). The collection of SRD data was undertaken on the Australian National Beamline Facility at the Photon Factory in Japan, operated by the Australian Synchrotron, Victoria, Australia. We acknowledge the Linkage Infrastructure, Equipment and Facilities Program of the Australian Research Council for financial support (proposal number LE0989759) and the High Energy Accelerator Research Organisation (KEK) in Tsukuba, Japan, for operations support.

REFERENCES

[1]   M. W. Barsoum, *Prog. Solid State Chem.* **28,** 201 (2000).

[2]   M. W. Barsoum and T. El-Raghy, *Am. Sci.* **89,** 334 (2001).

[3]   T. El-Raghy, M. W. Barsoum, A. Zavaliangos and S. R. Kalidindi, *J. Am. Ceram. Soc.* **82,** 2855 (1999).

[4]   J. Emmerlich, D. Music, P. Eklund, O. Wilhelmsson, U. Jansson, J. M. Schneider, H. Högberg and L. Hultman, *Acta Mater.* **55,** 1479 (2007).

[5]   N. F. Gao, Y. Miyamoto and D. Zhang, *Mater. Lett.* **55,** 61(2002).

[6]   E. H. Kisi, J. A. A. Crossley, S. Myhra and M. W. Barsoum, J. Phys. Chem. Solids **59,** 1437 (1998).

[7]   I. M. Low, *J. Eur. Ceram. Soc.* **18,** 709 (1998).

[8]   I. M. Low, E. Wren, K. E. Prince and A. Atanacio, *Mater. Sci. Eng. A* **466,** 140 (2007).

[9]   M. Radovic, M. W. Barsoum, T. El-Raghy, J. Seidensticker and S. Wiederhorn, *Acta Mater.* **48,** 453 (2000).

[10]   Z. Sun, J. Zhou, D. Music, R. Ahuja and J. M. Schneider, *Scripta Mater.* 54 (2006) 105.

[11]   B. J. Kooi, R. J. Poppen, N. J. M. Carvalho, J. T. M. De Hosson and M. W. Barsoum, *Acta Mater.* **51,** 2859 (2003).

[12]   C. Racault, F. Langlais and R. Naslain, *J. Mater. Sci.* **29,** 3384 (1994).

[13]   T. El-Raghy, P. Blau and M. W. Barsoum, *Wear* **238,** 125 (2000).

[14]   J. Emmerlich, D. Music, P. Eklund, O. Wilhelmsson, U. Jansson, J. M. Schneider, H. Hoberg and L. Hultman, *Acta Mater.* **55,** 1479 (2007).

[15]   Z. Oo, I. M. Low and B. H. O'Connor, *Phys. B Cond. Matt.* **385-386,** 499 (2006).

[16]   R. Radhakrishnan, J. J. Williams and M. Akinc, *J. Alloys Compd.* **285,** 85 (1999).

[17]   M.I. Avazov and T.A. Stenashkina, *Neorgani. Mater. Izv. Akad.Nauk SSSR* **11,** 1223 (1965).

STUDY OF STRUCTURAL DISCONTINUITY IN (Ce,Y)PdAl COMPOUNDS AT LOW AND HIGH TEMPERATURES

Martin Rusňák,[1,a] Jiří Prchal,[1,b] Hideaki Kitazawa,[2,c] Zdeněk Matěj[1,d], Toshiaki Furubayashi[2, e]

[1]*Department of Condensed Matter Physics, Faculty of Mathematics and Physics, Charles University, Ke Karlovu 5, 121 16 Prague 2, The Czech Republic*
[a] *rusnak.martin@gmail.com;* [b] *prchal@karlov.mff.cuni.cz;* [d] *matej@karlov.mff.cuni.cz*
[2]*National Institute for Materials Science (NIMS), 1-2-1 Sengen, Tsukuba, Ibaraki 305-0047, Japan*
[c] *KITAZAWA.Hideaki@nims.go.jp;* [e] *furubayashi-tky@umin.ac.jp*

ABSTRACT

The $Ce_{1-x}Y_xPdAl$ pseudoternaries belong to RTX compounds. An anomalous evolution of lattice parameters (a, c) and their ratio (c/a) was observed when varying the substitution parameter x between $0.2 < x < 0.8$ at room temperature. We have studied lattice of the samples from the (Ce,Y)PdAl series in the temperature range from 8 to 1073 K. Existence of a gap of forbidden values of c/a ratio was confirmed. The compounds with $x = 0.2$, 0.3 and 0.8 are single phased with values of c/a above- and below the forbidden gap, respectively, whereas the compounds with $x = 0.4$ and 0.5 exhibit coexistence of both the phases containing c/a values belonging to the range above- and below the forbidden gap over all the range of studied temperatures. New estimation of our work leads to conclusion that the values of the gap vary in dependence on composition and temperature, what is the main difference from the $RTAl$ and $RT_{1-x}T'_xAl$ compounds.

Keywords: CePdAl; Forbidden c/a ratio; Rare-earth; X-Ray diffraction; YPdAl; ZrNiAl

INTRODUCTION

The ternary RTX compounds (where $R \sim$ rare-earth element, $T \sim$ transition metal, $X \sim p$-metal) belong to a large group of 1:1:1 compounds. Quite large part of this group members crystallizes in the hexagonal ZrNiAl-type structure consisting of two types of layers alternating along the c-axis in the ABAB... sequence. These compounds exhibit interesting electronic and magnetic properties. While investigating these properties (Merlo *et al.*, 1998; Jarosz *et al.*, 2000; Talik *et al.*, 2001; Ehlers *et al.*, 1996; Dönni *et al.*, 1999) an unusual structural behaviour (abrupt change of lattice parameters a and c) was observed. Later, compounds with substitutions on the transitional-metal position were studied (Prchal *et al.*, 2004a; Ehlers *et al.*, 1997). There was discovered an existence of a forbidden c/a ratio in the range from 0.565 to 0.575. Using the substitution, it was tried to reach the forbidden ratio values. None of the compounds achieved these values. The compounds were either in the "low c/a" or in the "high c/a" state. Transition from one state to the other one is connected with coexistence of both types of the phases. Both - temperature and composition influence the position of the transition (Prchal *et al.*, 2008). New type of compounds which exhibit such phenomenon is $Ce_{1-x}Y_xPdAl$ (substitution on the rare-earth element position) (Kitazawa *et al.*, 2006; Prchal *et al.*, 2007). Unlike the other ones, its forbidden ratio values are not strictly following the range 0.565 to 0.575. Aim of this work is to investigate lattice parameters evolution of the $Ce_{1-x}Y_xPdAl$ in dependence on temperature (in the range from 8 to 1073 K).

EXPERIMENTAL PROOCEDURE

Five samples with nominal composition $x = 0.2$, 0.3, 0.4, 0.5, 0.8 ($Ce_{1-x}Y_xPdAl$) were prepared by arc-melting from the initial stoichiometric composition of pure elements (with purity for Ce, Y, Pd $\sim$ 3N; for Al $\sim$ 5N) in the monoarc furnace under protection of argon atmosphere.

The samples were checked for their quality by X-Ray powder diffraction at room temperature. The X-ray diffraction (XRD) experiment at low temperatures was performed using Rigaku diffractometer equipped by a low-temperature chamber, where cooling is provided by a closed-cycle cryocooler; installed at National Institute for Materials Science, Tsukuba. The range of temperatures varied from 8 K to 300 K with ~30K step. The high-temperature XRD experiment was performed on an X-Pert Pro MPD diffractometer with a MRI high temperature chamber; installed at Department of Condensed Matter Physics, Prague. The sample is put on a tantalum heating strip providing one heating option, but a platinum radiant heater is used as a main heating source to ensure good temperature homogeneity. The temperature range in this case was 300 K - 1073 K with ~100 K step. At temperatures higher than 773 K, the samples undergo an irreversible transformation to unknown, probably amorphous phase, what prevent us

CP1202, *Neutron and X-Ray Scattering in Advancing Materials Research: International Conference – 2009*
edited by A. Saat, H. A. Kassim, M. H. H. Jumali, J. M. Saleh, M. R. Othman, A. Ibrahim, F. M. Idris, and M. H. A.-R. M. Ahmad
© 2009 American Institute of Physics 978-0-7354-0739-8/09/$25.00

from further studies of the ZrNiAl-type phase at higher temperatures.

Both devices work with Bragg-Brentano geometry. The obtained diffraction patterns were measured with resolution from 0.02° to 0.03° in 2 θ.

Both, the sample and sample holder, were covered by a double-layer and single-layer cover for the low-temperature and the high-temperature experiment, respectively. The covers were evacuated to reduce the thermal exchange between the sample and the outer space and to prevent the sample oxidation at enhanced temperatures. The obtained diffraction patterns were refined using program FullProf (Rodriguez-Carvajal, 1993) with respect to corrections on instrumental deviations.

RESULTS AND DISCUSSION

We have studied the $Ce_{1-x}Y_xPdAl$ compounds with x = 0.2, 0.3, 0.4, 0.5, 0.8 by means of low-temperature *XRD* and, in addition, the samples with x = 0.3, 0.5 were studied by the high-temperature XRD experiment. All the samples were found to crystallize in the hexagonal ZrNiAl-type of structure (space group P6-2m; group number 189). At room temperature the samples with composition x = 0.2, 0.3, 0.8 exhibit single phase diffraction patterns. On the other hand the x = 0.4, 0.5 samples show the coexistence of two different phases (for $Ce_{0.5}Y_{0.5}PdAl$ see Figure 1). Both of the phases have the same symmetry (ZrNiAl-type), the difference is given by change in values of lattice parameters c and a, in accordance with previous studies at room temperature (Prchal *et al.*, 2007). Fitting with this model gives a good agreement as can be seen in Figure 2). The values of the lattice parameters a, c, c/a and R_{Bragg} (defined in Rodriguez-Carvajal, 1993) at room temperature are listed in Table 1.

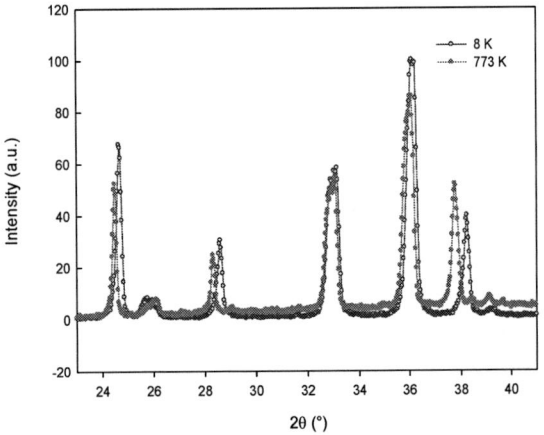

Figure 1: Part of the X-ray diffraction pattern of $Ce_{0.5}Y_{0.5}PdAl$ measured at 8 K and 773 K.

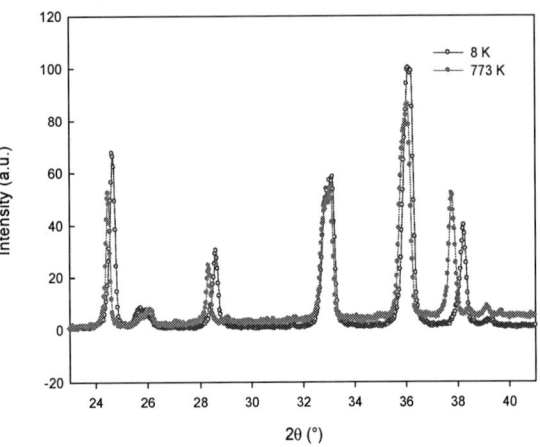

Figure 2: The X-ray diffraction pattern of Ce0.5Y0.5PdAl sample measured at 150 K. The points display the observed relative intensities. Black line passing through them is the calculated refinement.

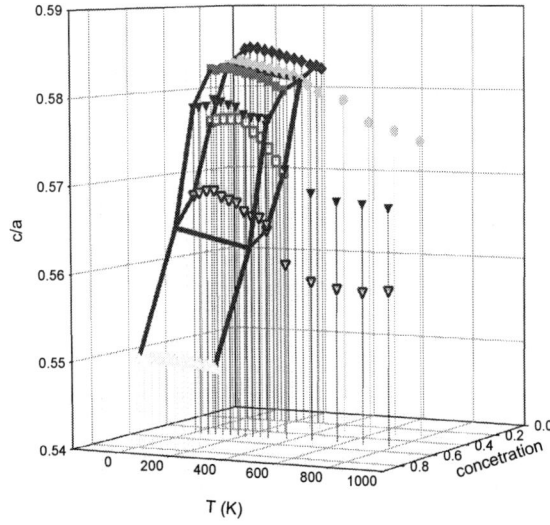

Figure 3: The temperature and concentration evolution of the c/a ratio of the lattice parameters a and c of the $Ce_{1-x}Y_xPdAl$ compound. The thick lines set bounds to the zone of forbidden values of c/a.

The existence of the two phases is connected with existence of a gap of forbidden c/a values. These critical values were already studied for $RTAl$ and $RT_{1-x}T'_xAl$ compounds (Prchal *et al.*, 2008) with conclusion of a criticality of the lattice when the c/a ratio crosses the values around 0.565-0.575. Our study confirms an existence of a forbidden gap, whereas in the case of (Ce,Y)PdAl compounds the gap varies in values of c/a that are not constant within both - temperature and composition. In these compounds the critical c/a ratio depends on temperature and constitution (see Table 2 and Figure 3). Varying the Y-concentration parameter from 0 to 1 at room temperature the forbidden gap changes from (0.570-0.579 for $x = 0.4$) to (0.555-0.566 for $x = 0.6$ (Prchal *et al.*, 2007)). For temperature evolution of the sample $x = 0.5$ the critical values of c/a can be found in the range from (0.568 - 0.578 for $T = 8$ K) to (0.558 - 0.567 at 773 K). Both - the chemical composition (different atomic radii) and temperature (change of the lattice parameters due to atomic thermal movement) influence the critical values of c/a in the $Ce_{1-x}Y_xPdAl$ compounds which form a quasi plane in the phase space.

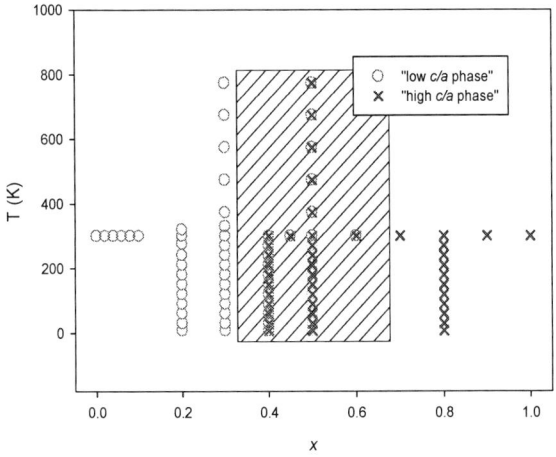

Figure 4: The phase diagram for the $Ce_{1-x}Y_xPdAl$ compound. Filled part marks the area of coexistence of both of the phases. It is obvious that border between the phases is parallel to the temperature. Additional data for compounds with $x = 0.0$ to 0.1 and 0.9 to 1.0 were taken from Prchal *et al.* (2008).

In contrast to previously studied systems with substitution on transition-metal position, where boundary between lattices of compounds having c/a above- and below the forbidden values (0.565 - 0.575) is crossing the "vector of temperature direction" (see Fig. 10 in Prchal *et al.* (2008)), so that the transition can be observed by varying the temperature; in the case of $Ce_{1-x}Y_xPdAl$ the border between the phases is parallel to the temperature direction and position of the boundary is almost temperature independent (see Figure 4).

The difference between the constant range of forbidden critical values - as in previously studied systems - and temperature- and concentration dependent ranges - as in the present studied system - are caused by the fact that substitution in the $Ce_{1-x}Y_xPdAl$ series is made on the position of rare-earth element. Since the rare-earth element in the ZrNiAl-type structure is present only in one type of the layers alternating along the c-axis.

Contrary to the transition metal, which is situated in both types of the layers), thus also the substitution takes place only within one of the two types of layers in $Ce_{1-x}Y_xPdAl$. This is the main difference from the previously studied $RT_{1-x}T'_xAl$ compounds. The R-Pd layer is thus directly affected by the substitution. The Pd-Al layer is affected by the Ce-Y substitution indirectly through the interlayer bonding. Because of conserving the crystallographic symmetry, also after the substitution the lattice parameter a has to be equal in both types of the layers. The volume of the unit cell shifts correspondingly to the difference of the volumes of the atoms substituted. In the layered type structure of $Ce_{1-x}Y_xPdAl$ it is more preferred to change the lattice parameter c, which is mainly affected by the weaker interlayer bonding, than to vary a in the Pd-Al layer, which is mainly bounded by the stronger interplanar bonding (the nearest neighbouring atoms are can be found within the planes). This leads to the conclusion that the governing parameter for the transition in our case is c. This statement is supported by the value of change in c compared to the change in a. In $Ce_{0.5}Y_{0.5}PdAl$ c increases from 406.83 pm to 414.67 pm at 300 K, i.e. by 2 % what is 7 times more than change in a, which at the same point changes from 720.93 pm to 719.20 pm. Contrary to that, in previously studied systems, the relative change of the parameters a and c was 0.6% and 1%, respectively for TbNiAl at 110 K (Prchal *et al.*, (2008)); also 1.3% in a and 2.8% in c for $ErNi_{1-x}Cu_xAl$ (Prchal *et al.*, (2004b)) between $x = 0.5$ and 0.6 at room temperature, what means only about

Table 1: The structural parameters of the $Ce_{1-x}Y_xPdAl$ compounds at room temperature

		a(pm)		c(pm)		c/a		R_{Bragg}
Low-T experiment								
$Ce_{0.8}Y_{0.2}PdAl$		720.17	(12)	419.79	(09)	0.58290	(22)	14.8
$Ce_{0.7}Y_{0.3}PdAl$		718.18	(20)	417.53	(14)	0.58140	(35)	13.2
$Ce_{0.6}Y_{0.4}PdAl$	phase 1	718.97	(37)	410.28	(27)	0.57060	(67)	9.7
	phase 2	718.26	(33)	416.81	(24)	0.58030	(60)	10.3
$Ce_{0.5}Y_{0.5}PdAl$	phase 1	720.93	(36)	406.83	(25)	0.56430	(63)	12.3
	phase 2	719.20	(45)	414.67	(34)	0.57660	(84)	13.1
$Ce_{0.2}Y_{0.8}PdAl$		722.79	(50)	397.15	(31)	0.54950	(81)	24.0
High-T experiment								
$Ce_{0.7}Y_{0.3}PdAl$		719.07	(72)	417.67	(47)	0.5808	(12)	23.5
$Ce_{0.5}Y_{0.5}PdAl$	phase 1	721.20	(59)	407.44	(42)	0.5649	(10)	17.5
	phase 2	719.59	(78)	415.42	(53)	0.5773	(14)	16.8

Table 2: The c/a ratio of the structural parameters a and c of the $Ce_{1-x}Y_xPdAl$ compounds at low- and high temperature

T (K)	Ce0.8Y0.2PdAl		Ce0.7Y0.3PdAl		Ce0.6Y0.4PdAl				Ce0.5Y0.5PdAl				Ce0.2Y0.8PdAl	
					phase 1		phase 2		phase 1		phase 2			
8 K	0.58482	(41)	0.58305	(38)	0.57646	(82)	0.58269	(82)	0.56808	(82)	0.57844	(78)	0.55005	(80)
30 K	0.58500	(39)	0.58325	(38)	0.57673	(81)	0.58259	(76)	0.56840	(87)	0.57853	(81)	0.55014	(82)
60 K	0.58494	(37)	0.58363	(38)	0.57682	(86)	0.58284	(78)	0.56869	(86)	0.57865	(81)	0.55022	(78)
90 K	0.58469	(38)	0.58329	(37)	0.57684	(87)	0.58279	(77)	0.56873	(81)	0.57943	(74)	0.55022	(93)
120 K	0.58450	(36)	0.58299	(37)	0.57687	(78)	0.58252	(70)	0.56807	(82)	0.57919	(73)	0.55025	(89)
150 K	0.58419	(35)	0.58290	(35)	0.57653	(81)	0.58231	(74)	0.56768	(80)	0.57884	(74)	0.55014	(86)
180 K	0.58388	(31)	0.58268	(36)	0.57537	(78)	0.58214	(74)	0.56738	(78)	0.57862	(74)	0.55011	(89)
210 K	0.58371	(34)	0.58244	(36)	0.57473	(77)	0.58176	(79)	0.56640	(74)	0.57771	(75)	0.55012	(90)
240 K	0.58341	(32)	0.58212	(35)	0.57343	(77)	0.58144	(74)	0.56586	(66)	0.57743	(71)	0.54988	(94)
270 K	0.58296	(28)	0.58185	(36)	0.57201	(64)	0.58077	(59)	0.56573	(65)	0.57737	(73)	0.54974	(93)
300 K	0.58289	(22)	0.58138	(35)	0.57064	(67)	0.58030	(60)	0.56432	(63)	0.57657	(84)	0.54948	(81)
300 K			0.5808	(12)					0.5649	(10)	0.5773	(14)		
373 K			0.5800	(13)					0.5604	(12)	0.5714	(13)		
473 K			0.5791	(13)					0.5585	(11)	0.5687	(12)		
573 K			0.5765	(13)					0.5578	(11)	0.5679	(13)		
673 K			0.5756	(12)					0.5576	(09)	0.5676	(10)		
773 K			0.5744	(14)					0.5579	(11)	0.5673	(13)		

2-times difference between the changes of the respective lattice parameters. We assume that the different character of the substitution of the compounds is the main reason for the temperature and composition dependence of the forbidden values of the c/a.

CONCLUSIONS

The $Ce_{1-x}Y_xPdAl$ samples with composition $x = 0.2$, 0.3, 0.8 exhibit single phase, whereas the compounds with $x = 0.4$, 0.5 exhibit coexistence of two different phases over the whole studied range of temperatures (8-773 K). The coexistence of the phases is connected with a gap of forbidden values of c/a which evolves with temperature and composition. This is the main difference from previously studied systems, where the substitution on the position of transition-metal element causes changes in both of the layers, whereas in our case of (Ce,Y)PdAl compounds the substitution occurs only in one of the hexagonal layers altering along the c-axis.

ACKNOWLEDGMENTS

This work is a part of a research plan MSM0021620834 financed by the Ministry of Education of the Czech Republic.

REFERENCES

Dönni, A., Kitazawa, H., Fischer, P., Fauth, F., (1999), Evidence for an isostructural phase transition in the metastable high-temperature modification of TbPdAl, *Journal of Alloys and Compounds* 289: 11-17.

Ehlers, G., Maletta, H., (1996), Magnetic order in TbNiAl and TbCuAl intermetallic compounds, *Zeitschrift für Physik B* 99: 145-150.

Ehlers, G., Ahlert, D., Ritter, C., Miekeley, W., Maletta, H., (1997), Anomalous transition from antiferromagnetic to ferromagnetic order in the pseudoternary series TbNi1-xCuxAl, *Europhysics Letters* 37: 269-274.

Kitazawa, H., Prchal, J., Tsujii, N., Imai, M., Kido, G., (2006), Magnetic properties of $Ce_{1-x}Y_xPdAl$ Kondo-lattice system, *Physica B* 378-380: 803-804.

Jarosz, J., Talik, E., Mydlarz, T., Kusz, J., Böhm, H., Winiarski, A., (2000), Crystallographic, electronic structure and magnetic properties of the GdTAl; T = Co, Ni and Cu ternary compounds, *Journal of Magnetism and Magnetic Materials* 208: 169-180.

Merlo, F., Cirafici, S., Canepa, F., (1998), Structural anomaly in GdNiAl: a crystallographic, electric and magnetic investigation, *Journal of Alloys and Compounds* 266: 22-25.

Prchal, J., Javorský, P., Daniš, S., Jurek, K., Dlouhý, J., (2004a), Structural study of the DyNi$_{1-x}$Cu$_x$Al system, *Czechoslovak Journal of Physics* 54: D315-D318.

Prchal, J., Javorský, P., Sechovský, V., Dopita, M., Isnard, O., Jurek, K., (2004b), Development of magnetic order in the pseudo-ternary series ErNi$_{1-x}$Cu$_x$Al, *Journal of Magnetism and Magnetic Materials* 283: 34-45.

Prchal, J., Kitazawa, H., Suzuki, O., (2007), Structural anomaly in the Ce$_{1-x}$Y$_x$PdAl pseudo-ternary series, *Journal of Alloys and Compounds* 437: 117-119.

Prchal, J., Javorský, P., Rusz, J., de Boer, F., Diviš, M., Kitazawa, H., Dönni, A., Daniš, S., Sechovský, V., (2008), Structural discontinuity in the hexagonal RTAl compounds: Experiments and density-functional theory calculations, *Physical Review B* 77: 134106.

Rodriguez-Carvajal, J., (1993), Recent advances in magnetic structure determination by neutron powder diffraction, *Physica B* 192: 55-69.

Talik, E., Skutecka, M., Kusz, J., Böhm, H., Jarosz, J., Mydlarz, T., Winiarski, A., (2001), Magnetic properties of GdPdAl single crystals, *Journal of Alloys and Compounds* 325: 42-49

The Commissioning Of A Three Dimensional Depolarized Neutron Beamline For Studying The Magnetic Correlation Length Of A Magnetic Material At Tsing Hua Open-Pool Reactor

Hui-Chia Su[1], Chih-Hao Lee[1,2*], Hsin-Ho Chang[1], Yu-Han Wu[1,2] Chih-Wei Hu[1], L. J. Chang[2], A. Ioffe[3], and H.T. Kraan[4]

[1]Department of Engineering and System Science, National Tsing Hua University, Hsinchu, Taiwan 30013. * Corresponding author: Chih-Hao Lee, chlee@mx.nthu.edu.tw, tel:886-3-5715131 ext. 34281

[2]Nuclear Science and Technology Development Center, National Tsing Hua University, Hsinchu, Taiwan 30013.

[3]Jülich Centre for Neutron Science (JCNS) Forschungszentrum Jülich Outstation at FRM-II, Lichtenbergstrasse 1, D-85747 Garching, Germany

[4]Delft University of Technology, Delft, Netherlands.

ABSTRACT

A new three dimensional neutron depolarization beamline was commissioning at W3 beam port of Tsing Hua Open Pool Reactor (THOR, 2 MW) located in Hsinchu, Taiwan. The beamline consists of a pair of Heusler crystals as neutron polarizer and analyzer. Two three-dimensional rotators were placed before and after the sample position. The sample chamber was enclosed in a μ-metal enclosure with a gradient field at incident and exit ports. Typically, 0.237 nm of monoenergetic neutron was selected for experiments. A flipping ratio of 23 can be obtained which corresponds to a polarization ratio of 92% of this beamline. The typical intensity is 400 n/cm^2s at 1 MW operation. Magnetic correlation lengths of Ni-ferrite powders with different packing densities were measured to understand the characteristics of this beamline. The magnetic correlation lengths were observed to be 2 μm at virgin state and about 3.1 μm at remanent state. This magnetic correlation length in the virgin state is similar than the particle size. No significant change of domain size at packing density up to 60% implies that the domain wall motion is hindered by the porosity of the sample.

Keywords: neutron beamline, neutron depolarization, Ni-ferrite, magnetic correlation length, polarized neutron

PACS No.: 61.12.-q, 61.12.Ha, 77.22.Ej

INTRODUCTION

Three dimensional neutron depolarization (ND) is a powerful tool to measure the magnetic correlation length (magnetic domain) of a magnetic material. It is complementary to the polarized small angle neutron scattering (SANS). However, the neutron intensity needed for a ND experiment is much smaller than SANS because the ND setup is a transmission type experiment. Therefore, ND experiment is suitable for a medium or low flux neutron reactor. In the following, we will describe a commission job of an ND experiment at Tsing Hua Open Pool Reactor (THOR, 2 MW). In 1941, the one dimensional ND principle was delivered firstly by Halpern and Holstein [1], and then extended to three-dimension (3D) by Rekveldt, Maleev and Ruban [2]. An ND technique can show the statistical average of the magnetic correlation length in the interior of a bulk magnetic material non-destructively [3]. The static magnetic correlation length with mean magnetization aligned to a mean orientation can be obtained by collecting the transmitted polarized neutrons pass through the magnetic material [4-6]. The depolarization matrix $D = P/P_0$ is an experimental parameter for analyzing the experimental data, where P_0 and P are the polarization vectors before and after neutrons transmitting through the sample. The change of polarization vector during neutrons transmitting through the samples is due to the influence of fluctuation in B(r) in the local magnetic induction ΔB around the mean magnetic induction <B>. ΔB is denoted as $\Delta B = B(r) - <B>$ [7-8]. The neutron intensity, the neutron polarization ratio, and the neutron wavelength affect the precision of the experimental data and the measurable magnetic domain sizes. The range of magnetic domains that can be measured by ND is from 10 nm up to mm's theoretically, if a good statistical data can be collected

CP1202, Neutron and X-Ray Scattering in Advancing Materials Research: International Conference – 2009
edited by A. Saat, H. A. Kassim, M. H. H. Jumali, J. M. Saleh, M. R. Othman, A. Ibrahim, F. M. Idris, and M. H. A.-R. M. Ahmad
© 2009 American Institute of Physics 978-0-7354-0739-8/09/$25.00

Table 1 The characteristics of the components of THOR-W3 beamline

Component	Position	Function	Neutron flux (neutrons/cm^2-s)
Filter 18 cm sapphire single crystal + 10 cm bismuth block	Inside the biology shielding, before the shutter	To filter γ-ray and fast neutrons from the reactor core	10^9 at the reactor face
Monochromator (Highly Oriented Pyrolitic Graphite) (HOPG (0002)) (Mosaic angle 0.4°)	1 m from the reactor face	To select the neutron wavelength (0.16, 0.237, and 0.45 nm were chosen)	$10^7 – 10^8$
Polarizer (Heusler crystal) (Cu$_2$MnAl (111))	1 m from the monochromator	To align the neutron spins in the same direction	- - - - - - -
Magnetic guide	Along the path of polarized neutrons	Set along the neutron path between polarizer and analyzer to make sure the direction of neutron spin would not be fluctuated.	- - - - - - -
3D spin rotator	Before/after the sample stage	To rotate the neutron spins to a specific direction.	- - - - - - -
Sample stage	1.5 m from the polarizer	For samples, minimum motion step of the motors is about 0.005°	4×10^3 at λ= 0.237 nm
Analyzer (Heusler crystal) (Cu$_2$MnAl (111))	0.45 m from the sample stage	To analyze the direction of neutron spins	- - - - - - - -
^{3}He proportional detector, which is enclosed by a $25 \times 25 \times 100$ cm^3 B$_4$C shielding house	0.05 m after the sample stage, and also after the analyzer	Detector: counting neutrons Shielding house: to prevent the stray scattering neutrons from striking the surrounding materials	4×10^2 at λ= 0.237 nm

[9]. The statistical errors coming from the neutron intensity can be reduced by extending the data

collecting time. The ND experiment was setup at W3 port of THOR [10]. To verify the correctness of ND experiment, a Ni-ferrite power was prepared and test. The final result is also double checked at PANDA beamline at Delft University at Netherlands [4-9,11-12]. A consistent result run at PANDA was also obtained.

INSTRUMENTATION SETUP AND EXPERIMENTAL

Figure 1 shows the schematic diagram of THOR-W3 beamline. Thermal neutrons diffused from the reactor core are monochromated by highly oriented pyrolitic graphite (HOPG (0002)) with a mosaic angle of 0.4° and three neutron wavelengths, 0.16 nm, 0.237 nm, and 0.4 nm can be selected. Usually, concrete plugs are inserted into the beam port of the monochromator tank and left only one port for the selected energy during the experiment. A filter combining with a 10 cm bismuth and an 18 cm sapphire single crystal is set right before the shutter, to filter out the γ-rays and fast neutrons from the reactor core. The neutron flux is counted by a ^{3}He proportional detector which is enclosed by a 25 x 25 x 100 cm^3 B$_4$C shielding house to prevent the stray

scattering neutrons from the surrounding materials. The neutron flux at each component is listed in Table 1. The background is about 0.3 – 0.4 counts/s; therefore, the signal-to-noise ratio is about 1000. For polarized neutron beamlines, a pair of Heusler (Cu$_2$MnAl (111)) crystals were used as a polarizer and an analyzer. The polarizer is usually set before the sample stage to align the neutron spins in the same direction. After reflecting from the polarizer, the neutron flux at sample stage is about 4 x 10^3 counts/cm^2-s. The analyzer is set after the sample stage. After reflecting from the analyzer, the neutron flux collected by the detector is about 400 counts/cm^2-s. Two 3D adiabatic spin rotators were installed to rotate the spin orientations of the neutrons. The polarization ratio currently is about 92%.

The nickel ferrite samples were prepared from a mixture of powders containing with the composition of mainly γ-Fe$_2$O$_3$ (78%) and NiO (20.5%), together with slight additives, such as Co (0.5%) and MnO (1%). The powders are sintered at 1350 °C and then ball milling 24 h to have a sample with particle size of about 2-5 μm. The low packing density nickel ferrite was prepared by mixing the powder with polymer (PDMS); the high density samples were prepared by compression under high pressure. With a pressure of 100 bars, the maximum packing density is about 60% of the bulk value. To do the neutron depolarization experiments, samples of 20 mm x 20 mm x 2 mm were prepared. The sample was mounted in a solenoid closed by a soft magnetic yoke. The sample is measured at $\lambda = 0.237$ nm.

$\lambda = 0.237$ nm) and the uncertainty under there three selected neutron wavelengths at THOR-W3 beamline were calculated and plotted in Fig. 2.

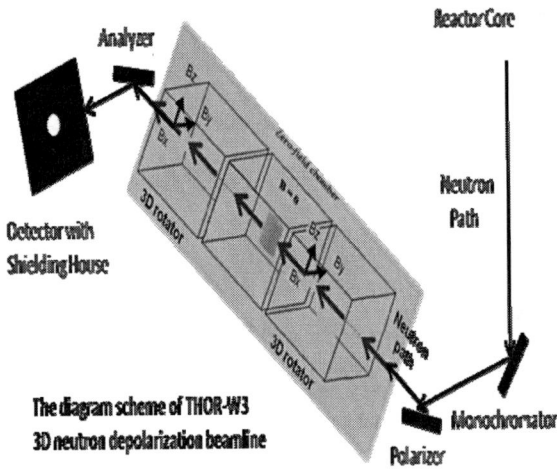

Fig. 1 The schematic diagram of THOR-W3 neutron beamline for 3D neutron depolarization experiments.

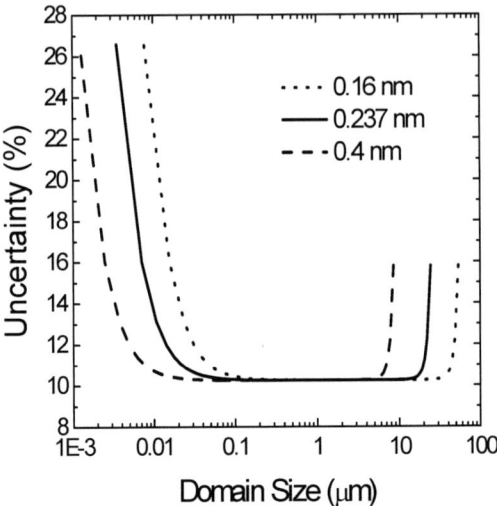

Fig. 2 The uncertainty of the measurable magnetic domain size under the different selected neutron wavelengths, 0.16, 0.237, and 0.4 nm. The measurable magnetic domain ranges corresponding to these selected wavelengths are 0.008 – 7 μm ($\lambda = 0.16$ nm), 0.04 ; 20 μm ($\lambda = 0.237$ nm), and 0.1 ; 50 μm ($\lambda = 0.4$ nm), respectively.

RESULTS AND DISCUSSION

The remanent magnetization was obtained from the rotation of the polarization of the transmitted beam assuming the model of small magnetic domains ($\omega t \ll 1$, where the ω is the neutron angle velocity in each magnetic domain and t is the time for a neutron passed through the small domain). The remanence can be obtained from the rotation angle Φ by:

$$\mu_0 M = \Phi/(\sqrt{c}L_w), \qquad (1)$$

where $\Phi = \tan^{-1}(D_{xy} - D_{yx})/(D_{xx} + D_{yy})$, c is 2.18 x 10^{29} $T^{-2}m^{-4}$. The applied magnetic field, H_{app} is perpendicular to beam direction which is parallel to the x-axis and parallel to the z-axis. From the rotation angle measured, the magnetic filling fraction = 0.9 can be calculated from the magnetization of the sample and the bulk material. From the depolarization matrix, the correlation length of the fluctuation of the local induction $\mu_0 M$ can be calculated as following equation [11]:

$$\zeta = \frac{-\ln(\det D)}{c\lambda^2 (\mu_0 M)^2 \varepsilon L_W}, \qquad (2)$$

where L_w is the effective thickness of the sample [12]. Typically, the error of the neutron wavelength is about 1%. The statistical error of the local induction and the effective thickness are both about 5%. Therefore, the measurable range of the magnetic correlation length (0.04 – 20 μm at

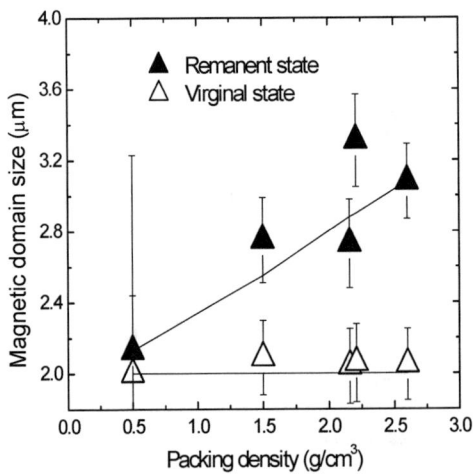

Fig. 3 The magnetic correlation length of Ni-ferrite at different densities measured under the virgin state (open triangle) and the remanent state (solid triangle).

For the packing density of 60%, the measured magnetic correlation length of the virgin and the remanent states are about 2 μm and 3.1μm, respectively. The magnetic correlation lengths are slightly increased as the packing density increases (see Fig. 3) at the remanent state but independent of the packing density at the virgin state. The magnetic domain sizes are larger at remanent states. These domain sizes measured are comparable to the particle sizes, which indicates that no cross talk between two near particles and the domain wall motion is pinned by the porosity.

CONCLUSION

In summary, the THOR-W3 neutron beamline was constructed for 3D ND experiments to study the magnetic correlation length of a magnetic material. The depolarization phenomena can be observed clearly on the Ni-ferrite samples. The neutron beamline at THOR-W3 is ready for studying the magnetic correlation length of a bulk magnetic material.

ACKNOWLEDGE

The authors would like to thank all members of the reactor division in the Nuclear Science and Technology Development Center, National Tsing Hua University for their engineering work at THOR-W3 beamline and the financial support from National Science Council of Taiwan under the contact no. NSC96-2793-M007-001.

REFERENCES

[1] O. Halpern, T. Holstein, Phys. Rev. 59 (1941) 960.

[2] R. Rosman *et al.*, Z. Phys. B 79 (1990) 61.

[3] A. Hubert, and R. Schäfer, "Magnetic Domains: The analysis of magnetic microstructures" (Spriger-Verlag Berlin Heidelberg, 1998).

[4] R. Rosman, "Magnetic particles studied with neutron depolarization and small-angle neutron scattering", Ph.D. thesis, Technique University Delft.

[5] M. Th. Rekveldt, and F. J. van Schaik, J. Magn. Magn. Mater. 14 (1979) 325.

[6] M.Th. Rekveldt, Physica B 267-268 (1999) 60.

[7] S. V. Grigoriev, S. V. Maleyev, A. I. Okorokov, H. Eckerlebe, and N. H. van DijK, Phys. Rev. B 69 (2004) 134417.

[8] R. Rosman, and M. Th. Rekveldt, J. Magn. Magn. Mater. 95 (1991) 319.

[9] R. Rosman, and M. Th. Rekveldt, Phys. Rev. B 43 (1991) 8437.

[10] H. C. Su, C. W. Hu, J. J. Peir, and C. H. Lee, Chinese J. Phys. 45 (2007) 374.

[11] P. T. Por, W. H. Kraan, P. I. Mayo, M. Th. Rekveldt, and K. O. Grady, J. Magn. Magn. Mater. 162 (1996) 139.

[12] S. V. Grigoriev, S. V. Maleyev, A. I. Okorokov, and V. V. Runov, Phys. Rev. B 58 (1998) 3206.

Calculation of Ion Charge State Distributions After Inner-Shell Ionization in Xe Atom

Adel M. Mohammedein, Adel A. Ghoneim, Kandil M. Kandil, Ibrahim M. Kadad

Applied Sciences Department, College of Technological Studies P.O. Box 42325, Shuwaikh 70654, Kuwait.
e-mail : admohamed@yahoo.com, am.mohammedein@paaet.edu.kw

ABSTRACT

The vacancy cascades following initial inner-shell vacancies in single and multi-ionized atoms often lead to highly charged residual ions. The inner-shell vacancy produced by ionization processes may decay by either a radiative or non-radiative transition. In addition to the vacancy filling processes, there is an electron shake off process due to the change of core potential of the atom. In the calculation of vacancy cascades, the radiative (x-ray) and non-radiative (Auger and Coster-Kronig) branching ratios give valuable information on the de-excitation dynamics of an atom with inner-shell vacancy. The production of multi-charged ions yield by the Auger cascades following inner shell ionization of an atom has been studied both experimentally and theoretically. Multi-charged Xe ions following de-excitation of K-, L_1-, $L_{2,3}$-, M_1-, $M_{2,3}$- and $M_{4,5}$ subshell vacancies are calculated using Monte-Carlo algorithm to simulate the vacancy cascade development. Fluorescence yield (radiative) and Auger, Coster- Kronig yield (non- radiative) are evaluated. The decay of K hole state through radiative transitions is found to be more probable than non-radiative transitions in the first step of de-excitation. On the other hand, the decay of L, M vacancies through non-radiative transitions are more probable. The K shell ionization in Xe atom mainly yields Xe^{7+}, Xe^{8+}, Xe^{9+} and Xe^{10+} ions, and the charged X^{8+} ions are the highest. The main product from the L_1- shell ionization is found to be Xe^{8+}, Xe^{9+} ions, while the charged Xe^{8+} ions predominate at $L_{2,3}$ hole states. The charged Xe^{6+}, Xe^{7+} and Xe^{8+} ions mainly yield from $3s_{1/2}$ and $3p_{1/2,3/2}$ ionization, while Xe in $3d_{3/2,5/2}$ hole states mainly turns into Xe^{4+} and Xe^{5+} ions. The present results are found to agree well with the experimental data.

Keywords: Auger cascade, multiple charged ions, charge state distributions

INTRODUCTION

The vacancy cascades following initial inner-shell vacancies in a single and multi-ionized atom often lead to highly charged residual ions. The inner-shell vacancy produced by ionization processes may decay by either a radiative or non-radiative transition. In addition to the vacancy filling processes, there is an electron shake off process due to the change of core potential of the atom. In the calculation of vacancy cascades, the radiative (x-ray) and non-radiative (Auger and Coster-Kronig) branching ratios give valuable information on the de-excitation dynamics of an atom with inner-shell vacancy. The production of multi-charged ions yield by the Auger cascades following inner shell ionization of an atom has been studied both experimentally and theoretically. Krause et al. [1] and Carlson et al. [2-5] measured the multicharged ions formed as a result of relaxation of inner-shell vacancies in rare gases using characteristic x-ray. Holland et al. [6] studied the partial cross-section for multiple photoionization in rare gases using a time-of-flight (TOF) mass spectrometer and synchrotron radiation. The multiple photoionization of Xe in the L- and M- subshells regions using synchrotron radiation are measured [7-9].

There are two major theoretical methods used to calculate the vacancy cascades originating from initial inner-shell vacancies. The first model is based on straightforward construction of de-excitation decays [10-17]. The second method is based on Monte Carlo technique to simulate possible de-excitation pathways including the radiative and non-radiative transition processes [18-22].

In the present work, Mont Carlo technique is adapted to calculate the multiply charged ions following K-, L- and M- subshells vacancies creation in Xe atom. The radiative (x-ray) and non-radiative (Auger and Coster-Kronig) branching ratios for ionization are evaluated using Multiconfiguration Dirac Fock wave functions from Grant et al. [23] and Dirac Fock Slater wave

CP1202, *Neutron and X-Ray Scattering in Advancing Materials Research: International Conference – 2009*
edited by A. Saat, H. A. Kassim, M. H. H. Jumali, J. M. Saleh, M. R. Othman, A. Ibrahim, F. M. Idris, and M. H. A.-R. M. Ahmad
© 2009 American Institute of Physics 978-0-7354-0739-8/09/$25.00

functions from Lorenz et al. [24], respectively. The results of multicharged Xe^{i+} ions are compared with the available theoretical and experimental data.

CALCULATION METHOD

Monte- Carlo technique is used to simulate vacancy cascades following K- L- and M- subshells in Xe atom. Radiative and non- radiative transitions and electron shake off processes are considered. The details of the calculations are described in El-Shemi et al [21] and Abdullah et al. [22]. A brief description of the method is given in the following.

The simulation of each cascade begins with the implementation of atomic data for all possible radiative and non-radiative transitions for a single ionized Xe atom. The radiative and non-radiative branching ratios which give valuable information on the de-excitation dynamic of an atom with a core vacancy, are calculated as following:

fluorescence yield:

$$\omega(C_i \rightarrow C_f) = \frac{\Gamma_R(C_i \rightarrow C_f)}{\Gamma(C_i)} \quad (1)$$

And Auger yield:

$$a(C_i \rightarrow C_f) = \frac{\Gamma_A(C_i \rightarrow C_f)}{\Gamma(C_i)} \quad (2)$$

Where the initial configuration C_i will decay into final configurations C_f. The sum of partial widths $\Gamma(C_i)$ of all radiative $i \rightarrow j$ and non-radiative $i \rightarrow jk$ transitions is given by:

$$\Gamma(C) = \sum_{i,j} \Gamma_{ij}(C_i \rightarrow C_f) + \sum_{i,jk} \Gamma_{ijk}(C_i \rightarrow C_f) \quad (3)$$

The partial widths of radiative and non-radiative transitions for single ionized atom are calculated as follows:

$$\Gamma_R(n_i l_i \rightarrow n_f l_f) = \frac{4k^3}{3g_i} \sum_{\substack{L,S,J,M_J \\ L',S',J',M'}} \left| \left\langle n_i l_i LSJM_J |D| n_f l_f L'S'J'M'_J \right\rangle \right|^2 \quad (4)$$

$$\Gamma_A(n_i l_i \rightarrow n_j l_j, n_k l_k) = \frac{2\pi}{g_i} \sum \left| \left\langle n_i l_i, \varepsilon_A l_A, L'S'J'M' |H^{ee}| n_j l_j, n_k l_k, LSJM \right\rangle \right|^2 \quad (5)$$

where $n_i l_i$, $n_j l_j$ and $n_k l_k$ denote the atomic subshell involved in the transition. g_i is the statistical weight of the initial state $n_i l_i$, the value k is equal to the x-ray transition energy divided by light velocity c, D is the dipole operator, and H^{ee} is the operator of electron-electron interaction operator.

An analysis of each cascade started with consideration of all possible electron transitions, which may fill an initial core vacancy, then one transition is selected. The selection is random in the interval (0,1) based on the branching ratios of all possible transitions. In the second step, a new configuration of vacancies appears after the occurrence of the selected transition.

The consideration of this configuration is similar to that at the first step. The random selection of the next transitions in the given cascade has been realized with allowance for the relative probabilities of possible transitions in all available vacancies. Successive vacancy configurations appear until all the vacancies reach the outer shell leading to the production of highly charged ions have been produced. After finishing with one cascade, the same initial hole is created again in the inner -subshell and the cascade is simulated again.

Finally after multiple simulations 10^5 times, the final charge state distribution (CSD) in outer shells and average charge state,$< i >$, of ions are calculated. The probabilities of charged ions state distributions, $p(i)$, and the average charged ions, $< i >$, are calculated for ith charged ions as:

$$p(i) = \frac{\sum_{1}^{n} i}{10^5} \quad (6)$$

The mean charge state of ions is given by:

$$< i >= \sum p(i) i \quad (7)$$

Where i denote the number of ejected electrons.

Radiative (x-ray) and non-radiative (Auger and Coster-Kronig) transitions rates for single ionized atoms are calculated using Multiconfiguration Dirac Fock wave function and Dirac Fock Slater wave functions respectively. The radiative and non-radiative transition rates for multi-ionized atoms are evaluated using a statistical weighting procedure from Jacobs et al. [25]. The transition probabilities of multi-ionized atoms change proportional to the number of electrons in the subshells. The radiative transition rates for multi-ionized atoms are given by:

$$a_r(n_1 l_1^{N_1}, n_2 l_2^{N_2} \rightarrow n_1 l_1^{N_1-1}, n_2 l_2^{N_2+1}) =$$
$$N_1 \frac{(4l_2 + 2 - N_2)}{(4l_2 + 2)} A_r(n_1 l_1 \rightarrow n_2 l_2) \quad (8)$$

and the non-radiative transition rates are given by:

$$a_a(n_1 l_1^{N_1}, n_3 l_3^{N_3}, n_4 l_4^{N_4} \rightarrow n_1 l_1^{N_1-1}, n_3 l_3^{N_3+1}, n_4 l_4^{N_4+1}) =$$
$$N_1 \frac{(4l_3 + 2 - N_3)}{(4l_3 + 2)} \frac{(4l_4 + 2 - N_4)}{(4l_4 + 2)} A_a(n_1 l_1 \rightarrow n_3 l_3, n_4 l_4) \quad (9)$$

If both of the final vacancies occur in the same principle shell, the non-radiative transition rates are given:

$$a_a(n_1 l_1^{N_1}, n_3 l_3^{N_3} \rightarrow n_1 l_1^{N_1-1}, n_3 l_3^{N_3+2}) =$$
$$\frac{N_1}{2} \frac{(4l_3 + 2 - N_3)}{(4l_3 + 2)} \frac{(4l_3 + 1 - N_3)}{(4l_3 + 1)} A_a(n_1 l_1 \rightarrow n_3 l_3^2) \quad (10)$$

where A_r and A_a are the radiative and non-radiative transition rates for single ionized atom, respectively, a_r and a_a are the transition rates for atom with various spectator inner-shell vacancies, $n_1 l_1$ is initial state and $n_3 l_3, n_4 l_4$ are a the final state, and N_i is the number of vacancies in the $n_i l_i$ sub-shell.

RESULTS AND DISCUSSIONS

The fluorescence yield (radiative branching ratios) of an atomic shell or subshell configurations is defined as the probability that a vacancy in that shell or subshell is filled through a radiative transition. The Auger yield (non-radiative branching ratios) is the probability that a vacancy in the atomic shell and subshell is filled through a non-radiadive transition by an electron from a higher shell or subshell. Table 1 presents the radiative (fluorescence) yield and non-radiative (Auger and Coster - Kronig) yield for single ionized Xe atom. The present results are compared with those published by other authors [10,26]. The decay for K shell vacancy in Xe atom through radiative transitions is more probable than the non-radiative transitions. The total branching ratio of radiative yield is 88.8% while the branching ratio of non-radiative yield is 11.2% as shown in Table 1. On the other hand, the decay of L- and M-subshells is more probable through non-radiative transitions (Auger and Coster- Kronig transitions).

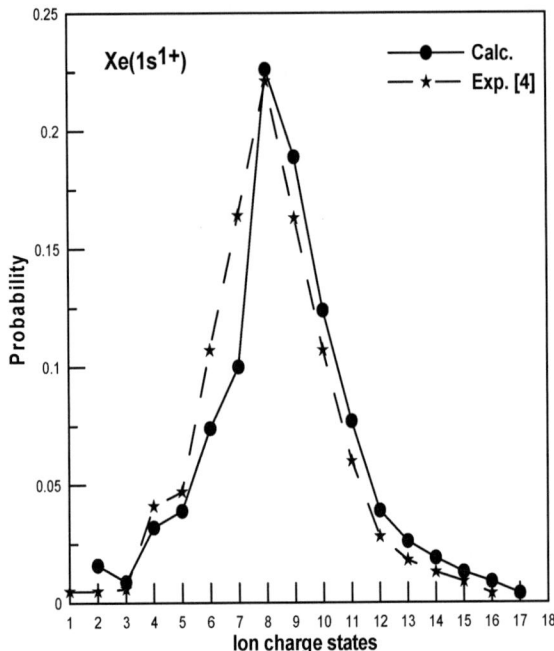

Table 1: Total radiative and non-radiative branching ratios of xenon

	K	L_1	L_2	L_3	M_1	M_2	M_3
RADIATIVE YIELD Radiative yield[10] Radiative yield[26]	88.8	4.97 5.43	8.79 7.97	8.80 7.31	0.05 0.05 0.05	0.01 0.01 0.01	0.05 0.07
C K YIELD CK yield[10] CK yield[26]		49.80 43.25	16.10 12.37 15.40		95.20 94.83 94.30	89.60 89.84 89.20	88.10 87.61
Auger yield Auger yield[10] Auger yield[26]	11.2	45.20 51.32	75.10 79.65 76.90	91.20 92.69	4.80 5.11 5.70	10.75 10.70 10.70	11.80 12.32

Figure 1 shows the probability of final charge state distributions for Xe^{i+} ions after de-excitation of K vacancy.

Figure1. Final charge state distribution of Xe^{i+} following de-excitation decay of initial K vacancy in Xe atom.

The calculation and experimental Xe^{i+} ion charge states have a maximum at i=8 and a shoulder at i= 4. This may be attributed to the initial K-L$_{2,3}$ radiative transitions which have a probability (88.8%) as shown in Table 1. The decay through radiative transition K-L$_{2,3}$ replace the vacancy to the L$_{2,3}$ subshells as a first decay step in the cascade. Then the resulted L$_{2,3}$ vacancies may decay through L-NN, L-NO and L-OO Auger transitions leading to the production of Xe^{4+} ions. Ions mainly produced from Xe in K shell vacancy state are found to be Xe^{7+}, Xe^{8+}, Xe^{9+} and Xe^{10+}. The number of ejected electrons result from de-excitation decay of inner-shell vacancies in xenon forms an asymmetric peak. In comparison with the available experimental data [4] the present calculations agree well.

The charge state distribution of Xe^{i+} ions yield after de-excitation decay of an initial L$_1$-shell vacancy is shown in Figure 2. The charged Xe^{9+} ions predominate in the charge state distributions. In the L$_1$ vacancy state, the intensity of Xe^{8+} ions decreases gradually, while that of Xe^{9+} ions increases. The spectrum of ejected electrons after an initial L$_1$ shell vacancy forms asymmetric peak. The spectrum of charged Xe^{1+}, Xe^{2+} and Xe^{3+} ions is weak and has no significant effect on the charge state distributions. The results of ion charge state distributions after de-excitation of an initial L$_1$- shell vacancy are very close to Carlson and Krause [4] experimental results.

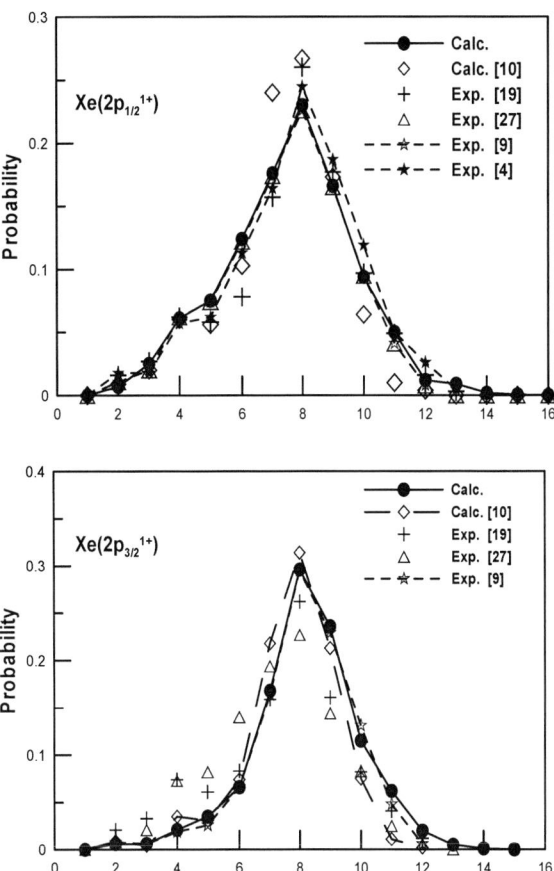

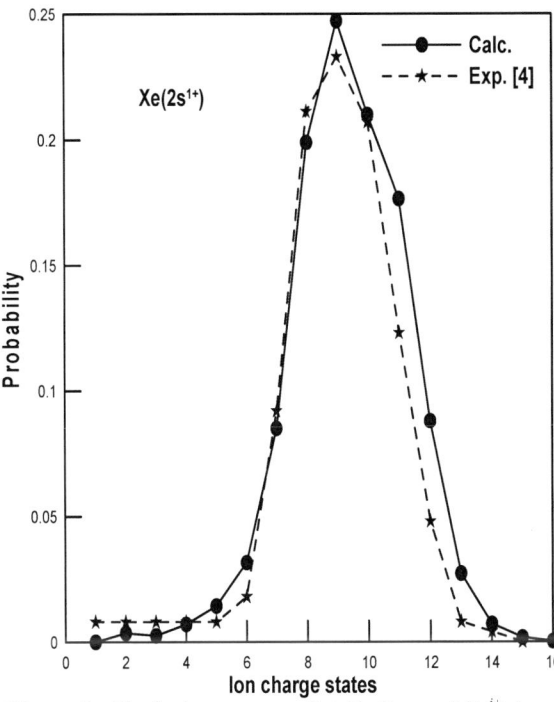

Figure 2. Final charge state distributions of Xe^{i+} ions following after de- excitation decay of initial L$_1$ vacancy in Xe atom

Figure 3. Final charge state distributions of Xe^{i+} ion result after de-excitation decay of initial L$_2$ and L$_3$ vacancies in Xe atom

Figure 3 shows the experimental and the theoretical charge state distributions of ions after de-excitation decay of an initial L$_2$- and L$_3$ subshell vacancies. The charged Xe^{8+} ions predominate in the charge state distributions. The shoulder appears at i=4 in both experimental data and calculations. This behaviour is previously observed at the K hole state. The reason of appearance of this shoulder is related to the first decay step of L$_2$ and L$_3$ subshell vacancies that may occur through Lα radiative transitions. These transitions lead to vacancy movement to a higher M$_{4,5}$ subshells without changing the number of vacancies producing a maximum abundance of Xe^{4+} ions. The probability of charge state distributions are maximum in the middle of the distribution, while it decreases at the lower and higher values of ion charge states. The present results are compared with theoretical calculations [10] and experimental data [4,9,19, 27] and the agreement is reasonable as indicated in Figure 3.

Figure 4 shows the probabilities of multiple Xe^{i+} ions yield after de-excitation decay of initial M$_1$, M$_{2,3}$ and M$_{4,5}$ vacancies. The distributions of Xe^{i+} ions are compared with the available theoretical [10] and experimental data [4,8]. The decay of the M subshell holes mainly proceeds through Auger and Coster-Kronig transitions producing electrons from N and O subshells. The charged ions mainly yield from Xe in 3s$_{1/2}$ and 3p$_{1/2,3/2}$ hole states are found to be X^{6+}, Xe^{7+} and Xe^{8+} ions. The 3s and 3p vacancies are presumed to decay mainly into M$_{4,5}$N hole states (Coster- Kronig transitions) which further de-excite through sequential Auger transitions to form Xe^{i+} ($6 \leq i \leq 8$). The single and double charged Xe^{i+} ions are yield from direct N - or O- subshells. As shown in Figure 4, Xe in 3d$_{3/2,5/2}$ hole states mainly turns into Xe^{4+} and Xe^{5+} ions, and the charged Xe^{3+} formed from de-excitation decay through N$_{4,5}$O Auger transition. After production Xe^{2+} ions the N$_{4,5}$ vacancies de-excites into OO sushells through Auger transitions and produce Xe^{4+} ion. The Xe^{5+}, Xe^{6+} ions formed from N$_1$N$_{4,5}$ de-excitation through Auger and /or Coster -Kronig transitions. The present calculations agree well with the experimental data as shown in Figure 4.

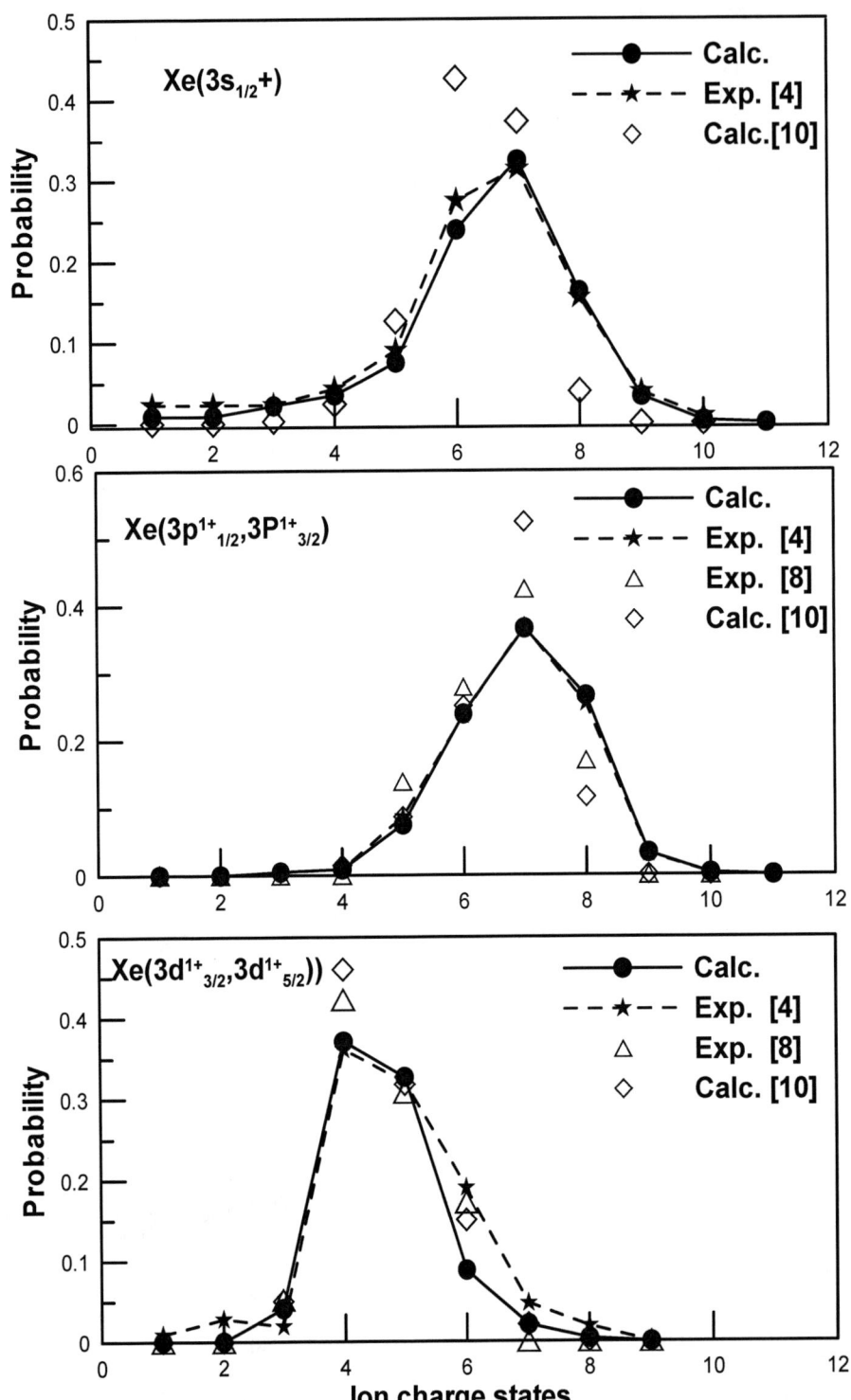

Figure 4. Final charge state distributions of Xe^{i+} following after de-excitation decay of $M_1, M_{2,3}$ and $M_{4,5}$ vacancies in Xe atom.

CONCLUSIONS

The charge state distributions of Xe ions produced after de-excitation decays of inner shell vacancies are calculated using Monte- Carlo (MC) simulation technique. At K and $L_{2,3}$ hole states, the yield of Xe^{8+} ions are predominate. The L_2 hole state mainly turns to Xe^{9+} and the M_1 and $M_{2,3}$ hole states mainly yield Xe^{7+} ions. The charged Xe^{4+} ions are found mainly after de-excitation decay of $M_{4,5}$ vacancies. The present results of charge state distributions of ions agree well with the experimental and theoretical data.

ACKNOWLEDGEMENT

The authors would like to thank Kuwait Foundation for the Advancement of Sciences (KFAS) for financial support of the present work.

REFERENCES

[1] M. O. Krause, M. V. Vestal, W. H. Johnson, T. A. Carlson; Phys. Rev. **133**, A385 (1964).

[2] T.A. Carlson and M.O. Krause; Phys. Rev. **137**, A1655 (1965).

[3] T.A. Carlson and M.O. Krause; Phys. Rev. Letters **14**, 390 (1965).

[4] T.A. Carlson, W. E. Hunt, and M.O. Krause; Phys. Rev. **151**, 41(1966).

[5] T.A. Carlson and M.O. Krause; Phys. Rev. **140**, A1054 (1965).

[6] D.M. P. Holland, K. Coding, J. B. West, and G. V. Marr; J. Phys. B: At. Mol. Phys. **12**, 2465 (1979).

[7] T. Tonuma, A. Yagishita, H. Shibata, T. Koizumi, T. Matsuo, K. Shima, T. Mukoyama and H. Tawara; J. Phys. B: At. Mol. Phys. **20**, L31 (1987).

[8] N. Saito and I. H. Suzuki; J. Phys. B: At. Mol. Opt. Phys. **25**, 1785 (1992).

[9] H. Tawara, T. Hayaishi, T. Koizumi, T. Matsuo, K. Shima and A. Yagishita; J. Phys. B: At. Mol. Opt. Phys. **25**, 1476 (1992).

[10] A. G. Kochur, A. I. Dudenko, V. L. Sukhorukov and I. D. Petrov; J. Phys. B: At. Mol. Opt. **27**, 1709 (1994).

[11] A. G. Kochur, A. I. Dudenko, V. L. Sukhorukov and I. D. Petrov; J. Phys. B: At. Mol. Opt. **28**, 387 (1995).

[12] A.G.Kochur and V.L.Sukhorukov; J.Phys.B:At. Mol. Opt. **29**, 3587 (1996).

[13] A. Kochur; J. Synchrotron Rad. **8**, 218 (2001).

[14] J. W. Cooper, S. H. Southworth, M. A. MacDonald and T. LeBrun Phys. Rev. A **50**, 405 (1994).

[15] F.von Buch, J. Doppelfeld, C. Gunther and E. Hartmann; J. Phys. B: At. Mol. Opt. Phys. **27**, 2151 (1994).

[16] G. Omar and Y. Hahn; Phys. Rev. A **43**, 4695 (1991).

[17] G. Omar and Y. Hahn; Phys. Rev. A **44**, 483 (1991).

[18]. T. Mukoyama; Bull. Inst. Chem. Res. Kyoto Univ. **63**, 373 (1985).

[19] T. Mukoyama, T. Tonuma, A. Yagishita, H. Shibata, T. Matsuo, K. Shima and H. Tawara; J. Phys. B: At. Mol. Phys. **20**, 4453 (1987).

[20] N.Mirakhmedov and E.S.Parilis;J. Phys. B:At. Mol. Opt. Phys. **21**,795 (1988).

[21] A. M. El-Shemi , A. A. Ghoneim and Y. A. Lotfy; Turk. J. Phys. **27**, 51 (2003).

[22] A. H. Abdullah, A. M. El-Shemi and A. A. Ghoneim; Rad. Phys. and Chem. **68**, 697 (2003).

[23] I. P. Grant, B. J. Mckenzie, P. Norrington, D. F. Mayers and N. C. Pyper; Comput. Phys. Commun. **21**, 207 (1980).

[24] M. Lorenz, E. Hartmann; Report ZFI-109, Leipzig, 27 (1985).

[25] V. L. Jacobs, J. Davis, F. B. F. Rozsnyai and J. W. Cooper; Phys. Rev. A **21**, 1917 (1980).

[26] O.Keski-Rahkonen and M.O.Krause; At. Nucl. Data.Tables **14**,139 (1974).

[27] S. Drees; Diploma thesis University of Bonn (1993) (Bonn-IR-93-50).

Computation of Ion Charge State Distributions After Inner-Shell Ionization In Ne, Ar And Kr Atoms Using Monte Carlo Simulation

Adel M. Mohammedein, Adel A. Ghoneim, Jasem M. Al-Zanki, Ashraf H. El-Essawy

Applied Sciences Department, College of Technological Studies, P.O. Box 42325, Shuwaikh 70654, Kuwait.
e-mail : ghoneim2000@hotmail.com, aa.ghoniem@paaet.edu.kw

ABSTRACT

Atomic reorganization starts by filling the initially inner-shell vacancy by a radiative transition (x-ray) or by a non-radiative transition (Auger and Coster-Kronig processes). New vacancies created during this atomic reorganization may in turn be filled by further radiative and non-radiative transitions until all vacancies reach the outermost occupied shells. The production of inner-shell vacancy in an atom and the de-excitation decays through radiative and non-radiative transitions may result in a change of the atomic potential; this change leads to the emission of an additional electron in the continuum (electron shake-off processes). In the present work, the ion charge state distributions (CSD) and mean atomic charge ions produced from inner–shell vacancy de-excitation decay are calculated for neutral Ne , Ar and Kr atoms. The calculations are carried out using Monte Carlo (MC) technique to simulate the cascade development after primary vacancy production. The radiative and non-radiative transitions for each vacancy are calculated in the simulation. In addition, the change of transition energies and transition rates due to multi vacancies produced in the atomic configurations through the cascade development are considered in the present work. It is found that considering the electron shake–off process and closing of non-allowed non-radiative channels improves the results of both charge state distributions (CSD) and average charge state. To check the validity of the present calculations, the results obtained are compared with available theoretical and experimental data. The present results are found to agree well with the available theoretical and experimental values.

Keywords: highly charged ions, de-excitation Auger processes, vacancy cascades

INTRODUCTION

The inner-shell vacancy in an atom can decay by two independent transition processes, i) the emission of photons (radiative transitions) and ii) the ejection of Auger electrons (non-radiation transitions). Radiative transitions are due to displacement of vacancies to higher shells, while non-radiation transitions are accompanied by ejection of an electron from the atomic sub-shells instead of the initial vacancy in the inner–shell. The de-excitation of this inner-shell hole via successive radiative (x-ray) and non-radiative (Auger and Coster-Kronig) transitions leads to highly charged ions. The production of inner-shell vacancies causes a sudden change of atomic potential, that lead to additional monopole ejection of outer–shell electrons (electron shake–off). The emission of electrons in the course of de-excitation vacancy cascade are accompanied by shift of energy levels, which may induce the closing of some Coster-Kronig channels during the vacancy cascade development in an atom. The decay of inner-shell

vacancies corresponds to transitions with emission of photon or electrons for each branch in the de-excitation cascade tree. As such, strong Auger satellite spectra arises from these transitions between various configurations, with additional vacancies (spectator) through the de-excitation cascades.

Early measurements on the charge state distribution (CSD) of ions produced by the inner-shell photoionization of rare gas atoms were performed at some restricted energies of the photons that could be obtained from use of an x-ray tube [1-4]. Monochromatic synchrotron radiation sources are now being used for detailed studies of the ion charge state distributions as a function of the exciting photon energy [5]. The multiply charged ions from K and Ca targets have been measured in the L-shell ionization region using a time of flight (TOF) mass spectrometer [6]. The production of multiple ion yields following the photoionization at an energy just above the K- threshold in the argon atom have been measured by Lindle et al. [7], Levin et al. [8], Ueda et al.[9], and Dopplfeld et al. [10]. The same measurements were performed for the xenon atom in the L region by Tawar et al. [11] and in

CP1202, *Neutron and X-Ray Scattering in Advancing Materials Research: International Conference – 2009*
edited by A. Saat, H. A. Kassim, M. H. H. Jumali, J. M. Saleh, M. R. Othman, A. Ibrahim, F. M. Idris, and M. H. A.-R. M. Ahmad
© 2009 American Institute of Physics 978-0-7354-0739-8/09/$25.00

M-shells region by Saito et al. [12], and Koizumi et al. [13]. The Auger satellite spectrum arising from multiple vacancy states (existence of spectator vacancies) during the de-excitation cascades for rare gases have been studied by Kochur [14,15], a von Buch [16], Cooper [17], and Alkemper et al.[18].

The calculation of vacancy cascades produced in excited atoms can be performed by two different methods. The first method is based on analytic description of vacancy cascade following inner-shell ionization [19,20]. Omar and Hahn [21,22] used the radiative and non-radiative transitions in cascade (RAC) to calculate the yields of multiply charged ions produced after de-excitation vacancy cascades in neutral Ne and Ar atoms with one or more initial inner-shell vacancies. Kochur et al. [23-25] suggested a method of straightforward construction of de-excitation trees to describe the vacancy cascades resulting from inner-shell ionization in rare gas atoms. In the second method, Monte Carlo simulations have been carried out on vacancy cascades following inner-shell ionization in rare gases and some other elements [26-30].

In the present work the multiply charged ions and mean charge state distributions which result from the creation of various inner-shell vacancies in neutral Ne, Ar, and Kr are calculated using Monte Carlo technique. The calculation is performed by evaluating the radiative and non-radiative branching ratios and electron shake–off probabilities. The change of transition energies and transition rates due to multi vacancies produced in the atomic configurations through the cascade development are considered in the present work. In addition, the electron shake-off probabilities and the closing of forbidden Coster-Kronig channels are included in the calculation. The results are compared with available theoretical and experimental values.

METHOD OF CALCULATION

The description of the de-excitation vacancy cascades resulting from inner–shell ionization in atoms and ground state ions is discussed in detail in the following section.

The Monte Carlo technique is employed to simulate the de-excitation vacancy cascade originating from the configuration with a single vacancy. Each cascade starts with the implementation of atomic data for all possible x-ray, Auger and Coster-Kronig channels, and electron shake-off probabilities. To realize a Monte Carlo selection of the actual de-excitation channel, the probabilities of all de-excitation channels were normalized to 1. Then a random number generated in the interval [0,1] selects the next de-excitation step including vacancy transfer and ionization.

N is used to denote the number of vacancies created in nl sub-shells of an atomic configurations via radiative or non-radiative transitions. As an example, the initial C_i configuration $1s\ 2s^2\ 2p^6\ 3s^2\ 3p^6$ represents a distribution with a single K-shell vacancy (N=1) in Ar; and the final

C_f configuration $1s^2\ 2s^2\ [2p^5]\ 3s^2\ [3p^5]$ represents a distribution that could be formed through an Coster-Kronig transition [N=2]. The square brackets are used to indicate spectator holes during the de-excitation cascade. After creating a new spectator vacancy in an actual configuration, the program first controls whether an electron shake-off takes place or not. If the random number generated is smaller than the sum of all normalized shake-off probabilities of the preceding vacancy configuration, then a shake-off process takes place, i.e. an additional electron will be ejected in a higher sub-shell. The channel whose sub-shell shake–off probability value coincides with the random number generated will be activated. After the decision concerning the occurrence of shake–off processes, the program selects the next de-excitation trees by generating a new random number. First, a comparison of the value of random number and the fluorescence yield checks whether radiative or non-radiative transitions take place. The actual de-excitation channel after this decision is chosen in analogy with the determination of the shake-off channels. Each new configuration is analyzed to see if further decays are possible. If further decays are possible, then the program code goes back to the first step mentioned above. The generation of new vacancy configurations continues until all vacancies reach the outer shells or no further decays are possible.

Finally the degree of charge state ions (Z_f^i) in outer shells are recorded. After finishing the de-excitation tree, the same inner shell vacancy will be simulated again. The probabilities of charged ions state distributions $p(Z_f^i)$ and the mean charge state of ions $\langle Z_f \rangle$ are recorded after 10^5 histories. The probability of charged ions state distributions $p(Z_f^i)$ is given by :

$$p(Z_f^i) = \frac{\sum_{1}^{10^5} Z_f^i}{10^5} \qquad (1)$$

The mean charge state of ions $\langle Z_f \rangle$ is expressed as :

$$\langle Z_f \rangle = \sum p(Z_f^i) Z_f^i \qquad (2)$$

The scheme of various decay modes of the $3s^{1+}$ primary vacancy in Kr atom is presented in Figure 1. The possible vacancy states generated by radiative and /or non-radiative processes are shown. The electron shake–off transitions accompanying either the initial ionization or the subsequent decay steps are included in the scheme. The shake–off shifts the subsequent part of the cascade towards increased numbers of holes. As shown in Figure 1, the last transition in the de-excitation cascade leads to a stable N^{-n} hole state.

The calculations of x-ray emission rates for neutral atoms are based on Multiconfiguration-Dirac-Fock (MCDF) wavefunctions [31,32]. The non-radiative transition rates for neutral atoms are computed with a code [33] using Dirac-Fock- Slater (DFS) wave functions [34]. The radiative and non-radiative branching ratios which give valuable information on the de-excitation dynamic of an atom with a core hole, are computed. The radiative branching ratios (fluorescence yields) is defined as the probability that the vacancy in an initial state C_i is filled by an electron from final state C_f through x-ray transitions (photon emission) and is given by :

shells. The electron shake-off probabilities are calculated using a code developed by El-Shemi [35].

The changes of radiative and non-radiative transitions are considered in the present work. These changes arise from the photons and electrons emissions in the course of atomic rearrangement following inner–shell vacancy production. The calculation of the transitions for intermediate configurations with multiple vacancies

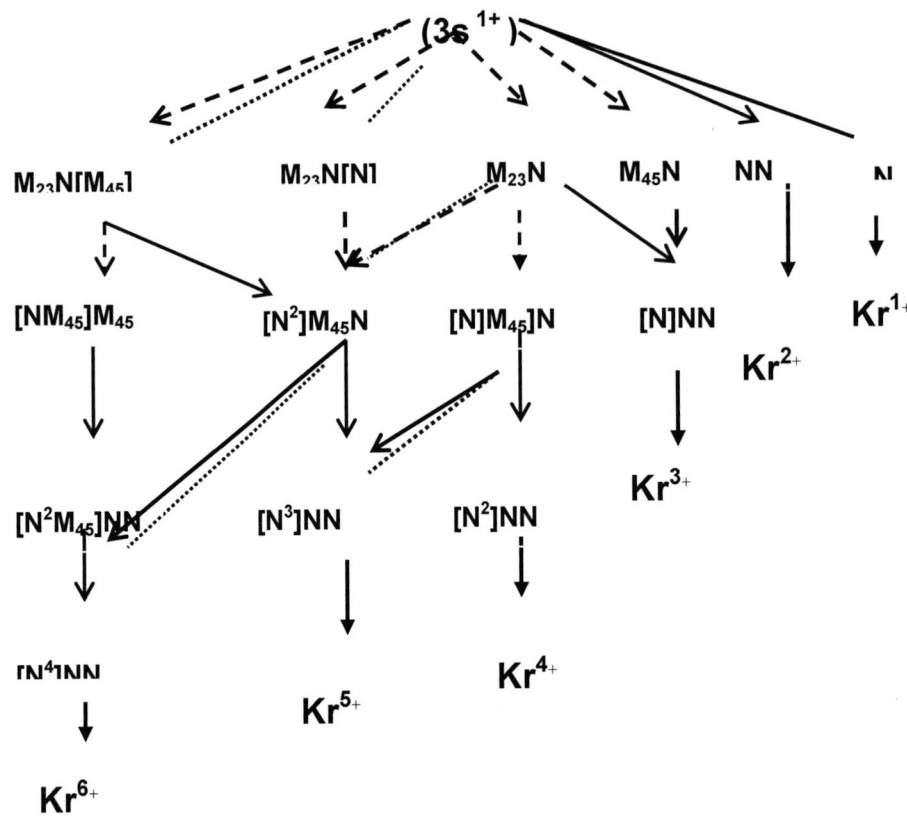

Figure 1. Decay branches pathway after 3s vacancy production in Kr atom. The solid arrow denotes to Auger transition, dotted one denotes to Coster-Kronig, the solid line is X-ray transition and the dotted line is shake off processes

$$\omega(C_f \rightarrow C_i) = \frac{\Gamma_{if}^R}{\Gamma} \qquad (3)$$

The non-radiative branching ratios (Auger yields), is the probability that the vacancy is filled through Auger or Coster-Kronig transitions, and is given by:

$$a(C_f \rightarrow C_i) = \frac{\Gamma_{if}^A}{\Gamma} \qquad (4)$$

where Γ is the sum of the total radiative width (Γ^R) and non-radiative width (Γ^A). The relationship between fluorescence yield ω and Auger yield a satisfies:

$$\omega + a = 1 \qquad (5)$$

The electron shake-off process due to sudden change of atomic potential during vacancy cascade development leads to the ejection of additional electrons from atomic

during the cascade propagation is based on the radiative and non-radiative transitions for single ionized atom [20]. The radiative transitions are given by :

$$a_r(n_1l_1^{N_1}, n_2l_2^{N_2} \rightarrow n_1l_1^{N_1-1}, n_2l_2^{N_2+1}) = N_1 \frac{(4l_2+2-N_2)}{(4l_2+2)} A_r(n_1l_1 \rightarrow n_2l_2)$$

(6)

and the non-radiative transitions are given by :

$$a_a(n_1l_1^{N_1}, n_3l_3^{N_3}, n_4l_4^{N_4} \rightarrow n_1l_1^{N_1-1}, n_3l_3^{N_3+1}, n_4l_4^{N_4+1}) =$$

$$N_1 \frac{(4l_3+2-N_3)}{(4l_3+2)} \frac{(4l_4+2-N_4)}{(4l_4+2)} A_a(n_1l_1 \rightarrow n_3l_3, n_4l_4)$$

(7)

66

$$a_a(n_1l_1^{N_1}, n_3l_3^{N_3} \rightarrow n_1l_1^{N_1-1}, n_3l_3^{N_3+2}) =$$

$$\frac{N_1}{2} \frac{(4l_3+2-N_3)}{(4l_3+2)} \frac{(4l_3+1-N_3)}{(4l_3+1)} A_a(n_1l_1 \rightarrow n_3l_3^{2})$$

$$(8)$$

where A_r and A_a are the radiative and non-radiative transition rates for single ionized atom, respectively, a_r and a_a are the transition rates for atom with various intermediate vacancies, n_1l_1 is initial state, n_3l_3, n_4l_4 are final states, and N_i is the number of vacancies in the n_il_i sub-shell.

The emissions of electrons in the course of vacancy cascade decays are accompanied by shifts of energy levels. So some initially allowed Coster–Kronig channels may be become energetically forbidden. That is because most Coster – Kronig energies are low and the rates are sensitive to the transition energy. The Auger and Coster–Kronig transition energies are computed in the vacancy cascade simulation using "Z+1 rule". This rule specifies that the effect of vacancy in nl sub- shell on the binding energy of the electron in n'l' sub-shell can be approximated by taking the nl binding energy in the neutral atom of next–higher atomic number [36]. It should be noted that ignoring the shifts in energy levels during vacancy cascade calculation leads to a discrepancy between the calculation results of the ion charge state distributions and the experimental data.

RESULTS AND DISCUSSIONS

The radiative and non-radiative branching ratios give valuable information on the de-excitation dynamic of an atom with a core vacancy. The branching ratios for possible radiative and non-radiative transitions from K-shell ionization in Ne, Ar and Kr atoms are shown in Figure 2.

The radiative transitions (x-ray transitions) are generally much weaker than non-radiative transitions for K-shell ionization in Ne. The dominant non-radiative channels are found to be K-L_2L_3 and K-L_3L_3 transitions. Consequently, the probability of filling the K- shell vacancy through Auger transitions is high. The decay of a K-shell vacancy in the Ne atom through non-radiative transitions leads to doubly charged ions. For Ar K-shell ionization, the K-L_2L_3 Auger channel is the strongest one. In the figure 2 it is shown that the radiative K-L transitions (x-ray transitions) are less than 20%, and the Auger electron emission is clearly the dominant mechanism for the production of the highly charged ions.

The radiative transitions K-L are the dominant transitions for K-shell vacancy relaxation in Kr atoms. The K-L_3 ($K_{\alpha 1}$) transition is the strongest one after K-shell ionization in Kr atoms. This transition results from the fact that the inner-shell vacancies can be filled by successive Auger and Coster-Kronig transitions. These

Auger and Coster-Kronig channels cause the emission of many Auger electrons and electron shake-off.

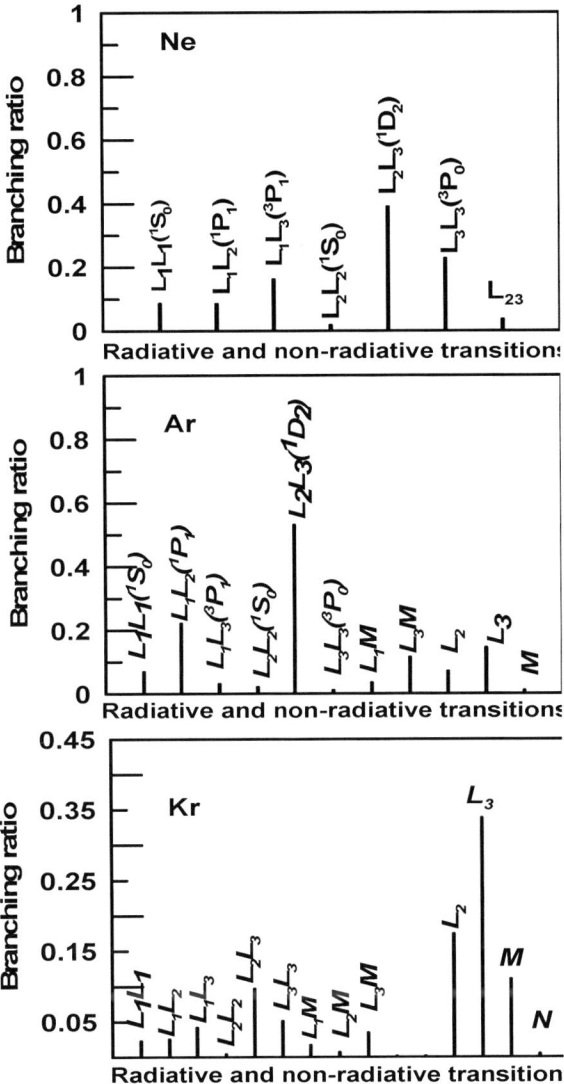

Figure 2. Radiative and non-radiative branching rations for the decay of Ne, Ar, and Kr atoms after K-shell ionization.

So, Coster-Kronig transitions are expected to play an important role in the formation of multiply charged ions following K- shell vacancy production in Kr atoms. The radiative and non-radiative branching ratios give valuable information on the de-excitation dynamic of an atom with a core vacancy.

Electron emission in the course of the cascade development causes shifts of the electron energy levels in atoms. Consequently some initially allowed Coster-Kronig channels may become energy forbidden, thus influencing further de-excitation decay. To demonstrate the importance of considering forbidden Coster-Kronig channels, the calculations are performed with and without considering this effect. Figure 3 shows the results of calculation of ion charge state distributions (Z_f).

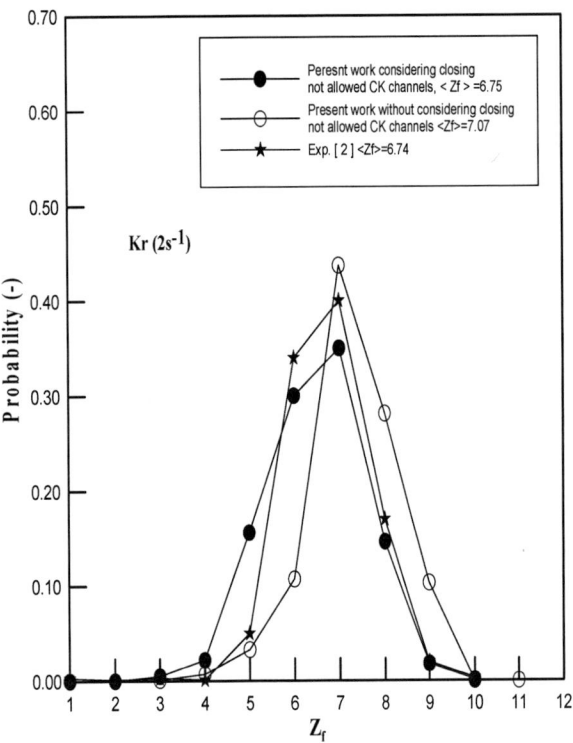

Figure 3. *Ion charge state probabilities* Z_f *following* L_I *sub-shell vacancy production in Kr atom with and without consideration of closing the forbidden Coster- Kronig (CK) channels.*

Figure 3 presents the calculation of ion charge state distributions (Z_f) following L_I inner sub-shell vacancy production in neutral Kr atom with and without considering closing of the energy forbidden Coster-Kronig channels together with the corresponding experimental data [2].

The Coster-Kronig LLM and LLN transitions are the most probable processes in the course of the de-excitation vacancy cascade after L_I vacancy production. As shown from the figure, ignoring the closing of forbidden Coaster-Kronig channels in the course of vacancy cascades leads to a discrepancy between the calculated ion charge state distributions (Z_f) and the corresponding experimental data. In addition, the mean atomic charge $<Z_f>$ values are 7.07, 6.75 and 6.74 for calculation without closing, calculation with closing and the experimental data respectively.

Figure 4 presents the results of Kr^{i+} ion yields in the K-shell de-excitation process obtained by performing the calculation with and without considering the electron shake-off transitions. The difference in the results for ion charge state, calculated with and without inclusion of the shake-off process demonstrates the importance of this process in the study of vacancy cascades in atoms and ions. As shown in the figure, the highest final charge state distribution from K-shell ionization in calculations without considering skake-off is Kr^{4+} while it is Kr^{5+} for calculations that include electron shake-off, the latter agreeing with the experimental data. In addition, the

mean atomic charge $<Z_f>$ is 5.80, 6.15, and 6.16 for calculations that ignore the shake-off process, for calculations that consider the shake-off process and experimental data respectively.

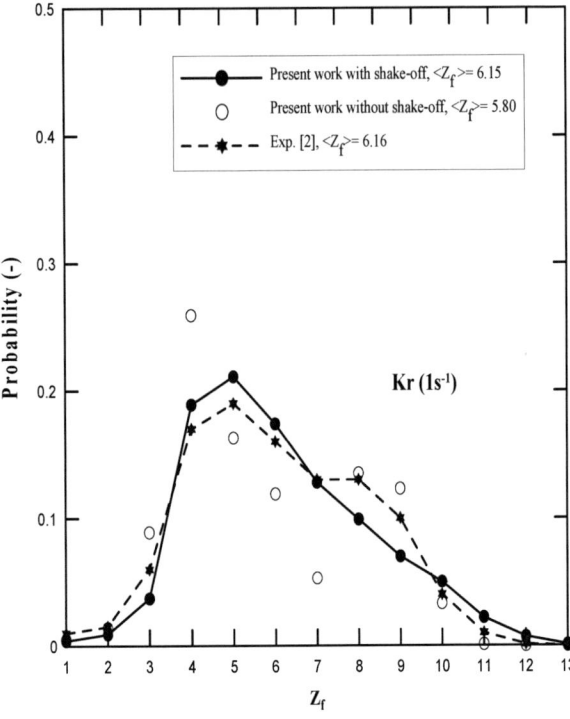

Figure 4. *Ion charge state Distributions (CSD)* Z_f *following K- vacancy production in neutral Kr atom with and without inclusion electron shake –off probabilities.*

The experimental data in Figure 4 shows a bimodal distribution at Kr^{8+} and Kr^{9+} because the emission of electrons starts from $3d_{3/2}$ and $3d_{5/2}$ in the course of the cascade development. This bimodal distribution at this range has not been observed in our calculation. This may be attributed to the change in the radiative and non-radiative transitions for multiply ionized configurations through cascade propagation. Consequently consideration of electron shake–off effects improves agreement between the calculated ion charge state distributions (Z_f) and the mean charge state ions $\langle Z_f \rangle$ with experimental data [2].

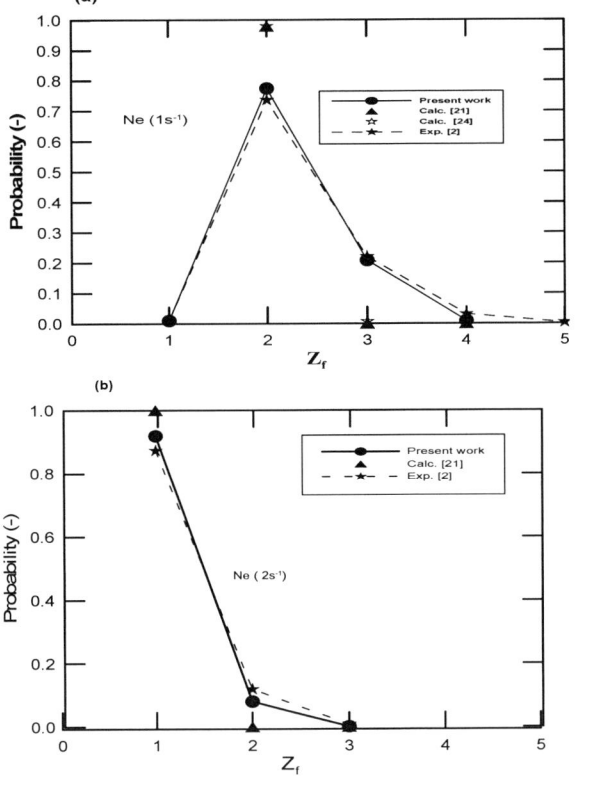

Figure 5. Probabilities of Z_f for Ne after K and L_1 vacancies production compared with experimental data by Carlson et al. [2] and with calculations by Omar and Hahn [21] and by Kochur et al. [24].

The ion charge state distributions (Z_f) following K-shell vacancy production in Ne atom are represented in Figure 5(a). The filling of the K vacancy in the Ne atom through K-LL Auger transitions is most probable because the value of Auger yield is very high ($a_K = 96.45\%$). Decay of the K shell vacancy via K-LL Auger transitions generates a double charged ion Ne^{2+} with a probability of 77.5 %. The L shell vacancies in Ne atom do not further decay by the Auger processes, because the Auger transitions for L shells in Ne are impossible. Therefore the production of Ne^{3+} and Ne^{4+} charged ions is mainly due to electron shake-off processes. Figure 5(b) shows the charged ions Ne^{i+} resulting from L_1 vacancy de-excitation decay in Ne.

Since the non-radiative transitions are impossible, the L_1 do not decay further by Auger channels. Production of the Ne^{1+} charged ion is dominant (92.5 %). The higher charged ions are produced by electron shake–off transitions. The calculations are compared with another theoretical calculation [21,24] and experimental data [2]. As shown from Figure 5, the present results are in good agreement with the experimental data.

The ion charge state distributions (Z_f) of Ar^{i+} after de-excitation decay of K, L_1 and L_{23} shells vacancies in neutral Ar atom are illustrated in Figure 6. In Figure 6(a) the results produced after K vacancy are compared with available theoretical predictions [22, 24, 26, 28] as well

as available experimental values [2, 4, 5, 7, 9, 10]. The production of ion charge state varies from Ar^{1+} to Ar^{8+}. The main product from the K-shell ionization is found to be Ar^{4+} with a value of 39.4%. This is because more K-LM and L-LM Auger and Coster- Kronig channels contribute to the vacancy cascades. On the other hand, the contributions of charged Ar^{7+} and Ar^{8+} after K shell ionization appear weakly with values less than 2% and 1% , respectively.

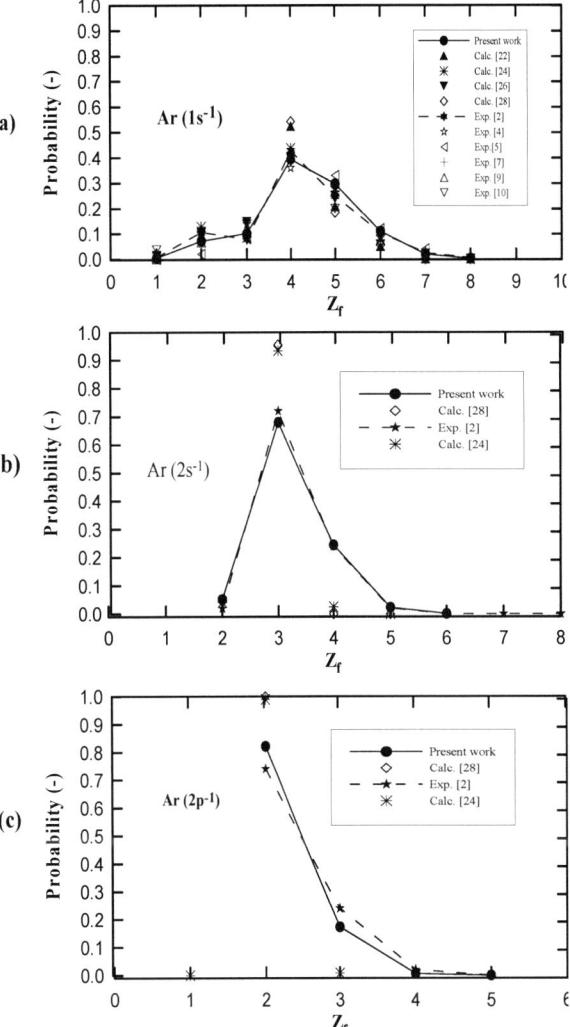

Figure. 6 . Probabilities of final ion charge states Z_f for Ar after K, L_1, and L_2 vacancies production compared with experimental data Carlson et al. [2] and with calculations by Omar and Hahn [21] and by Kochur et al. [24].

Ar^{3+} ions arise from L, and M vacancies which are generated through K ionization followed by K-LM Auger transitions. The L-MM emission (final Auger step) produces Ar^{3+} ions with a value of 10%. The calculations and experiments reproduce the same shoulder behaviour at $Z_f =2$. This behaviour may be attributed to the KL fluorescence yield which occurs with a probability of 7.29% for K-L_2 and with 14.44% for K-L_3. If radiative decay of the K hole occurs with the emission of K_α, a new vacancy will be generated in the

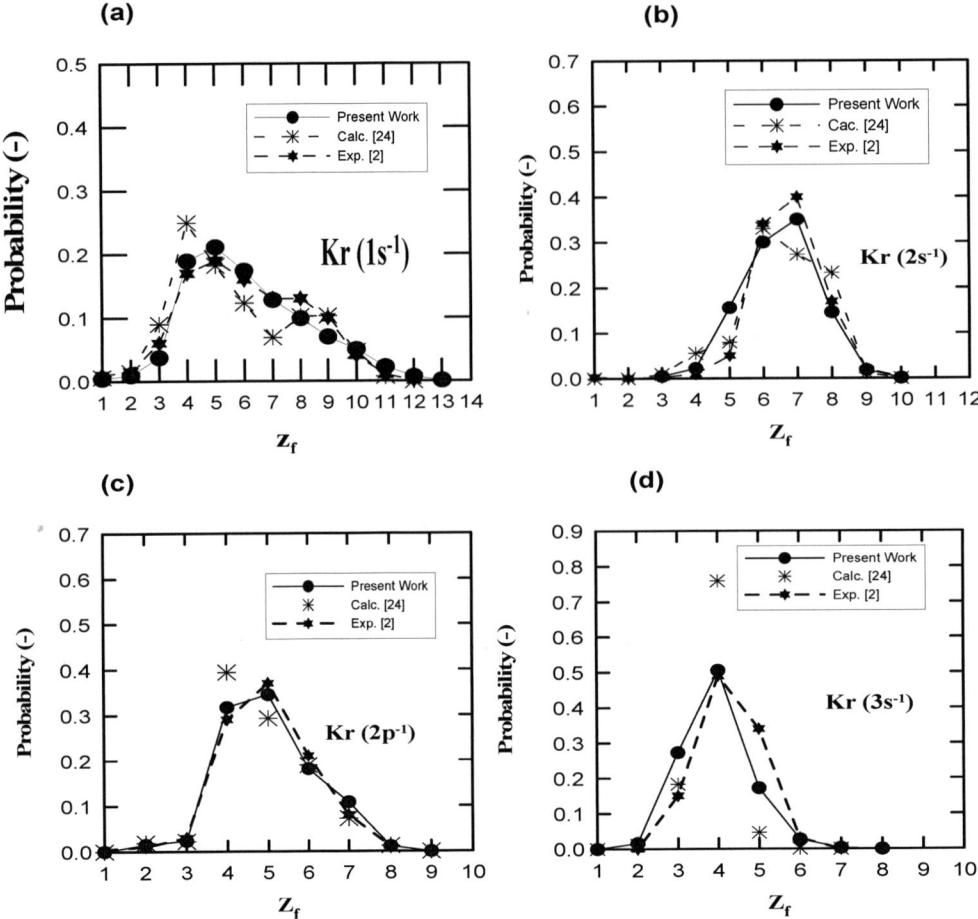

(2p) sub-shells. Therefore L-MM Auger transitions would follow the K_α emission and the ion produced through this de-excitation cascade would be Ar^{2+} with intensity of 7%. The Ar^{2+} ions cannot be produced through the K-shell Auger processes because the K-MM Auger yields are less than 1%.

In Figures 6(b) and 6(c), the results of ion charge state distributions (Z_f) produced after L sub-shell ionization in Ar atom are compared with available experimental and predicted values. The main product from the K-shell ionization is found to be Ar^{3+} with 68% ions. These ions arise from initial L_1 vacancies that are transferred by L_1-LM Coster–Kronig transitions. The de-excitation of the L_1 hole produces ion charge states from Ar^{2+} to Ar^{6+} ionization. As shown in figure 5 (c), the highest final charge is Ar^{2+} which is generated from L–MM Auger transitions. Figure 6 clearly indicates good agreement between the present predictions and the experiments.

Calculated Z_f of multiple Kr^{i+} ions produced by de-excitation of the primary K, L_1, L_{23} and M_1 vacancies are compared with experimental and theoretical values in Figure 6.   Figure 7(a) shows the ion charge state distribution following K shell hole in Kr. Threshold

ionization in the K-shell induces the multiply charged ions from Kr^{2+} to Kr^{13+}.   The main productions are Kr^{4+}, Kr^{5+} and Kr^{6+} ions. One can see from figure 7 (a), that the highest final charge state from K-shell ionization is found to be Kr^{5+}.   If the K-shell hole decays through K_α x-ray transition (K-L_{23}), then the L_{23} holes decays further successive Auger and Coster–Kronig transitions producing the final charge states from Kr^{4+} to Kr^{7+} ions. On the other hand, if the K vacancy decays through K-LL Auger processes and Shake off transitions (which accompanies the decay transitions), the most probable final charge states will be from Kr^{8+} to Kr^{13+} ions.

The charge state distributions of ions (Z_f) produced by de-excitation of an L_1 vacancy in Kr atom are shown in Figure 7(b). The main production of Kr^{i+} is i = 7. For the de-excitation of L_1 vacancies L-LM, Coster–Kronig processes are the most probable transitions. Figure 7(c) shows Z_f production by L_{23} vacancy de-excitation cascades. In this case, Kr^{5+} is the main production. The decay of L_{23} vacancies produces the multiple charged ions up to Kr^{10+}. In the krypton atom, the highest Z_f is found to be equal to 5 for K and L_{23} de-excitation decay, while it equals 7 for L_1 de-excitation decay. The L_1 de-excitation leads to a higher ionization degree than the K

70

and L_{23} de-excitation due to the appearance of L-LM, and L-LN Coster-Kronig processes. Figure 7(d) shows the charge state distributions (Z_f) following M_1 vacancy decay. The highest charge state production is Kr^{4+} ion. The multiple charged ions produced by M_1 vacancy decay are from Kr^{1+} to Kr^{6+}. As shown from Figure 6, the results agree reasonably well with the experimental data.

The average charge state $\langle Z_f \rangle$ for Ne, Ar and Kr atoms obtained from equation (2) are in right corner of each figure. Generally, the mean ion charge state after the de-excitation of inner-shell vacancies increases with increasing the atomic number. This is mainly due to the increase of the number of possible Auger and Coster-Kronig transitions. An exception to this rule was predicted, the mean ionization degree produced from K-shell de-excitation being found to be smaller than the one produced from L_1 de-excitation. This behaviour in K vacancy cascade is caused by x-ray transitions (K_α and K_β), which shift the K vacancy into higher atomic sub-shells (L_3 and M_3) without creation of further vacancies. On the other hand, the de-excitation of L_1 vacancies through L-LM and L-LN Coster–Kronig processes are more probable. The mean charge values $\langle Z_f \rangle$ are compared with available experimental and theoretical values.

CONCLUSIONS

The charge state distribution (Z_f) and mean charge state $\langle Z_f \rangle$ produced by the de-excitation of inner-shell vacancies in Ne, Ar and Kr atoms have been calculated. The Monte Carlo (MC) technique was employed to simulate the de-excitation vacancy cascades following inner-shell ionization. The calculation considered the effects of the x-ray, Auger and Coster-Kronig transitions. The electron shake–off probabilities and the closing of forbidden Coster– Kronig channels in cascade development were considered in the present calculations. Consideration of electron shake-off and the closing of forbidden Coster-Kronig channels improve the calculations of vacancy cascades following inner shell vacancy production in atoms. The results have been compared with available theoretical and experimental values and found to be in favorable agreement with experiments.

ACKNOWLEDGEMENT

The authors would like to thank the Public Authority for Applied Education and Training (PAAET), Kuwait for financial support of the present work under the project No. (TS-08-04).

REFERENCES

[1] T. A. Carlson, and M. O. Krause; Phys. Rav. A 137 (1965) 1655-62.

[2] T. A. Carlson, W.E. Hunt, and M. O. Krause; Phys. Rav. A 151 (1966) 41.

[3] M. O. Krause, and T. A. Carlson; Phys. Rav. A 158 (1967) 18.

[4] T. A. Carlson, M. O. Krause and W. H. Johnsten; Phys. Rev. A 133 (1964) 385

[5] D. A. Church et al. ; Phys. Rev. A 36 (1987)2487.

[6] T. Matsuo , T. Hayaishi, Y. Itoh, T. Koizumi, T. Nagata, Y. Sato, E. Shigemasa, A. Yagishita, M. Yashino and Y. Itikawa; J. Phys. B: At. Mol. Opt. Phys. 25 (1992) 121-133.

[7] D. W. Lindle, W. L. Manner, L. Steinbeck, E. Villalobos, J. C. Levin and I. A. Sellin; J. Electron Spectrosc. 67 (1994) 373-85

[8] J. C. Levin, C. Biedermann, N. Keller, L. Liljeby, R. T. Short, and T. A. Sellin; Phys. Rev. Lett. 65 (1990) 988.

[9] K. Ueda, E. Shigemasa, T. Sato, A. Yagishita, M. Ukai, H. Maezawa, T. Hayaishi and T. Sasaki; J. Phys. B: At. Mol. Opt. Phys. 24 (1991) 605.

[10] J. Doppelfeld, N. Anders, B. Esser, F. Von Busch, H. Scherer and S. Zinz; J. Phys. B: At. Mol. Opt. Phys. 26 (1993) 445.

[11] H. Tawara, T. Hayaishi, T. Koizumi, T. Matsuo, K. Shima, T. Tonuma and A. Yagishita; J. Phys. B: At. Mol. Opt. Phys. 25 (1992) 1467-1473

[12] N. Saito and I. Suzuki; J. Phys. B: At. Mol. Opt. Phys. 25 (1992) 1785-1793

[13] T. Koizumi, T. Hayaishi, T. Matsuo, K. Shima, H. Tawara, T. Tonuma, and A. Yagishita; J. Phys. Soc. Japan 58 (18990 1.

[14] A. G. Kochur, Ye. B. Mitkina and V. L. Sukhorukov; J. Phys. B: At. Mol. Opt. Phys. 31 (1998) 5293-5300

[15] A. G. Kochur, V. L. Sukhorukov and Ye. B. Mitkina; J. Phys. B: At. Mol. Opt. Phys. 33 (2000) 2949-2953.

[16] F. von Busch, U. Kuetgens, J. Doppelfeld and S. Fritzsche; Phys. Rev. A 59 (1999) 2030.

[17] J. W. Cooper, S. H. Southworth, M. A. MacDonld, T. LeBrun; Phys. Rev. A 50 (1994) 405.

[18] U. Alkemper, J. Doppelfeld and F. von Busch; Phys. Rev. A 56 (1997) 2741.

[19] J. C. Weisheit; Astrophys. J. 190 (1974) 735.

[20] V. L. Jacobs, H. Davis, B. F. Rozenyai and J. W. Cooper; Phys. Rev. A 21 (1980) 1917.

[21] G. Omar, and Y. Hahn; Phys. Rev. A 43 (1990) 4695.

[22] G. Omar, and Y. Hahn; Phys. Rev. A 44 (1990) 483.

[23] A. G. Kochur, A. I. Dudenko, V. L. Sukhorukov and I. D. Petrov; J. Phys. B: At. Mol. Opt. Phys. 27 (1994) 1709.

[24] A. G. Kochur, V. L. Sukhorukov, A. I. Dudenko and P. V. Demekhin; J. Phys. B: At. Mol. Opt. Phys. 28 (1995) 387.

[25] A. G. Kochur and Ye. B. Mitkina; J. Phys. B: At Mol. Opt. Phys. 32 (1999) L41 - L43.

[26] T. Mukoyama; J. Phys. Soc. Japan 55 (1986) 3054.

[27] M. N. Mirakhmed and E. S. Parilis; J. Phys. B: At. Mol. Opt. Phys. 21 (1988) 795.

[28] M. G. Opendak; Astrophys. Space Sci. 165 (1990) 9.

[29] A. El- Shemi, Y. A. Lotfy and G. Zschornack; J. Phys. B: At. Mol. Opt. Phys. 30 (1997) 237.

[30] A. El- Shemi, A. A. Ghoneim, Y. A. Lotfy; Turk. J. Phys. 27 (2003) 1-9.

[31] I. P. Grant, B. J. Mckenzie, P. Norrington, D. F. Mayers and N. C. Pyper; Comput. Phys. Commun. 21 (1980) 207.

[32] B. J. Mckenzie , I. P.Grant and P.H. Norrington; Comput. Phys. Commun. 21 (1980) 233.

[33] M. Lorenz, and E. Hartmann (1985) report ZfI −109, Leipzig p 27.

[34] J. P. Desclaux; Comput. Phys. Commun. 9 (1975) 31

[35] A. El-Shemi; Egypt. J. Phys. 27 (1996) 231.

[36] M. F. Chung and L. H. Jenkins, Surf. Sci. 22 (1970) 479.

PARTICLE SIZE DISTRIBUTION MODELS OF SMALL ANGLE NEUTRON SCATTERING PATTERN ON FERROFLUIDS

Sistin Asri Ani [a], **Darminto** [a], **Edy Giri Rachman Putra** [b]

(a) Department of Physics, Faculty on Mathematics and Natural Sciences, Sepuluh Nopember Institute of Technology
Kampus ITS Sukolilo, Surabaya 60111, Indonesia

(b) Neutron Scattering Laboratory, National Nuclear Energy Agency of Indonesia (BATAN)
Gedung 40 Kawasan Puspiptek Serpong, Tangerang 15314, Indonesia

e-mail: sistin.aa_16@yahoo.com

ABSTRACT

The Fe_3O_4 ferrofluids samples were synthesized by a co-precipitation method. The investigation of ferrofluids microstructure is known to be one of the most important problems because the presence of aggregates and their internal structure influence greatly the properties of ferrofluids. The size and the size dispersion of particle in ferrofluids were determined assuming a log normal distribution of particle radius. The scattering pattern of the small angle neutron scattering measurement were fitted by the theoretical scattering function of two limitation models are log normal sphere distribution and fractal aggregate. Two types of particle are detected, which are presumably primary particle of 30 Å in radius and secondary fractal aggregate of 200 Å with polydispersity of 0.47 up to 0.63.

Keywords: ferrofluids, fractal structure, particle size distribution.

INTRODUCTION

Ferrofluids (magnetic liquids) are stable colloidal dispersions of nano-sized particle of ferro- or ferrimagnetic particles in a carrier liquid[1]. Colloidal stability of biocompatible water-based ferrofluids is particularly important for biomedical application such as magnetic cell separation, drug delivery system, hyperthermia contrast enhancement in magnetic resonant imaging. One of the fundamental problems of physics of ferrofluids (magnetic liquids) is the determination of their macroscopical characteristic as a function of their inner composition, such as the shape, size distribution, physical properties and concentration of the magnetic particles as well as the properties of the carrier liquid[2].

Since the particles in typical ferrofluids are small (mean diameter is about 10 nm), they are subject to intensive translational and rotational Brownian motion. Any inner structure in a ferrofluid is a result of a competition between thermal motion of the particles, their magnetic interactions and hydrodynamical forces.

In order to achieve stable colloidal dispersions reproducibly, it is of paramount important to monitor the particle size for each preparation. It is an obvious advantage to have a narrow particle-size distribution since a large particle, which may adversely affect the performance of the fluid, are absent. For most application it is absolutely essential that the ferrofluids has to be stable (isolated) with regard to temperature and in the presence of magnetic fields. The presence of agglomeration of particles must be avoided at all cost.

In this research, we report an analysis concerning the particle size distribution of ferrofluids Fe_3O_4 in several concentrations using non-polarized SANS technique.

MATERIALS AND METHODS

Sample preparation

The water based of ferrofluids has been synthesized by co-precipitation methods. The natural magnetite Fe_3O_4 was extracted directly from iron-sands taken from several rivers in East Java, Indonesia, using a magnetic separator. The synthesis was started from a co-precipitation (~70 °C) of magnetite from chloride acid solution, Fe^{3+} and Fe^{2+} ions in the presence of concentrated NH_4OH solution, after continuous stirring for an hour. A homogeneous solution could be obtained. The products were filtered and washed several times using distilled water. Finally these magnetite particles were then coated and stabilized in water using tetra-methyl ammonium hydroxide as surfactant. In order to find out more detail on aggregation phenomena and particle-size distribution in magnetic nanoparticles, ferrofluids samples were prepared as a function of concentration i.e. 0.5, 1, 2, and 3M. The samples were characterized by means of small angle neutron scattering spectrometer and magnetic force microscopy (AFM-MFM). Detail preparation of this ferrofluids samples was described elsewhere[3].

CP1202, Neutron and X-Ray Scattering in Advancing Materials Research: International Conference – 2009
edited by A. Saat, H. A. Kassim, M. H. H. Jumali, J. M. Saleh, M. R. Othman, A. Ibrahim, F. M. Idris, and M. H. A.-R. M. Ahmad
© 2009 American Institute of Physics 978-0-7354-0739-8/09/$25.00

Small angle neutron scattering measurement

Small angle neutron scattering (SANS) is a technique that allows characterizing structures or object on the nanometer scale, typically in the range between 1 nm and 150 nm. The information one can be extracted from SANS is the average size of a primarily particle, particle size distribution and spatial correlation on nanoscale structures as well as shape and internal structure of particles such as core-shell structure. Further, the scattering intensity on an absolute scale contains the product of scattering contrast of the investigated structures in surrounding medium and number of volume density. SANS also widely use in many fields, like to characterize nanoparticle (in solution or in bulk), clusters, void, precipitates, etc in the nanometer size range[4]. The objective of SANS experiment is to determine the differential cross section, since it is this which contains all the information on the shape, size and interaction of the scattering bodies in the sample.

The differential cross section per unit volume in general can be given by

$$\frac{d\Sigma}{d\Omega}(Q) = \frac{1}{V}\left\langle \left| \sum_j b_j \exp(-iQr_j) \right|^2 \right\rangle \quad (1)$$

where $\partial\Sigma/\partial\Omega(Q)$ has dimension of (length)$^{-1}$ is normally in unit of cm^{-1}, b is the neutron scattering length, V is the volume containing the n atom, Q is the scattering vector and $\exp(-iQr_j)$ is the spatial arrangement of material.

In case SANS, we can replace the sum over atoms with integral over scattering length density

$$\sum_j b_j = \int_V \rho(r)dr \quad (2)$$

Then the equation (1) becomes

$$\frac{d\Sigma}{d\Omega}(Q) = \frac{1}{V}\left\langle \left| \int \rho(r)\exp(-iQr)dr \right|^2 \right\rangle \quad (3)$$

For two phase system, the equation is given by

$$\frac{d\Sigma}{d\Omega}(Q) = \frac{1}{V}\left\langle \left| \begin{array}{l} (\rho_p - \rho_m)\int_{V_P}\exp(-iQr)dr + \\ \rho_m\left\{ \int_{V_m}\exp(-iQr.dr + \int_{V_P}\exp(-iQr)dr \right\} \end{array} \right|^2 \right\rangle \quad (4)$$

The second term is the total scattering amplitude from solvent or matrix. It is negligible at low Q. Then the equation (4) becomes

$$\frac{d\Sigma}{d\Omega} = (\rho_P - \rho_m)^2 V^2 \left\langle \left| \sum_k F_k(Q)\exp(-iQR_k) \right|^2 \right\rangle \quad (5)$$

where

$$F(Q) = \frac{1}{V}\int_V \exp(-iQ.r)dr$$

The diferential cross section is gives by

$$\frac{\partial\Sigma}{\partial\Omega}(Q) = N_P V_P^2 (\Delta\rho)^2 P(Q)S(Q) + B_{inc} \quad (6)$$

with $\quad P(Q) = \left\langle |F(Q)|^2 \right\rangle$

where N_P is the number concentration of scattering bodies (given the subscript "p" for particles). V_P is the volume of one scattering bodies, $(\Delta\rho)^2$ is the square of difference in neutron scattering length density from system. $P(Q)$ and $S(Q)$ is shape factor and structure factor or interparticle structure, respectively. B_{inc} is a incoherent background signal. $\partial\Sigma/\partial\Omega(Q)$ has dimension of (length)$^{-1}$ is normally in unit of cm^{-1} [5].

The small-angle neutron scattering measurements using a 36 m SANS BATAN spectrometer (SMARTer) were carried out at the neutron scattering laboratory (NSL) in Serpong, Indonesia. All samples were exposed using a neutron wavelength λ of 3.90 Å with the detector position are 1.5, 4 and 13 m. These experimental settings cover the scattering vector of $0.005 < Q < 0.25$ Å^{-1} with $Q = (4\pi/\lambda)\sin 2\theta$, where 2θ is the angle between the incident and scattered beams.

The scattering intensity of noncorrelated particle becomes

$$I(Q) = I_e\left[\frac{(A-A_0)^2}{b_e^2} \right](4\pi/3)^2 \quad (7)$$
$$x\int_R f(R)R^6\Theta^2(Q,R,R_a,\kappa)dR$$

I_e and b_e denote the scattering intensity and the scattering amplitude of a free electron under the same condition.

For $f(R)$ is a log normal distribution of particle radius, the equation becomes

$$f(\mu,\sigma,R) = \frac{1}{(2\pi)^{1/2}\sigma R}\exp\left\{ -\frac{[\ln(R)-\mu]^2}{2\sigma^2} \right\} \quad (8)$$

with the moment $\langle R \rangle = \exp(\mu + \sigma^2/2)$ and $\Delta R = \exp(\mu)\left\{ \exp(\sigma^2)[\exp(\sigma^2)-1] \right\}^{1/2}$ is assumed[1].

For the model of log normal sphere and aggregate fractal can be describe by[6]

$$f(r) = \frac{N}{c_{LN}}\frac{1}{r^p}\exp\left(\frac{1}{2\sigma^2}(\ln r - \ln r_{med})^2 \right) \quad (9)$$

where

$$c_{LN} = \sqrt{2\pi}\, r_{med}^{1-p} \exp\left((1-p)^2 \frac{\sigma^2}{2} \right) \qquad (10)$$

The basic properties of log normal distribution were established long ago (Weber 1834, Fechner 1860, 1897, Galton, 1879), and it is not difficult to characterize the log normal distributions mathematically. Two parameters are needed to specify a log normal distribution. Traditionally, the mean μ and the standard deviation σ of log (X)[7].

AFM-MFM measurement

Particle-size distribution can be monitored very well by using a transmission electron microscopy. However, another method based on measurement of the magnetization curve of the fluid is particularly convenient. The method assumes that the particle-size distribution can be described by a log normal volume distribution. This distribution have been found to be a good representation for most system studied whether they be based on metallic or ferrite particles[2].

Magnetic force microscopy (MFM) is an extension of atomic force microscopy (AFM) that images magnetization patterns with sub-micron resolution. The measurement has been carried out using Universal QScope Scanning Probe Microscope with AFM and STM is included. The maximum sample size is 25mm(W) x 75mm (L) x 15mm(H) (USPM) / 150mm x 150mm x 67mm (Qscope). Scan head is content of easy laser alignment for AFM wit X-Y thumbweel control of the laser and the photodetector under a video microscope with pixel resolution at 1024x1024.

RESULTS AND DISCUSSION

The scattering pattern of ferrofluids in several concentrations and the positions of sample to detector distance from 1.5 m to 13 m are shown in Figure 1. It looks different scattering pattern as an effect of the concentration, even though the changes is inconsistently. The slope in the region between Porod area and Guinier area (low Q range) is appear indicating the aggregation of particle becomes a larger particle size. This is because of the molecular force of each magnetite particles. According to Figure 1, it showed that the ferrofluids is not monodisperse system. Here, both of the particle size and its distribution were analysed from the scattering pattern using SANS PSI SASfit program data analysis[7]. The data were analysed assuming a spherical particles shape. The model of log normal sphere and aggregate fractal is good approximation to fitting the scattering pattern from ferrofluids. The corrected scattering intensity data was obtained by substracted of each raw data samples by its background using a standart ILL GRASP programme[8]. The results are given in Table 1.

Table 1. Data analysis of ferrofluids samples was measured using SANS spectrometer at various concentrations.

No.	Sample	σ	R (Å)	R₀ (Å)
1.	0,5 M	0,63	28,53	93,86
2.	1 M	0,47	31,90	69,34
3.	2 M	0,48	30,17	114,88
4.	3 M	0,53	29,75	131,64

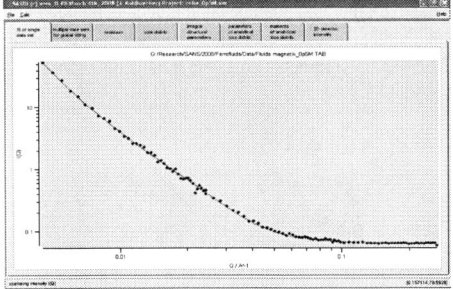
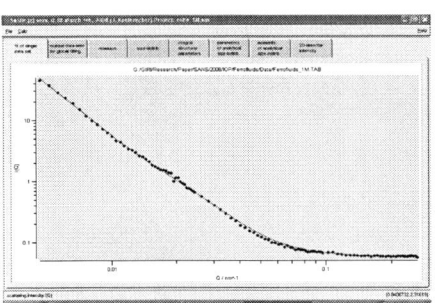
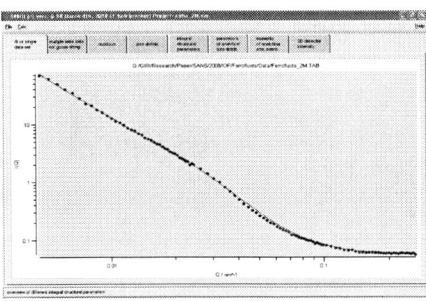
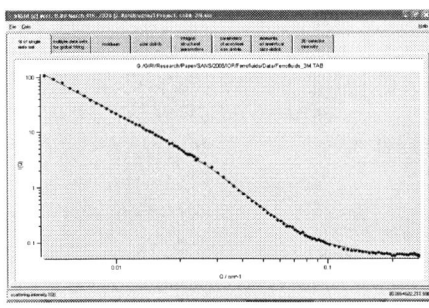

Figure 1. Fitting results from the neutron scattering pattern on ferrofluids at several concentrations from 0.5 M to 3 M using SASfit program. The red line one shows the fitting line and the black point shows the corrected scattering data from ferrofluids.

Nanosized particle of ferrofluids produced by a co-precipitation methods described almost invariably have a rough and very approximate spherical shape. While by a co-precipitation method the particle-size distribution is relatively wide that is usually about 0.4 in that the standard deviation of the log normal distribution[2].

75

Polydispersity is important factor that mainly determines the behaviour of magnetic colloids. In principle, specific features introduced by polydispersity to the process of chain aggregation formation can be monitored already for multi component system. Figure 2 shows the particle size distribution of ferrofluids at several concentrations. It is also important to observe that the particle and the cluster size distribution are fairly broad. In particular, the log normal distribution σ have the range value of 0.47 – 0.63 and have the radius of particle size of 30 Å.

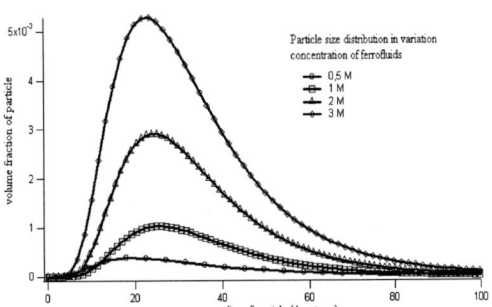

Figure 2. Particle size distribution of ferrofluids at several concentrations.

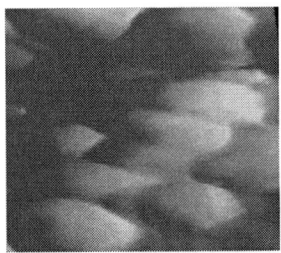

Figure 3. AFM-MFM images from ferrofluid samples shows that aggregate particles formed in the sample.

The particle size of ferrofluids which is determined from SANS measurement is in agreement with those obtained from AFM-MFM measurement. However, the particle size distribution which is observed by SANS technique is higher than the AFM-MFM measurement due to the limitation of coverage in scattering angles or momentum transfer. It gives a truncation error in measured correlation function and hence the distribution differs.

The difference value of R and R_0, Table 1 is clearly describing that many primary particles become the secondary particles (aggregated) on specified form. The specified form can be observed by a microscopy measurement, Figure 3 which shows the AFM-MFM image monitored the particle size distribution.

The average of particle diameter which is a secondary particle is 25 nm. A primary particle is likely growing in three dimensional spaces becomes the bigger size of spherical shape.

CONCLUSION

The Fe_3O_4 ferrofluids samples were synthesized by a co-precipitation method and the investigation of ferrofluids microstructures were carried out by means of the small angle neutron scattering and the AFM-MFM measurement. In particular, the radius of particle size of ferrofluids is about 30 Å with the log normal distribution σ range of 0.47 – 0.63. The particle size and its distribution which are obtained from SANS measurement are in a good agreement with those obtained from AFM-MFM measurement.

ACKNOWLEDGMENT

This work was supported in part by the program of Research Incentive in Nanoscience and Nanotechnology, sponsored by the State Ministry of Research and Technology (KMNRT), Republic of Indonesia, 2006 (Darminto as the Principle Investigator). S.A. Ani acknowledges the support from ICNX2009 committees for attending the ICNX2009 in Kuala Lumpur, Malaysia, June 29 – July 1, 2009.

REFERENCES

[1] Dietmar E., Jürgen Blasing, *J.Apply.Cryst.*(1999). 32. 273-280.

[2] S. Odenbach (Ed.), Hand Book, *Ferofluids, Magnetically Controllable Fluids and Their Application,* Springer,Bremen, Germany, 2002.

[3] S.A. Ani, S. Pratapa, S. Purwaningsih, Triwikantoro, Darminto, E.G.R. Putra, A. Ikram, Neutron and X-Ray Scattering in Materials Science and Biology, edited by A. Ikram et al., AIP Conference Proceeding 989, American Institute of Physics, Melville, NY, April 2008, 176-179.

[4] Yimei Zhu, Modern Technique for Characterizing Magnetic Materials, London, 2005, Chapter 2.

[5] Stephen M. King, hand out lectures, Small Angle Neutron Scattering, Chilton, U.K., 2003.

[6] J. Kohlbrecher, SASfit ver. 0.82, November 2007, PSI.

[7] Eckhard Limpert, Werner A., Stahel, Markus ABBT, May 2001, Vol.51 No.5 BioScience.

[8] C. Dewhurst, "GRASP: Graphical Reduction and Analysis SANSProgramforMatlab",http://www.ill.eu/fileadmin/users _files/Other_Sites/lssgrasp/grasp_main.html, Institut Laue Langevin 2001 – 2007.

A NEW METHOD FOR NEUTRON CAPTURE THERAPY (NCT) AND RELATED SIMULATION BY MCNP4C CODE

*[1]Mousavi Shirazi, Seyed Alireza; [2]Taheri, Ali

[1]*Ph.D Student, Department of Nuclear Engineering, Islamic Azad University, Science and Research Branch*

[2]*M.SC Graduated, Department of Nuclear Engineering, Islamic Azad University, Science and Research Branch, Tehran, Iran*

[1]Email: alireza_moosavi@yahoo.com; [2]Email: taheri60@gmail.com

ABSTRACT

Neutron capture therapy (NCT) is enumerated as one of the most important methods for treatment of some strong maladies among cancers in medical science thus is unavoidable controlling and protecting instances in use of this science. Among of treatment instances of this maladies with use of nuclear medical science is use of neutron therapy that is one of the most important and effective methods in treatment of cancers. But whereas fast neutrons have too destroyer effects and also sake of protection against additional absorbed energy (absorbed dose) by tissue during neutron therapy and also naught damaging to rest of healthy tissues, should be measured absorbed energy by tissue accurately, because destroyer effects of fast neutrons is almost quintuple more than gamma photons. In this article for neutron therapy act of male's liver has been simulated a system by the Monte Carlo method (MCNP4C code) and also with use of analytical method, thus absorbed dose by this tissue has been obtained for sources with different energies accurately and has been compared results of this two methods together.

Keywords: Monte Carlo method, absorbed dose, neutron, , analytical method.

INTRODUCTION

In this article has been simulated and shown a system by Monte Carlo method (MCNP4C code) for dosimetery act and determine of absorbed neutron spectrum for sources with different energies for naught absorption of unallowable dose [2] by a tissue such as male's liver during of neutron therapy. So, this tissue is putting in centre of spherical lacuna from polyethylene stuff with radius 20cm that has a thin layer from cadmium stuff with thickness 100μm and this layer is same of veneer on the polyethylene sphere surface. This cadmium layer has a large absorption cross section for thermal neutrons. Also for leakage decrease of emitted fast neutrons from the external source is used from a graphite reflector with radius 25cm and thickness 5mm.

(re to slowing down ratio means $\xi \dfrac{\Sigma_s}{\Sigma_a}$).

In fact with consider of emitted neutrons seepage from the external source and their passing through polyethylene sphere thickness and their coming to liver tissue, value of absorbed energy by this tissue is calculated from the Monte Carlo method (MCNP4C code) and analytical method and is compared together. Thus re to obtained values and consider to maximum of allowable absorbed energy value

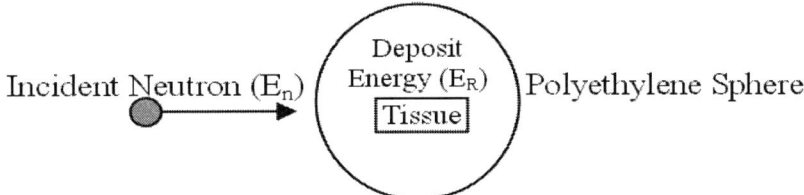

Figure 1: *Emitted neutron from the source and also spherical lacuna with tissue*

CP1202, *Neutron and X-Ray Scattering in Advancing Materials Research: International Conference – 2009*
edited by A. Saat, H. A. Kassim, M. H. H. Jumali, J. M. Saleh, M. R. Othman, A. Ibrahim, F. M. Idris, and M. H. A.-R. M. Ahmad
© 2009 American Institute of Physics 978-0-7354-0739-8/09/$25.00

(maximum of allowable absorbed dose), can perform neutron therapy act. In the neutron therapy is used usually from the D-Be source that generates fast neutrons almost in energy range 14 MeV.

So is better that be used from the single energy neutron source in neutron therapy, thus because Am-Be neutron source has energy pick and is not single energy, so is not suitable for the neutron therapy [1]. Liver tissue is inclusive of some components such as: water, glycogen and heavy molecules such as protein [3]. If a neutron arrives into a tissue, during of neutron therapy, is deposited energy in tissue.

By this simulation, can be obtaining absorbed dose in the tissue, moreover can be determine extent of absorbed energy in the rest of structure and other constitutive components of this system. Moreover in this article, calculation of absorbed dose in liver tissue under neutron therapy, has done by analytical method and use of pair random numbers generation and use of respective neutronic formulas, and has been compared with produced results from the MCNP4C code.

MATERIALS AND METHODS

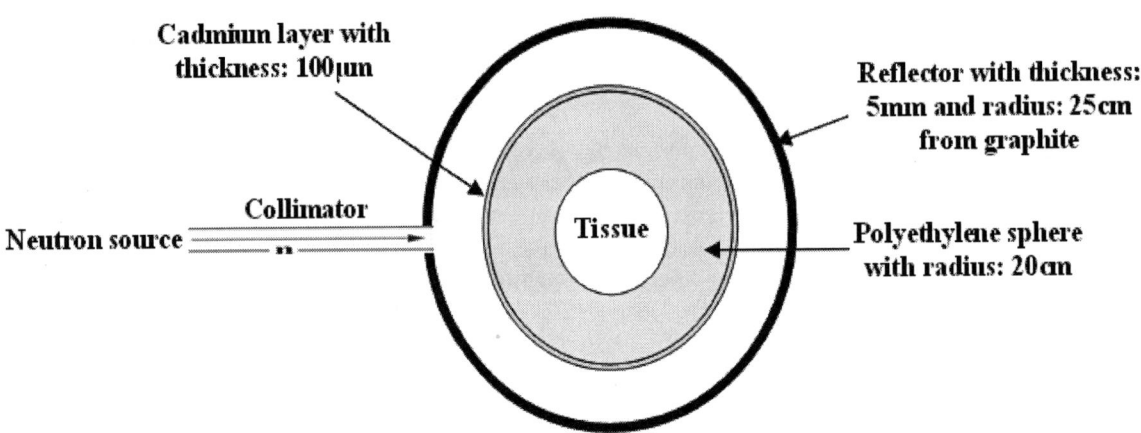

Figure 2: Spherical lacuna inclusive liver tissue for the neutron therapy

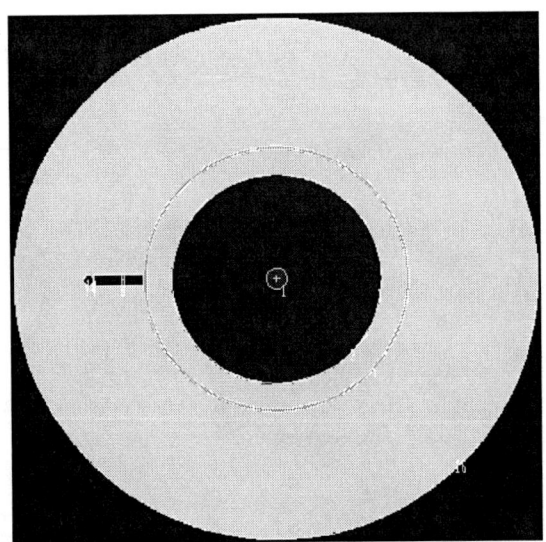

Figure 3: The produced 3 dimensions perspective from the Monte Carlo method for the respective system

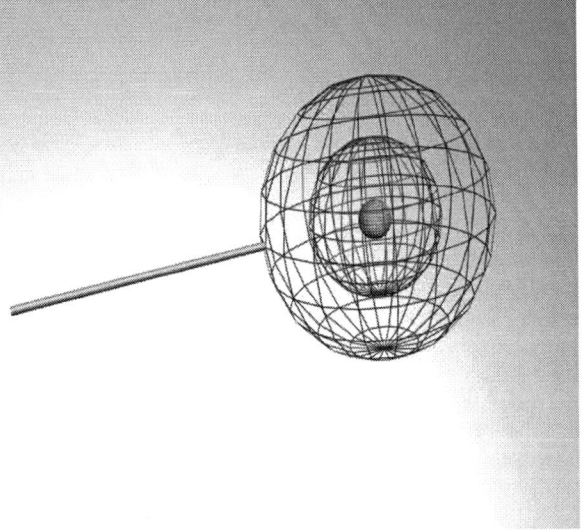

Figure 4: 3 dimensions perspective of the respective system

Simulation of the said system by MCNP4C code

For doing of the neutron therapy act and respective dosimetery, has been considered a system which exists liver tissue in its centre, see following figures:

With accurate calculation of structure component of a special tissue such as male's liver, has done simulation of said dosimetery system by MCNP4C code programming and in this programming, has been entered accurate information from the compound materials of liver tissue for the male sex and also accurate information from the system geometry as input to MCNP4C program.

Table 1: Component and materials of liver tissue for male sex [3]

Mass Percent	Material
69.69%	Water
0.35%	Glycogen ($C_{24}H_{42}O_{21}$)
29.9%	Protein and Glucose ($C_{44189}H_{71252}N_{12428}O_{14007}S_{321}$ and $C_6H_{12}O_6$)

So hereby, values of absorbed dose in liver and absorbed energy in structure of component rest of respective system has obtained for the external source with different energies, means 0.001ev-12MeV via respective tally (F_6).
Moreover in this simulation has been considered thermalization of the neutrons. Liver tissue has included from these materials according to below table:

Using analytical method

To obtaining of absorbed neutron energy information by target nucleus and also recoiled nucleuses angle, is requirement know of scattered neutrons angular distribution. Angular distribution can present relative probabilities for the scattered neutrons into different vectors, so is used from random sampling.
To obtaining of Cos and Sin of an angle which is distributed in interval of 0 until 2π isotropic (SinX, CosX). When: $f(X) = \dfrac{1}{2\pi}$ for $0 \leq X \leq 2\pi$ (1)

Referring to the sketch (R_1, R2) is any point in the rectangle given by the points (1,0), (-1,0), (1,1), (-1,1). [4] The test accepts any point inside the semicircle of unit radius in the range from 0 to 2π:

$$Cos\theta = \frac{R_1}{\sqrt{R_1^2 + R_2^2}} \quad (2) \text{ [4]}$$

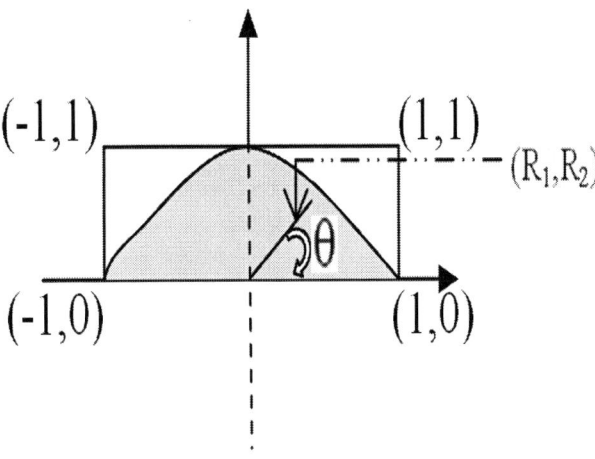

Figure 5: *Semicircle and test rectangle for generating of pair random numbers*

Alternatively a rejection technique yields the variables Cos and Sin as follows: choosing two random numbers R_1, R_2 where: $-1 \leq R_1 \leq +1$ and $0 \leq R_2 \leq +1$, if $(R_1^2 + R_2^2) > 1$ then this pair of two random numbers discarded and a fresh pair of two random numbers is chosen and the procedure is repeated again.

The efficiency of this method is: $\dfrac{\pi}{4}$ .If $\varphi = 2\theta$ then φ lies in the interval $(0, 2\pi)$ and

$$Cos\,\varphi = Cos\,2\theta = 2Cos^2\theta - 1 = \frac{R_1^2 - R_2^2}{R_1^2 + R_2^2} \quad (3) \text{ [4]},$$

$$Sin\,\varphi = \frac{2R_1R_2}{R_1^2 + R_2^2} \quad (4) \text{ [4]}.$$

Every one of the events e_1, e_2, ..., e_n which occur with probabilities p_1, p_2, ..., p_n respectively. Usually the events, e_i's are the types of interaction with any one of the constituents of the medium. With choose a random number R, any event e_j is randomly sampled if

$$(\sum_{i=1}^{j-1} P_i) \leq R \leq (\sum_{i=1}^{j} P_i) (5) \text{ where: } \sum_{i=1}^{n} P_i = 1 \text{ as an}$$

isotropic in the case of tracking neutrons in a hydrogenous medium like polyethylene (constituents are H and C) the e_i's are the various interactions such as elastic scattering with each one of the constituents and inelastic scattering to the two excite levels of carbon and absorptions in hydrogen and carbon.

The individual p_i's is the ratio of the respective macroscopic cross section to total macroscopic cross section.

e_1: Elastic scattering in hydrogen with a probability $$P_1 = \frac{\Sigma_H \, (Elastic\,)}{\Sigma_{Tot}} \quad (6),$$

e_2: Elastic scattering in carbon with a probability $$P_2 = \frac{\Sigma_C \, (Elastic)}{\Sigma_{Tot}} \quad (7),$$

e_3: Inelastic scattering (1st level 4.43MeV in carbon) with a probability $P_3 = \dfrac{\Sigma_C}{\Sigma_{Tot}}$ (8),

e_4: Inelastic scattering (2nd level 7.65MeV in carbon) with a probability $P_4 = \dfrac{\Sigma_C}{\Sigma_{Tot}}$ (9),

e_5: Absorptions in hydrogen and carbon with a probability $P_5 = \dfrac{[\Sigma_C + \Sigma_H (absorption)]}{\Sigma_{Tot}}$ (10)

In fact the main problem is obtaining and determination of E_R for different energies of emitted neutrons by too numbers tracing of emitted neutrons histories from the source that are changed to epithermal and thermal neutrons. If all directions in the 3 dimensions special angle be equally from the point of collision then there is possible without any preferred directions it is known as an isotropic distribution. The fractional number of emitted neutrons $(P(\Omega)d\Omega)$ in the direction Ω within a small cone of solid angle $d\Omega$, for isotropic scattering, is given by:

$$P(\Omega)d\Omega = \frac{d\Omega}{4\pi} = \frac{2\pi d(Cos\,\theta)}{4\pi} = \frac{1}{2}d(Cos\,\theta) \quad (11) \; [4],$$

where: θ is the scattering angle. Thus the probability $P(Cos\,\theta)d(Cos\,\theta)$ for the emitted neutrons between the angles θ and $(\theta + \Delta\theta)$ is given by:

$$P(Cos\,\theta)d(Cos\,\theta) = \frac{1}{2}[Cos\,\theta - Cos(\theta + \Delta\theta)] \quad (12)$$

[4], Because E_1, E_2 are emitted neutrons energies from the source respectively before and next of collision with target nucleus, thus the lost energy due to emission of scattered neutrons in the direction interval θ and $(\theta + \Delta\theta)$ is

$$E_1 - E_2(Cos\,\theta)P(Cos\,\theta)d(Cos\,\theta) \quad (13),$$

where: $E_2(Cos\theta)$ was seen early to be

$$E_2 = E_1 \cdot \frac{A^2 + 2A Cos\,\theta + 1}{(A+1)^2} \quad (14) \; [4],$$

where: $\overline{Cos\,\theta} = \displaystyle\int_{-1}^{+1} Cos\,\theta P(Cos\,\theta)d(Cos\,\theta) \quad (15)$

[4], and for isotropic scattering: $\overline{Cos\,\theta} = 0$.

So the mean last energy of neutron or mean acquired energy by recoiled nucleus $(\overline{E_R})$ is obtained by:

$$\overline{E_R} = \frac{2AE_1}{(A+1)^2} \quad (16) \; [4].$$ For energies below 14MeV, elastic scattering with hydrogen is known to be isotropic in the center of mass system. All values of $Cos\theta$ from -1 to +1 have equal probabilities. The angle between the

scattered neutron and recoil proton direction is $\dfrac{\pi}{2}$. Since the mass of carbon, nitrogen and oxygen are larger than the mass of the incident neutron; the scattering angle is approximately the same in both systems means laboratory and center of mass systems, and therefore the latter angle has almost the full range of values from 0 to π.

Inelastic interaction of neutrons with carbon is very important. Carbon has two excited levels of energy: 4.43MeV and 7.65MeV. The energy of recoiled nucleuses (E_R) is less than from the energy of inelastic scattering neutrons. So can be writing:

Deposited energy in target such as tissue = Energy of incident neutron – Energy of recoiled nucleuses

The transmitted energy to a recoiled nucleus (E_R) with mass number (A) in a collision by an incident neutron as a projectile with energy (E_n) is obtaining according to below formula:

$$E_R = \frac{2A}{(A+1)^2} E_n Cos^2\psi \quad (17) \; [4], \text{ where: } \psi \text{ is}$$

between angle of incident neutron way and recoiled nucleus in laboratory system. In tissue, collision with hydrogen has most importance, because a large value of tissue constituents are water and organic materials that all of these materials have significant percents from the hydrogen in their structures and 85-95% of neutron energy is transmitted through neutron collision with hydrogen. For $E_n > 10$MeV and superior energies, the (n,α) reactions will have more fraction for example in 14MeV, share of recoiled protons is about 2/3 and rest is respect to alpha particles and heavy recoiled particles. The extent of neutron seepage with consider of its passing through polyethylene sphere and its coming to liver tissue, is requirement to accurate know of interaction cross section and angular distribution of scattered neutrons. So with use of Turbo Pascal programming, for different values of incident neutron energy (E_n) in energy range: 0.001ev-12MeV and also mass numbers (A) affiliate to other constituents of said dosimeter system that are entered as input data's to program, so the absorbed dose in the liver tissue (E_R) and also absorbed energy in J/kg in other constituents of this system such as polyethylene sphere, collimator and exist air in system that have different mass numbers has been obtained in Gy and has been produced respective E_R-E_n graphs. This program on base of incident neutron energy and distribution of scattering angle (ψ) and mass number of target and also number of pair random numbers (R_1, R_2) which are as number of incident neutrons, can calculate value of absorbed energy in Gy (Kerma of absorbed dose in Gy).

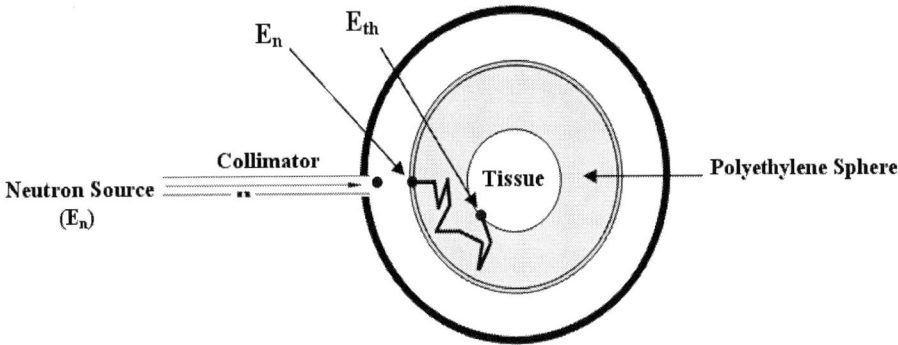

Figure 6: Collision of emitted neutron from the external source with polyethylene and slowing down and then thermalization in sequent collisions

Every pair of random number which is generated by program re to entered initial energy to program is as a emitted neutron from the source with that initial energy. Hereby with more generation of random numbers, can be surveyed more number of neutrons which are as nps in programming of MCNP4C code. The neutron when is passing from the thickness of polyethylene for coming to tissue, because during of passing its energy is decreased and is possible that its energy be too less than from the energy of initial emitted neutron, so should be calculated energy value of passed neutron from the polyethylene and prepared for collision with tissue, with consider to its emitted initial value from the external source (E_n).this new energy (decreased energy or E_{th}) is defined as a fresh input to program automatically.

Thus for obtaining of E_{th} value from E_n value, is requirement determine of done collisions number between emitted neutron from the source and polyethylene. So can be writing:

$$n = \frac{\ln(\frac{E_n}{E_{th}})}{\xi} \ (18)\ [5],\ E_{th} = E_n e^{-n\xi} \ (19)\ [5],$$

$$L_{sl}^2 = \frac{n(\lambda_{tr})^2}{3} \ (20),\ n = \frac{3L_{sl}^2}{\lambda_{tr}^2} = \frac{3L_{sl}^2}{\frac{1}{\Sigma_{tr}^2}} = 3L_s^2\Sigma_{tr}^2 =$$

$$3\frac{D_f}{\Sigma_{sl}}\Sigma_{tr}^2 \approx \frac{\Sigma_{tr}}{\Sigma_{sl}} = \frac{\Sigma_a + \Sigma_s(1-\frac{2}{3A})}{\Sigma_{sl}} ,\text{Thus:}$$

$$n \approx \frac{\Sigma_a + \Sigma_s(1-\frac{2}{3A})}{\Sigma_{sl}} \ (21),$$

so according to equation (19) can be writing:

$$E_{th} = E_n e^{-n\xi} = E_n e^{\left(-\frac{\Sigma_a+\Sigma_s(1-\frac{2}{3A})}{\Sigma_{sl}}\right)\left(\times\frac{2}{A+\frac{2}{3}}\right)} \ (22)$$

So hereby value of E_{th} is obtained with consider of input values to program, means: A, E_n, Σ_{sl} (Σ_a, Σ_s parameters are defined as constants), and by this method is obtained average of absorbed dose by liver tissue. Meantime absorbed energy in polyethylene has been obtained with consider to E_{th} which is in fact as passed neutron energy from totally of polyethylene thickness and E_n which is as emitted neutron and also with consider to defined small intervals into polyethylene thickness. Moreover these values of obtained energies into polyethylene thickness are averaged with obtained values of absorbed energy for E_n and last E_{th}, and then hereby mean absorbed energy value has been obtained by polyethylene accurately.

RESULTS AND DISCUSSION

The produced graphs of absorbed dose by liver tissue and absorbed energy by structure and other constituents of the said dosimeter system in Gy via simulating by Monte Carlo method (MCNP4C) and also analytical method (programming by Turbo Pascal software) are following graphs:

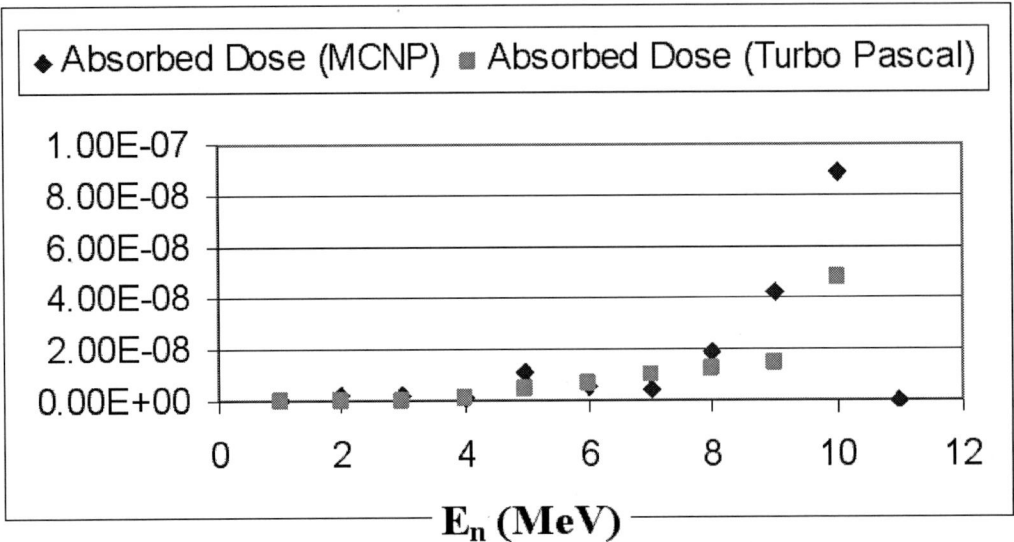

Figure 7: Absorbed dose in the liver tissue

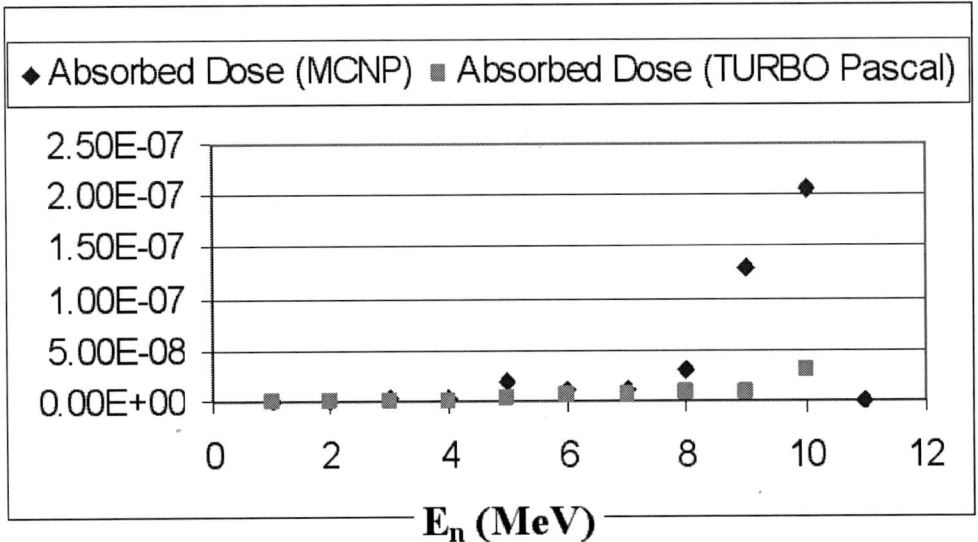

Figure 8: Absorbed energy in the polyethylene sphere with radius 20cm

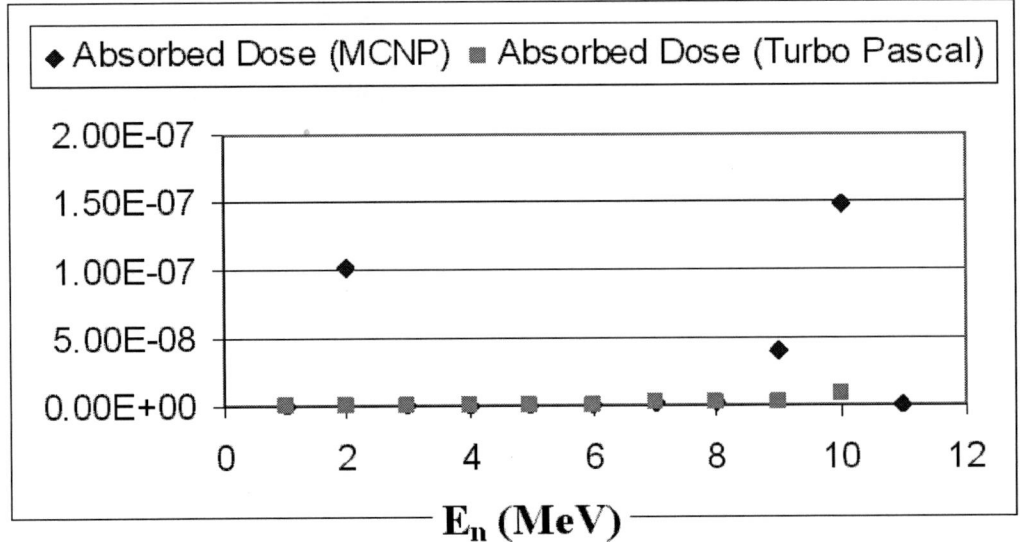

Figure 9: Absorbed energy in the cadmium layer

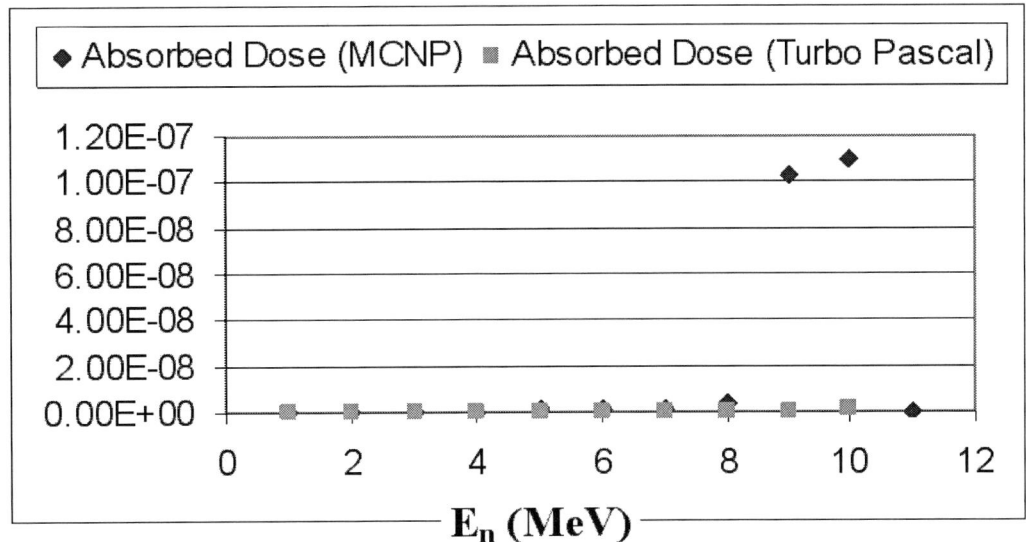

Figure 10: Absorbed energy in the air of spherical lacuna (dosimeter)

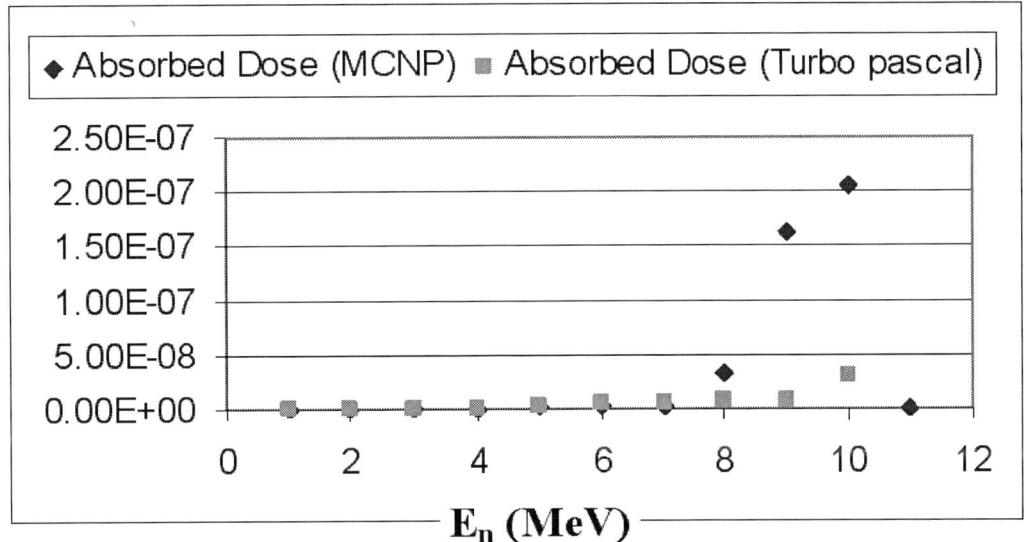

Figure 11: Absorbed energy in the collimator

From view of graphs is consequenced that absorbed dose in the liver tissue and also absorbed energy in other component of the said system such as polyethylene, with increase of the source energy is gotten more extent and for small extent of the neutron energy, absorbed dose and absorbed energy by tissue and other component are being too low and can be saying that produced graphs from the both methods have almost accordance together.

REFERENCES

[1] Annals of the ICRP, Recommendations of the International Commission Radiological Protection, ICRP.,Publication 26, Pergamon Press, New York (1977).

[2] National Committee on Radiation Protection, Protection against Neutron Radiation, NCRP Report No.39, National Bureau of Standards, Washington D.C (1971).

[3] J.J.McBride, M. MASON GUEST, and E.L.SCOTT, the Storage of the Major Liver Components; Emphasizing the Relationship of Glycogen to Water in the Liver and the Hydration of Glycogen, (February 1941).

[4] H.Kahn, Applications of Monte Carlo, AECU (1954).

[5] Stacy, Nuclear Reactor Physic, JW, (1999).

Presenting of the symbols

ξ: Lethargy, Σ_{sl}: Macroscopic cross section of slowing down, Σ_a: Macroscopic cross section of absorption, Σ_s: Macroscopic cross section of scattering, Σ_{tr}: Macroscopic cross section of transport, λ_{tr}: Average of free scanning for neutron transport, D_f: Diffusion constant, n: number of collisions.

Characterization of Titanium Silicide (TiSi$_2$) for Complementary Metal Oxide Semiconductor

Yee Mon Thu, Khin Maung Latt

Department of Physics, Yangon Technological University, Myanmar; (Corresponding author, e-mail: yeemonthu.2008@gmail.com).

Department of Physics, Mandalay Technological University, Myanmar. (e-mail: sayarkyee9@gmail.com)

ABSTRACT

TiSi$_2$ has been one of the most crucial silicides for complementary metal oxide semiconductor device applications such as contacts to source/drain actives and gate regions due to its low resistivity and good thermal stability. However, the polymorphic transformation, (C49–TiSi$_2$ ~60–70 µΩcm→C54–TiSi$_2$ ~5–20 µΩcm), required for self-aligned silicide process has become difficult on narrow, ~0.25 mm and thin lines, known as fine-line width effect. In the present paper, Titanium thin films of thickness 100 nm are deposited on 6" (100) single crystal Si wafers by ionized metal plasma deposition method. Then samples are annealed at different temperatures under high vacuum condition to study the for C49–TiSi$_2$ formation of TiSi. Transformation temperatures of C49 to C54 were studied using scanning electron microscopy (SEM), X-ray diffraction (XRD) and four point probe electrical method.

Keywords— TiSi$_2$, C49–TiSi$_2$, C54–TiSi$_2$, scanning electron microscopy SEM, X-ray diffraction XRD, four point probe method.

INTRODUCTION

TiSi$_2$ is widely used as a contact material in ultra-large scale integration (ULSI) circuit. The formation of TiSi$_2$ is of great importance not only for practical application in the microelectronic industry but also for the fundamental understanding of the phase formation of intermetallic compounds[1].Titanium disilicide (TiSi$_2$) has been widely applied as contact and gate electrodes for complementary metal-oxide-semiconductor (CMOS) transistor in ultralarge-scale integration (ULSI) technology as contacts to source/ drain actives and gate regions due to its low resistivity and good thermal stability.

It is generally accepted that TiSi$_2$ is a polymorphic material and may exist as an orthorhombic base-centered C49 phase or as an orthorhombic face-center C54 phase. In the SALICIDE process, a deposited Ti film is thermally reacted in a N$_2$ ambient to form a silicide in the exposed Si areas of the wafer(formation step).

In the Ti/Si reaction, nucleation and growth of the metastable, high resistivity C49 phase (60-90µΩ cm) has always been observed at about 600°C to precede the nucleation of the C54 phase (12-20 µΩ cm). As the low resistivity stable C54 phase (>700˙C) is desired for contact material, a high temperature annealing process is then necessary to transform the C49 phase to the thermo-dynamically stable, low resistivity C54 phase [2]. If the C54 phase formation is inhibited or fails to occur uniformly, the associated RC delay impact can be as much as 5-10% depending on the specific circuit layout [3]. To meet device performance requirements, it is essential that most of the silicide transforms to the low resistivity C54 phase.

In the current semiconductor industry, rapid thermal annealing (RTA) is used to fabricate TiSi$_2$ from Ti/Si wafers. The high resistivity (60-70 mΩ cm) metastable C49 phase forms first at RTA temperatures of 550-700°C and transforms to the low resistivity (15-20 mΩcm) stable C54 phase at higher temperatures of 750-850°C. C54 TiSi$_2$ is the preferred phase for integrated circuit (IC) fabrication due to its lower resistivity and better thermal stability as compared with C49 phase. Therefore, a complete C49 to C54 phase transition is important for IC fabrication. However, due to the "fine line effect", the transformation to C54 phase on narrow polysilicon lines requires very high annealing temperature, which causes severe problems such as agglomeration and "punch through" and results in device failure. Promoting C54 phase formation at low annealing temperatures is essential in terms of both industrial requirements and scientific interests.

EXPERIMENTS

Titanium thin films of thickness of 100nm were deposited on p-type (100) silicon wafers by the ionized metal plasma sputtering process. In this technique, the sputtered atoms, Ti, are ionized using inductively coupled Ar/N$_2$ plasma. Since the sputtered metal ions can be accelerated perpendicularly to the wafer by applying a simple DC potential in the plasma sheath at the sample's surface, the direction of the metal ions is well controlled and the step coverage for high aspect ratio contact/via

CP1202, *Neutron and X-Ray Scattering in Advancing Materials Research: International Conference – 2009*
edited by A. Saat, H. A. Kassim, M. H. H. Jumali, J. M. Saleh, M. R. Othman, A. Ibrahim, F. M. Idris, and M. H. A.-R. M. Ahmad
© 2009 American Institute of Physics 978-0-7354-0739-8/09/$25.00

structures can be improved. Samples of 20×20 mm^2 in size were cut from wafers with a Ti/Si bilayer structure. Following titanium deposition, a thermal annealing step is necessary to drive the Ti-Si reaction. In this experiment, a conventional tube furnace annealing which was set in a high vacuum condition was used to drive the silicidation. Reactions between titanium and silicon/ polysilicon through the furnace annealing (FA) have been extensively investigated and complete pictures of the overall silicidation sequence, although there still have been some controversies regarding the reaction products at low temperatures (before the C49-TiSi$_2$ forms). The observations can be summarized as follows [4-12]:

1) Raaijmakers *et al.*:
   Ti/Si $\rightarrow$ a-TiSi$_x$ $\rightarrow$ TiSi$_2$ (C49) $\rightarrow$ TiSi$_2$ (C54)
2) Wang and Chen:
   Ti/Si $\rightarrow$ Ti$_5$Si$_3$ $\rightarrow$ TiSi$_2$ (C49) $\rightarrow$ TiSi$_2$ (C54)
3) Muraka and Fraser:
   Ti/poly-Si $\rightarrow$ TiSi$\rightarrow$ TiSi$_2$ (C49) $\rightarrow$ TiSi$_2$ (C54)

The difference in these observations is not still well understood.

After the deposition was done then samples are annealed at different temperatures under high vacuum condition to study the for C49–TiSi$_2$ formation of Ti/Si bilayer. Transformation temperatures of C49 to C54 were studied using X-ray diffraction (XRD) in order to know phase changes and four point probe electrical method was used to monitor the change in sheet resistance and surface morphology examination was also done with scanning electron microscopy (SEM),

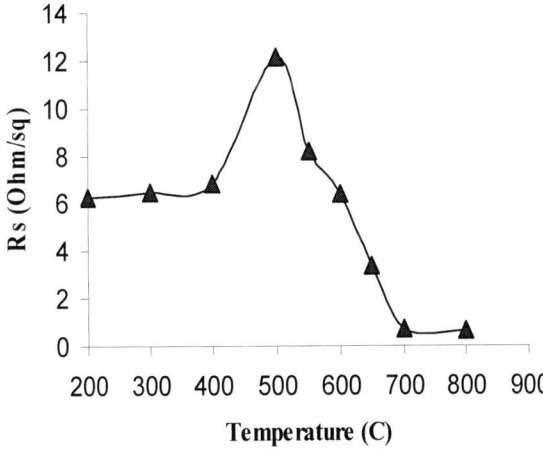

Fig. 1 Sheet resistance of the annealed 100nmTi/Si films as a function of annealing temperature.

The as-deposited sample can be considered as a Ti/Si bilayer system. During annealing, a reaction will occur at the interface between the Ti layer and Si substrate, resulting in the formation of the C49-TiSi$_2$/Si, and finally the C54-TiSi$_2$/Si system.

It is necessary to identify silicide phases formed in order to follow the Ti-Si reaction path during FA. Sheet resistance (Rs) of the titanium silicide varies with its composition, so the transformation sequence during the reaction can be easily tracked by measuring Rs. In this

work, *ex-situ* Rs measurements were carried out using a four-point probe. The number of measured points per sample was 30 and the average value was taken.

Another powerful technique to identify crystalline phases present in materials is X-ray Diffraction (XRD). In this work, XRD measurements were performed with a Rigaku Rint-2000 diffractometer using Cu K$_\alpha$ X-ray radiation ($\lambda = 1.542$Å at 50kV and 20mA. The radiation detection was done from $2\theta = 30$ ° to $2\theta = 55$ °. The scan step and speeds were 0.05° and 4°/min, respectively.

Annealing of the sample below 300°C caused little change in the sheet resistance. As shown in Fig. 1, the sheet resistance slightly increased with such a low temperature annealing. As the annealing temperature increased from 300°C to 500°C, Rs rapidly increased from 6.87Ω/sq. to 12.33Ω/sq. No crystalline phase was detected in XRD analysis, so the growth of amorphous-TiSi$_x$ is considered to be the main reason for the Rs increase since it has an extremely high resistivity value. However, a sudden decrease in Rs from 12.33Ω/sq. to 3.54Ω/sq. was observed when the annealing temperature increased from 400°C to 600°C. C49 peaks can be obviously found in Fig. 2 upon annealing at 600°C. The resistivity value of C49 phase is much lower than that of amorphous-TiSi$_x$. Thus, Rs greatly decreased with the formation of the C49 phase.

Increasing the temperature further above 650°C caused a final drop in Rs to a range of 0.62 to 0.86Ω/sq. XRD spectra show that most of the C49 phase transformed to the C54 phase as the temperature reached 650°C. The resistivity of the thermodynamically stable C54 phase (12 - 20$\mu\Omega$cm) is much lower than that of the C49 phase (60 - 90$\mu\Omega$cm) [13]. Thus, Rs kept decreasing with the growth of the C54 phase.

The C49-to-C54 phase transformation is considered to be a limiting case of massive transformation [14]. It's well known that the massive transformation is nucleation-controlled and often exhibits explosive growth [15]. As a result, the nucleation of C54 phase is difficult, and a certain thermal budget should be satisfied. Once the C54 nuclei form successfully at a certain temperature, the growth of the C54-TiSi$_2$ proceeds very fast. In the experiments, Rs dropped from 3.54Ω/sq. to 0.75Ω/sq. within 50°C (from 600°C to 650°C), while the variation of Rs was only about 0.1Ω/sq from 650°C to 900°C. The increase of the peak intensity values of the C54 phase, however, indicated that the C54 phase was forming more completely as the annealing temperature increased.

Surface morphology examination exhibit that TiSi$_2$ was found to agglomerate into islands after long vacuum FA at a high temperature [16]. Agglomeration starts with grain boundary grooving at the intersections between the silicide grain boundaries and the silicide film surface and the silicide/Si interface, followed by deepening of the grooves along the grain boundaries until the silicide grains eventually separate each other. The rough surface caused by agglomeration will lead to an increase in sheet resistance of the films as well as a spread of contact resistance [17].

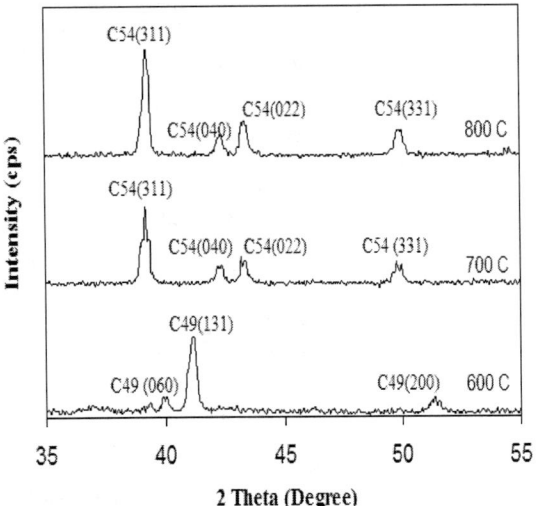

Fig. 2. XRD spectra of the 100nm IMP Ti/Si samples annealed at different temperatures (a) 600℃; (b) 700℃; and (c) 800℃

As a contact material, low resistivity is one of the most important properties. In order to fulfill this requirement, the C54 phase TiSi$_2$ should be formed completely and silicide agglomeration should be avoided. The process window of TiSi$_2$ is defined as the process temperatures at which a complete C54 phase formation can be achieved without the risk of agglomeration. Observations in this work indicates that scaling of the film thickness resulted in a narrower process window since a higher annealing temperature was needed for the C54 phase formation and a lower temperature was needed for agglomeration as the film thickness decreased.

Both annealing temperature and holding time can affect the C54 phase formation, while the former was found to be the dominant factor. If Ti thickness was scaled down from 100nm to 10nm (current deployment), higher temperatures were required to complete the C49-to-C54 transformation and the thermal stability was degraded. As a consequence, the process window for C54 formation shrunk significantly. An amorphous layer was detected (according to XRD) to form at the Ti/Si interface in the IMP sputtering process due to the heavy ions bombardment. This layer consists of a mixture of Ti and Si, and then quickly formed the distorted C49 layer during thermal treatment. It is believed that semi-coherent interface was formed between the C49 phase and the Si substrate. Such dislocations adjust the misfit between different lattices, and thus reduce the interface energy.

Recrystallization is defined as the nucleation of the new strain free grains and gradual consumption of the cold worked matrix by growth of the existing grains. The driving force for recrystallization also comes from the stored energy of the cold work.

On the one hand, the thermal mismatch between the Ti film and the Si substrate will generate some physical defects at the Ti/Si interface and affect the nucleation of the C49 phase. With the C49 grain size reduced, more C54 nucleation sites will be available in the C49 phase. This will make the C54 phase formation easier, leading to smaller C54 grains and smooth film surface. On the other hand, after the formation of the C49 phase, the thermal mismatch under the temperature change will cause plastic deformation in the C49 phase. Due to the significantly increased internal energy in the C49 phase as a result of heavy strain in the grains, the Gibbs free energy curve of this phase will shift up, and the equilibrium temperature of the C49 and C54 phases will decrease according to the thermodynamic theory. Thus, it is possible for the C54 phase to form at a lower temperature. Long time annealing in each thermal shock is not preferred since it might cause recrystallization of the C49 phase, eliminate the defects, and thus reduce the stored internal energy in the grains.

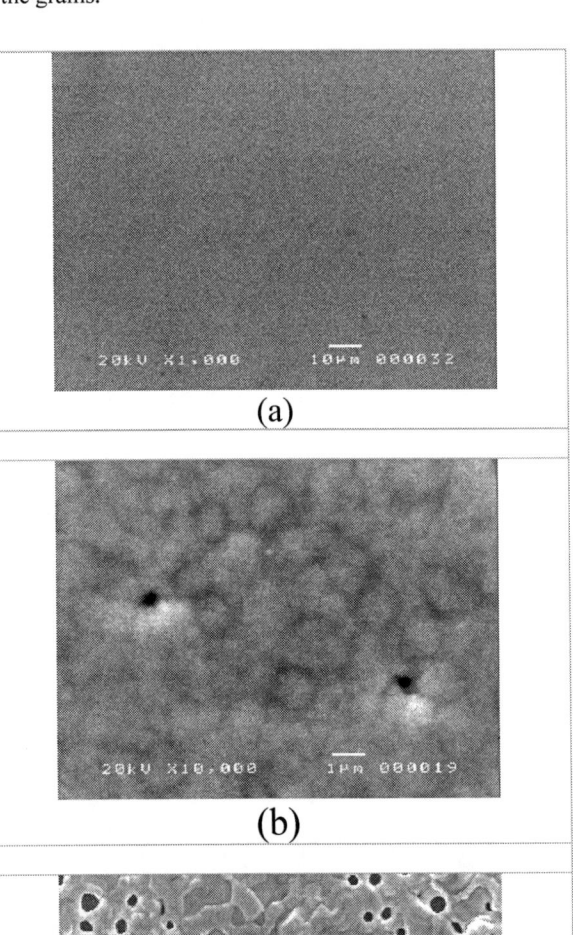

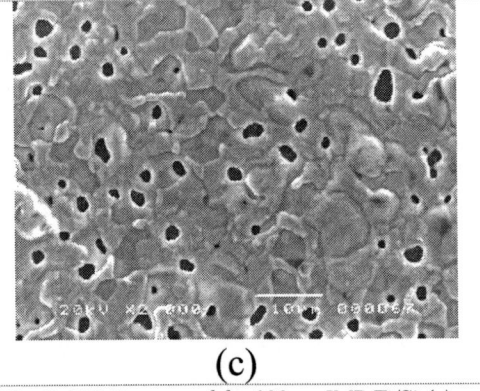

Fig. 3. SEM images of the 100nm IMP Ti/Si (a) as deposited samples (b) annealed at high temperatures 700℃; and (c) 950℃.

The pre-cooling treatment also induced thermal stress between the Ti and Si layers, leading to generation of defects like dislocations in the sample. It can be deduced that the more cooling cycle, the more thermal stress and thus the more defects at the interface.

CONCLUSION

The work presented in this paper can be summarized into two major parts. The first part focuses on the factors that may affect the C49-to-C54 phase transformation and the quality of the $TiSi_2$ films formed during the annealing process. Both the temperature and time have been found to affect the C54 phase formation, while the former played a more important role.

Compared with other deposition methods, the IMP sputtering process causes heavier bombardment damage on the Si substrate by the sputtered ions. An amorphous layer consisting of a mixture of Ti and Si atoms was observed to form accordingly.

The methods provided here are easy to implement so that they have the potential of reducing the complexity and cost associated with forming low resistivity titanium silicide on scaled structures for ULSI applications.

REFERENCES

Mourox A, ZhangS-L, Peterson CS,Enhancement of the formation of the C54 phase of $TiSi_2$ through the introduction of an interposed layer of tantalum. Phys Rev B 1997;56(16):10,614-20

Miles GL, Mann RW,Bertsch JE. $TiSi_2$ phase transformation characteristics on narrow devices. Thin Solid Films 1996;290-291:469-72.

Mann RW, Clevenger LA, Miles GL, Harper JME, Cabral Jr C, D'Heurle FM,Knotts TA, Rakowski DW. Reduction of the C54- $TiSi_2$ phase formation temperature using metallic impurities. Mater Res Soc 1996;402:95-100.

I.J.M.M. Raaijmarkers, L.J. van Ijzendoorn, A.M.L. Theunissen, and K.B. Kim, Mat. Res. Soc. Symp. Proc. 146, 267(1989).

M.H. Wang and L.J. Chen, Appl.Phys. Lett. 58, 463 (1991).

M.H. Wang and L.J. Chen, J. Appl. Phys. 71, 5918 (1992).

K. Holloway, and R. Sinclair, J. Appl. Phys. 61, 1359 (1987).

S. Ogawa, Y. Kouzaki, T. Yoshida, and R. Sinclair, J. Appl. Phys. 70, 827 (1991).

I.J.M.M. Raaijmarkers, and K.B. Kim, J. Appl. Phys. 67, 6255 (1990).

A. Kirtikar, and R. Sinclair, Mat. Res. Soc. Symp. Proc. 260, 227 (1992).

M.H. Wang and L.J. Chen, Appl. Phys. Lett. 59, 2460 (1991).

S.P. Muraka & D.B. Fraser, J. Appl. Phys. 51, 342, (1980)

A. Sabbadini, F. Cazzaniga, and T. Marangon, Microelectronic Eng. 50, 159 (2000).

A.K. Jena, and M.C. Chaturvedi, *Phase Transformations in Materials*, Prentic Hall, Englewood Cliffs, New Jersey, 1992.

T.B. Massalaki, *Phase Transformations*, ASM, Metals Purk, Ohio (1970).

E.G. Colgan, J.P. Gambino, and Q.Z. Hong, Mat. Sci. & Eng. R16(2), 43 (1996).

C.Y. Ting, E.M. d'Heurle, S.S. Iyer, and P.M. Fryer, J. Electrochem. Soc. 113(12), 2621 (1986).

POLARIZED NEUTRON REFLECTOMETRY AT THE PRESENCE OF SMOOTH INTERFACIAL POTENTIAL

Saeed S. Jahromi*, Seyed Farhad Masoudi

Department of Physics, K.N. Toosi University of Technology, P.O. Box 15875-4416, Tehran, Iran;

E-mail: Saeed S. Jahromi, saeed_s_jahromi@sina.kntu.ac.ir

ABSTRACT

Polarized neutron reflectometry (PNR) have developed theoretically and experimentally in the past decades. In order to resolve the phase problem in neutron reflectometry, several simulation methods have been proposed such as reference layer and the variation of surroundings. By considering some factors such as smoothness or roughness in simulations, it is tried to gain more compatible results with experiment. In this paper, by using four different functions "Linear, Airy, Eckart and Error function", we investigate the effects of the smoothness of the interfacial potential on reflectivity, polarization of reflected neutrons and determination of the SLD of the sample, using reference layer method with polarized neutrons. We have also proposed a solution in order to gain better results with less noise in output reflectivity data, at the presence of smoothness.

Keywords: polarized neutron reflectometry, linear potential, Eckart potential, Airy function, Error function

INTRODUCTION

Specular reflection of neutrons from nano-scale thin films can provide useful information about the physical and chemical properties of such layers and would help us to the study of surface structure [1]. By measuring the intensity of reflected neutrons, we can determine the scattering length density (SLD) of the sample along its depth. However the reconstruction of the surface profile has been hampered by the so-called phase problem. Like any scattering techniques in which only the intensities could be measured, in the absence of the phase, the least-squares fit methods could be used to determine the surface profile but generally more than one SLD could be find which corresponds with the same reflectivity data [1,2,4]. By knowing the phase, the real and imaginary parts of reflection coefficient, a unique result for the SLD profile could be retrieved with the help of the Gel'fan-d-Levitan integral equation [1,11-13]. In the past decades, several methods have proposed to find the phase of the reflection such as dwell time method and the reference layers which seems to be the best practical method [1]. The reference layer method which is based on transfer matrix method [2,7] was first achieved by Majkarzak et al. who also proposed and tested this method experimentally [1]. This method then developed by Leeb et al. by measurement of the polarization of reflected neutrons and a magnetic reference layer [3] and Masoudi et al. formulated the method in an straightforward way and enhanced the method [8-10].

In all of these works, it is supposed the interacting potential between neutrons and the sample at boundaries to be discontinuous and sharp. As we know from a realistic sample, there is some smearing at boundaries. Considering this smearing as a smoothness factor would cause some changes in the output reflectivity. In this paper by using four smooth varying functions; "Linear, Airy, Eckart and Error function", we have investigated the effects of the smoothness of interfacial potential on reflectivity, polarization of reflected neutrons and determination of the SLD of the sample, using reference layer methods with polarized neutrons. We have also proposed a solution in order to gain better results with less noise in output reflectivity data, at the presence of smoothness.

THEORY

In neutron specular reflectometry one is dealing with a one dimensional scattering problem with potential V proportional to the SLD perpendicular to the surface. This scattering problem is represented by a one dimensional schrödinger equation:

$$\left(\partial_x^2 + (q^2 - 4\pi\rho(x))\right)\Psi(q,x) = 0 \qquad (1)$$

where $\rho(x)$ is the scattering length density of the sample and q is the neutron wave vector in vacuum. The interacting potential between neutrons and the sample is represented by $v(x) = 2\pi\hbar^2\rho(x)/m$ [2].

CP1202, *Neutron and X-Ray Scattering in Advancing Materials Research: International Conference – 2009*
edited by A. Saat, H. A. Kassim, M. H. H. Jumali, J. M. Saleh, M. R. Othman, A. Ibrahim, F. M. Idris, and M. H. A.-R. M. Ahmad
© 2009 American Institute of Physics 978-0-7354-0739-8/09/$25.00

For a magnetic layer, when the magnetization is in the plane of the film, the SLD is defined by $(\rho_\pm = \rho(x) \pm \mu \cdot B)$ where $\rho(x)$ is the depth profile of none magnetized film, μ is the magnetic moment of the incident neutrons and B is the magnetic field due to the magnetization of the film. The plus and minus sign is proportional to the polarization of incident neutrons, parallel and anti parallel to the local magnetization, respectively [2,3]. Here we suppose the magnetic induction outside the ferromagnetic is small enough not to affect the neutron beam.

An alternative exact representation of reflection coefficient for such a layer, $r_\pm(q)$, is derived from the transfer matrix method of solving eq (1). The elements of the 2×2 transfer matrix, $A(q), B(q), C(q), D(q)$ for arbitrary surround can be expressed as[1,2]:

$$\begin{pmatrix} 1 \\ ih \end{pmatrix} t_\pm e^{ihqL} = \begin{pmatrix} A & B \\ C & D \end{pmatrix} \begin{pmatrix} 1+r_\pm \\ i(1-r_\pm) \end{pmatrix}$$

(2)

where $h = (1 - 4\pi\rho_s / q^2)^{1/2}$ is the refractive index of the substrate with SLD ρ_s.

The reflectivity, $R_\pm(q) = |r_\pm(q)|^2$, can be related to the elements of the transfer matrix in term of new quantity $\sum_\pm(q)$ [1,4]:

$$\sum_\pm(q) = 2\frac{1+R_\pm}{1-R_\pm} = h(A^2 + B^2) + \frac{1}{h}(C^2 + D^2)$$

(3)

The reflection coefficient in term of three new parameters, $\gamma^h, \beta^h, \alpha^h$, can be expressed as:

$$r(q) = \frac{\beta^h - \alpha^h - 2i\gamma^h}{\beta^h + \alpha^h + 2}$$

(4)

where

$$\alpha^h = hA^2 + h^{-1}C^2$$
$$\beta^h = hB^2 + h^{-1}D^2$$
$$\gamma^h = hAB + h^{-1}CD$$

(5)

Reference method

Suppose the sample is separated into two distinct regions. An unknown layer and the known magnetized reference film mounted between the substrate and the unknown film.

The total transfer matrix for such a sample is represented as [7]:

$$\begin{pmatrix} A & B \\ C & D \end{pmatrix} = \begin{pmatrix} w_\pm & x_\pm \\ y_\pm & z_\pm \end{pmatrix}\begin{pmatrix} a & b \\ c & d \end{pmatrix}$$

(6)

where $(w_\pm, ..., z_\pm)$ is transfer matrix of the known part and $(a, ..., d)$ describes the transfer matrix correspond to the unknown part. (+) and (−) denote the plus and minus magnetization with respect to the polarization of incident neutrons [3].

By using eq (6), the parameter $\sum_\pm(q)$ can be expressed as [8]:

$$\sum_\pm(q) = \beta^h_{k\pm}\tilde{\alpha}_u + \alpha^h_{k\pm}\tilde{\beta}_u + 2\gamma^h_{k\pm}\tilde{\gamma}_u$$

(7)

where the tilde represents the reversed unknown film; that is, the interchange of the diagonal elements of the corresponding transfer matrix $(a \leftrightarrow d)$ [1].

To derive the reflection coefficient of the sample as a function of polarization of reflected neutrons, we suppose the incident neutrons to be polarized in the direction normal to the reflection surface (x) and the reference layer is magnetized in a direction parallel to the reflecting surface (z). In this case the polarization of reflected neutrons is represented as follows [3]:

$$P_x + iP_y = \frac{2r_+^* r_-}{R_+ + R_-}$$

$$P_z = \frac{R_+ - R_-}{R_+ + R_-}$$

(8)

By using eq (5) to (8) for the polarization of reflected neutrons, we can write [8-10]:

$$P_x = 1 + \frac{2\zeta}{\sum_+ \sum_- - 4}$$

(9)

$$P_y = \frac{2\zeta}{\sum_+ \sum_- - 4}(c_{\gamma\beta}\tilde{\alpha}_u + c_{\alpha\gamma}\tilde{\beta}_u + c_{\alpha\beta}\tilde{\gamma}_u)$$

(10)

$$P_z = 2\frac{\sum_+ - \sum_-}{\sum_+ \sum_- - 4}$$

(11)

where

$$\zeta = 2(1 + \gamma^h_{k+}\gamma^h_{k-}) - (\alpha^h_{k+}\beta^h_{k-} + \beta^h_{k+}\alpha^h_{k-})$$

(12)

and

$$c_{ij} = i^h_{k+}j^h_{k-} - j^h_{k+}i^h_{k-}$$

(13)

for i and j = 'α', 'β' and 'γ'. The parameters c_{ij} and ζ are known from the elements of the transfer matrix of the know reference layer.

By knowing $\sum_+, \sum_-$ and P_y, we can determine the parameters $\gamma^h, \beta^h, \alpha^h$ for the unknown part of the sample, $\sum_\pm$ is determined directly from measurement or by using equation (9) to (13) as follows [8,10]:

$$\sum_\pm^2 \pm \frac{\zeta P_z}{2(P_x-1)}\sum_\pm -(4+\frac{\zeta}{(P_x-1)}) = 0$$

$$(14)$$

Eq (14) has two different solutions. The physical solution is selected from the fact that $\sum_\pm > 2$ [8]. Actually by knowing two of the parameters, $\sum_+$, $\sum_-$, P_x and P_z, the two others could be determined.

As an example, by using P_y, P_z and $\sum_-$, the three parameters are calculated as follows:

$$\begin{pmatrix} \tilde{\alpha}_u \\ \tilde{\beta}_u \\ \tilde{\gamma}_u \end{pmatrix} = M^{-1} \begin{pmatrix} \sum_- \\ \dfrac{4P_z\sum_- -2\sum_-}{P_z\sum_- -2} \\ \dfrac{2P_yP_z\sum_- -P_y\sum_-^2}{P_z\sum_- -2} -2P_y \end{pmatrix}$$

$$(15)$$

$$M = \begin{pmatrix} \beta_{k-}^h & \alpha_{k-}^h & 2\gamma_{k-}^h \\ \beta_{k+}^h & \alpha_{k+}^h & 2\gamma_{k+}^h \\ c_{\gamma\beta} & c_{\alpha\gamma} & c_{\alpha\beta} \end{pmatrix}$$

$$(16)$$

The parameters $\tilde{\alpha}_u$, $\tilde{\beta}_u$ and $\tilde{\gamma}_u$ denote the free reversed unknown sample.

Smooth interfacial potentials

As we mentioned in chapter 1, we are going to investigate the effects of the smoothness of interfacial potential at boundaries by choosing four continuous and smooth varying functions; 'Linear, Airy, Eckart and Error function'.

Linear potential

In this method the SLD of two adjacent layers with the thicknesses d_1 and d_2, is supposed to vary linearly from the value ρ_1 to ρ_2 at the interface. By considering the linear variation area to be a distinct layer with the thickness of $\varepsilon(d_1 + d_2)$, where ε is a parameter called smoothness factor which is $0 \le \varepsilon \le 1$, our sample is divided into three region; two rectangular film with the thicknesses $d_1(1 - \varepsilon)$, and $d_2(1 - \varepsilon)$ and SLD ρ_1 and ρ_2 respectively and the linear film. As the parameter ε is increased, the thickness of the linear layer increased too, and the interfacial potential is supposed to be smoother. $\varepsilon=0$ denotes the non-smooth potential.

The total transfer matrix for the entire sample is derived from the multiplication of the transfer matrix of the three parts. In order to determine the transfer matrix of the linear part, we can use one of the following methods; first, the step method and second, the direct solution of

the schrödinger for the linear potential which is the Airy function.

In step method, we divide the linear part into several small rectangular layers and multiply the transfer matrix of individual films, in this manner, the SLD of each layer increase or decrease like a step depending on the gradient of the linear function for plus and minus gradient, respectively.

The direct solution of the schrödinger equation, Eq (2), for the linear potential of previous chapter, with the smoothness factor ε, is presented as:

$$\psi(x) = C^+ Ai(\chi) + C^- Bi(\chi)$$

$$(17)$$

where Ai and Bi are the Airy function of the first and second kind, respectively and;

$$\chi = \frac{\eta x - \lambda}{(\eta)^{2/3}}$$

$$\eta = 4\pi m$$

$$\lambda = q^2 - 4\pi\rho_1$$

$$m = \frac{(\rho_2 - \rho_1)}{\varepsilon(d_2 + d_1)}$$

$$(18)$$

The elements of the transfer matrix for the linear layer are determined by using boundary conditions at the interfaces.

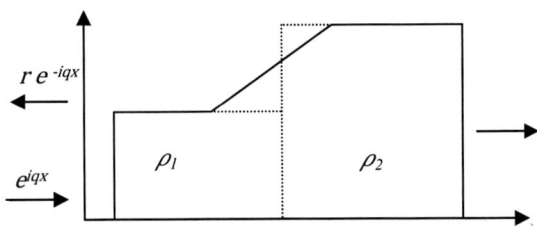

Figure 1. *The linear potential at the interface of two adjacent layer.*

As shown in fig 1, for an incoming from the left, we have:

$$A = \frac{Bi'(\chi)\big|_{x=x_0} Ai(\chi)\big|_{x=x_L} - Bi(\chi)\big|_{x=x_L} Ai'(\chi)\big|_{x=x_0}}{Ai(\chi)\big|_{x=x_0} Bi'(\chi)\big|_{x=x_0} - Bi(\chi)\big|_{x=x_0} Ai'(\chi)\big|_{x=x_0}}$$

$$(17\text{-}a)$$

$$B = \frac{q}{\lambda^{1/3}} \times \frac{Ai(\chi)\big|_{x=x_0} Bi(\chi)\big|_{x=x_L} - Bi(\chi)\big|_{x=x_0} Ai(\chi)\big|_{x=x_L}}{Ai(\chi)\big|_{x=x_0} Bi'(\chi)\big|_{x=x_0} - Bi(\chi)\big|_{x=x_0} Ai'(\chi)\big|_{x=x_0}}$$

$$(17\text{-}b)$$

$$C = \frac{\lambda^{1/3}}{q} \frac{Ai'(\chi)\big|_{x=x_L} Bi'(\chi)\big|_{x=x_0} - Ai'(\chi)\big|_{x=x_0} Bi'(\chi)\big|_{x=x_L}}{Ai(\chi)\big|_{x=x_0} Bi'(\chi)\big|_{x=x_0} - Bi(\chi)\big|_{x=x_0} Ai'(\chi)\big|_{x=x_0}}$$

(17-c)

$$D = \frac{Ai(\chi)\big|_{x=x_0} Bi'(\chi)\big|_{x=x_L} - Bi(\chi)\big|_{x=x_0} Ai'(\chi)\big|_{x=x_L}}{Ai(\chi)\big|_{x=x_0} Bi'(\chi)\big|_{x=x_0} - Bi(\chi)\big|_{x=x_0} Ai'(\chi)\big|_{x=x_0}}$$

(17-d)

By knowing the total transfer matrix, the reflectivity for the whole sample is known.

Error function

In this method, the smooth variation of the SLD of two adjacent films in term of layers depth is represented by [6]:

$$\rho(x) = \rho_1(x) + \frac{\rho_2(x) - \rho_1(x)}{2}\left[1 + erf\left(\frac{x-\Delta}{\sqrt{2}\,\sigma}\right)\right]$$

(18)

where σ is a parameter, which denotes the smoothness and Δ is the turning point of the error function.

Eckart potential

The smooth variation of the potential at boundaries can also be expressed by the so-called Eckart potential which is one of the most applicable functions in nuclear physics. The SLD variation corresponding with the Eckart potential is as follows [5]:

$$\rho(x) = \rho_1(x) + [\rho_2(x) - \rho_1(x)]/[1 + \exp(\frac{x-b}{\Delta})]$$

(19)

where b is the smoothness factor and Δ is the turning point of the Eckart potential.

EXAMPLE

To investigate the effects of the smoothness of the potential on reflectivity, polarization of reflected neutrons and the SLD, we consider a two layer sample composed of 20 nm thick Copper over a 15 nm thick gold with the SLD of 6.52 and 4.46 $\times 10^{-4}$ nm^{-2} for Copper and gold, respectively and a magnetic 20 nm Cobalt film with the SLD of 6.44 and -1.98 $\times 10^{-4}$ nm^{-2} for plus and minus magnetization, respectively as reference layer, which is mounted between the sample and a silicon substrate. It is supposed the incident neutron beam is polarized in the x direction and the magnetic field of the Cobalt reference is along the z axis.

*Figure 2. Arrangement of a sample for investigating the smoothness of the potential. The dashed line represents the effective potential experienced by neutrons parallel and anti parallel to the magnetic field **B***

Figure 3-a, shows the polarization of the reflected neutrons form the sample of figure 2 for the Eckart potential with smoothness factor $b=5A^o$. The real and imaginary parts of the reflection coefficient for the Error function with $\sigma=5A^o$, is illustrated in Figure 3-b. The reflectivity and phase of the reflection for the linear interfacial potential is depicted in Figure 4, **a** and **b**, using step method and airy function with smoothness factor $\varepsilon=0.05$ and 0.02, respectively.

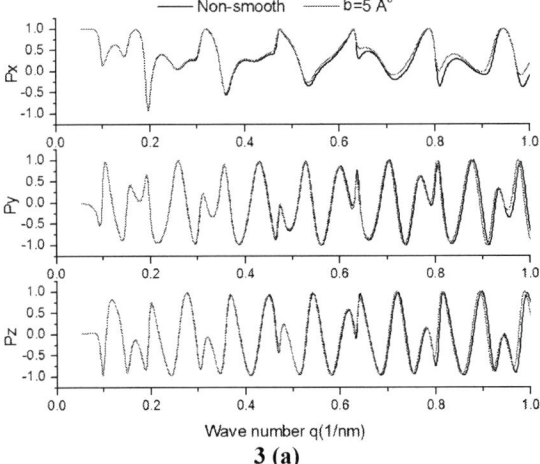

3 (a)

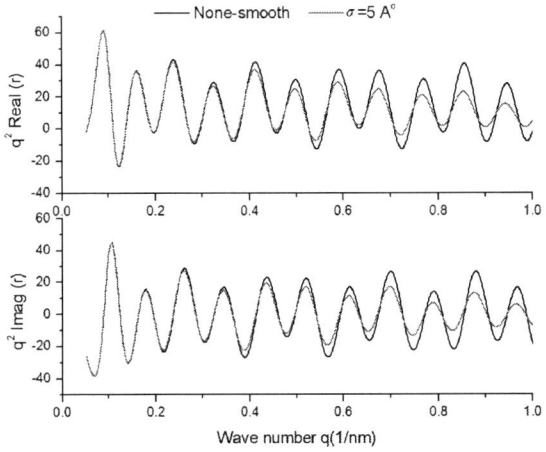

3 (b)

Figure 3. a) Polarization of the reflected neutrons for the Eckart potential with smoothness factor; b=5Aᵒ. b) Real and imaginary parts of the reflection coefficient for the Error function with smoothness factor σ=5Aᵒ.

whole sample, M (Eq. (16)). As the noises are existed even on the real and imaginary parts of the reflection coefficient graphs, we can't use them to reconstruct the SLD of the sample.

As the total transfer matrix depends on the reference layers, we can diminish or eliminate the noises or shift them to the q value larger than 0.6, by finding a suitable thickness for the reference layers. The rest of the noises can be removed by extrapolation.

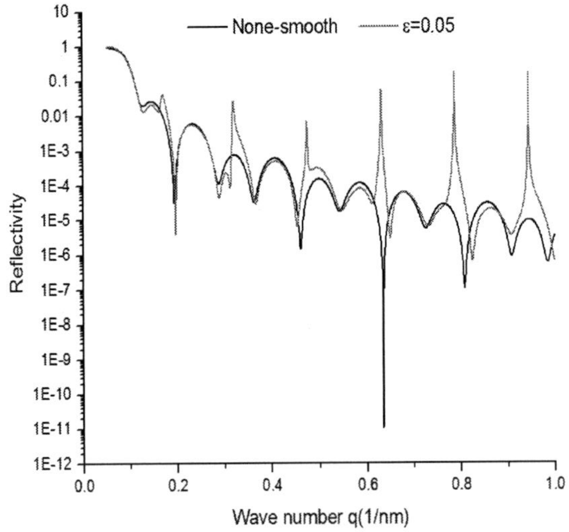

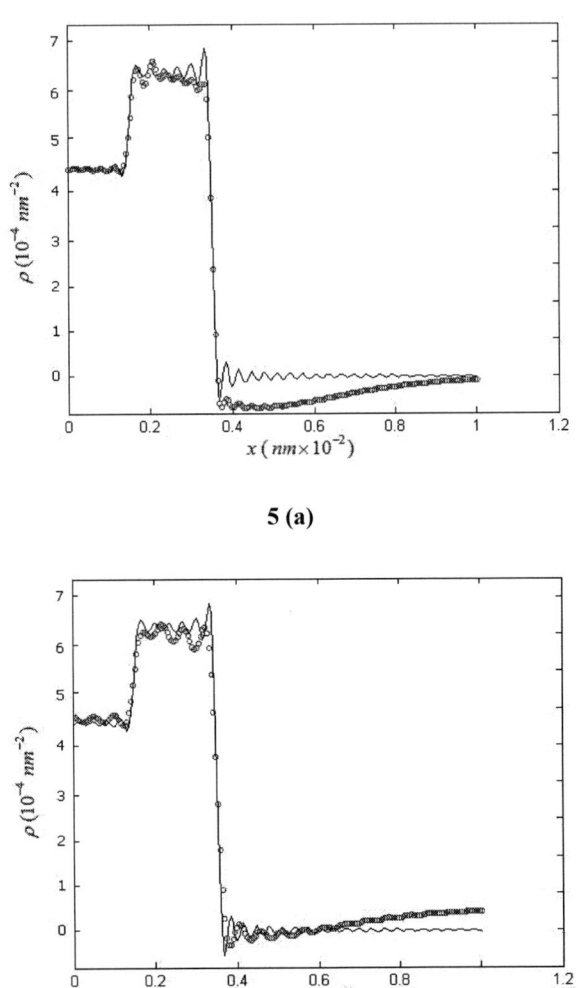

4 (a)

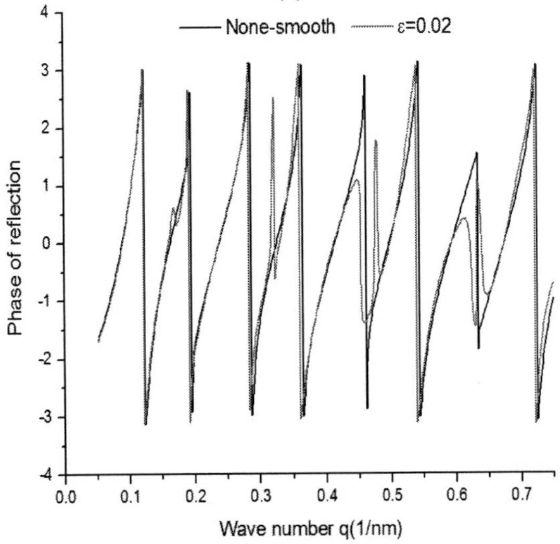

5 (a)

4 (b)

5 (b)

Figure 4. a) Reflectivity curve for the linear potential (step method) with $\varepsilon=0.05$. b) Phase of the reflection for the linear potential (Airy function) with $\varepsilon=0.02$.

Figure 5. Reconstructed SLD of the free reversed unknown sample; a) Error function, b) Linear potential (step method). The blue circled curves demonstrate the reconstructed SLD of the sample of Figure 2 at the presence of smoothness.

As it is presented in Figure 3 and 4, the effects of the smoothness of the interfacial potential on output results, is clear in all of the figures specially at large wave numbers. The results for the small wave numbers, $q \leq 0.6$, truly correspond with none-smooth data.

Figure 4-a,b demonstrate some noises in reflectivity and phase of the reflection even in small wave numbers for the linear potential. These noises are due to the abrupt changes in the elements of the transfer matrix of the

In order to retrieve the SLD of the unknown sample, the data for the real and imaginary parts of the reflection coefficient are needed as input data for some useful codes like the one which is developed by P. Sacks [11] based on Gel'fan-d-Levitan integral equation [11-13].

To show the stability of the reconstruction process of the SLD at the presence of the smoothness, the SLD of the unknown sample was retrieved for the Error function and linear potential (step method). As there were no noises in the data of Error function, they could be used directly as input data for Sacks code, however the data of linear potential needed to get clear from noises. The best result with less noises, obtained for the thickness of 10 nm for Cobalt reference. In order to remove the noises at large wave numbers, the data of q's larger than 0.6 was also neglected. Then these purified data was used as input for Sacks code.

The retrieved SLD of the free reversed unknown sample is illustrated in Figure 5-**a** and **b** for the Error function and linear potential, respectively. The 15 nm thick Copper with the SLD of 6.52 and 20 nm thick gold with SLD of 4.46, is clearly demonstrated in the Figure.

CONCLUSION

In most of simulation method in PNR, it is supposed the potential of two adjacent layers to be discontinuous and sharp. In this paper by introducing the reference method for retrieving the phase of the reflection for polarized neutrons, we investigated the effects of the smoothness of the interfacial potential on, reflectivity, polarization of reflected neutrons and the SLD of the sample for four smooth varying functions (Linear, Airy, Eckart and Error function).

The output results show that; depending on the range of the neutron wave numbers, consideration of the smoothness is very important. The output data for small wave numbers is clearly correspond with non-smooth data, nevertheless at large values of q, the data deflects from the none-smooth curve.

We can decrease the effects of smoothness on output data by selecting a suitable thickness for the reference layers and using the data of wave numbers smaller than 0.6.

It was shown that the reconstruction process of the SLD is yet stable at the presence of the smoothness.

REFERENCES

[1]- Xiao-Lin Zhou, Sow-Hsin Chem, (1995) "Theoretical foundation of X-ray and neutron reflectometry", Phys. Rep. 257, 223-348.

[2]- C.F. Majkrzak, N.F. Berk, and U.A. Perez-Salas, (2003) "Phase sensitive neutron reflectometry" Langmuir, 19, 7796-7810.

[3]- J. Kasper, H. Leeb, and R. Lipperheide, (1998)"Phase Determination in Spin-Polarized Neutron Specular Reflection" Phys. Rev. Lett. 80 2614.

[4]- C.F. Majkrzak, N.F. Berk, (1996) "Exact determination of the neutron reflection amplitude and phase" Physica B: Condensed Matter, Volume 221, Issues 1-4, 2 April, Pages 520-523.

[5]- M. R. Fitzsimmons, Los Alamos National Laboratory, C.F. Majkrzak, National Institute of Standards and Technology, "Application of polarized neutron reflectometry to studies of artificially structured magnetic materials".

[6]- D.A. Korneev, V.K. Ignatovich_, S.P. Yaradaykin, V.I. Bodnarchuk, "Specular reflection of neutrons from potentials with smooth boundaries", Physica B 364 (2005) 99–110

[7]- C.F. Majkrzak, N.F. Berk, (2003) "Phase sensitive reflectometry and the unambiguous determination of scattering length density profiles" Physica B: Condensed Matter, 336, 27-38.

[8]- S. Farhad Masoudi and Ali Pazirandeh, "Retrieval of the reflection coefficient in spin-polarized neutron specular reflectometry", J. Phys.: Condens. Matter 17 (2005) 475–484

[9]- S. Farhad Masoudi, A. Pazirandeh, "Application of polarized neutron in determination of the phase in neutron reflectometry" , Physica B 356 (2005) 21–25

[10]- S.F. Masoudi, "Exact determination of phase information in spin-polarized neutron specular reflectometry" , Eur. Phys. J. B 46, 33–39 (2005)

[11]- M.V. Klibanov, P.E. Sacks, (1994) J. Comput. Phys. 112, 273

[12]- K. Chadan, P.C. Sabatier, (1989) "Inverse Problem in Quantum Scattering Theory", 2nd edn. (Springer, New York)

[13]- T. Aktosun, P. Sacks, (1998) Inverse Probl. 14, 211; T.Aktosun, P. Sacks, (2000) SIAM (Soc. Ind. Appl. Math.) J. Appl. Math. 60, 1340; T. Aktosun, P. Sacks, (2000) Inverse Probl. 16, 821

USING SYNCHROTRON X-RAY DIFFRACTION (SXRD) FOR STUDYING THE BASO₄ FORMATION KINETICS AND THE EFFECT OF INHIBITORS ON BARITE FORMATION

Eleftheria Mavredaki, Anne Neville, Ken S. Sorbie*

menema@leeds.ac.uk
School of Mechanical Engineering, University of Leeds UK
*Institute of Petroleum Engineering, Heriot Watt University, Edinburgh UK

ABSTRACT

Synchrotron X-Ray Diffraction (SXRD) was used *in-situ* to investigate the formation of barium sulphate on a stainless steel surface at high temperatures. For the first time *in-situ* SXRD measurements of BaSO₄ formed in the presence of foreign ions (Sr^{2+}, Ca^{2+} etc) are presented. The formation kinetics of BaSO₄ on the surface has been determined and the crystallographic nature of the barite was investigated. In addition the effect of Poly – phosphinocarboxylic acid (PPCA) as chemical scale inhibitor on barium sulphate was examined at two temperatures. The barite crystal faces present on the surface after the treatment with inhibitors were detected. The lattice planes recorded with the *in-situ* SXRD measurements revealed that the presence of Sr^{2+} in the initial formation water resulted in the co-precipitation of Sr^{2+} within the barite lattice. The *in-situ* SXRD measurements allowed the assessment of information on the kinetics and crystallography of the formed scale in the absence and presence of inhibitor. The crystallography of the barite revealed high sensitivity to temperature and inhibition effects.

Keywords: Barium sulphate, Crystallography, Inhibitors, SXRD

INTRODUCTION

The precipitation of barium sulphate from aqueous supersaturated solutions has been widely studied; it comprises one of the most well known flow assurance problems in the oil industry. Formation of barite and other insoluble scale types (metal sulphates, calcium carbonate and silicates) occurs due to the incompatibility of the injected seawater with the formation water during their mixing in the reservoirs. The driving force of the scale formation is the supersaturation index which dominates the mixed solution. The inorganic salts deposit and adhere to production tubing and equipment parts leading to serious flow assurance issues and further to reduction in the performance of the petroleum facilities. Due to its extremely low solubility barium sulphate scale resists many chemical methods of removal including treatments with strong acids. Inhibition of barium sulphate by chemicals is generally recognized as the most appropriate approach for flow assurance.

So far the number of techniques used for barium sulphate studies with main interest on the precipitation process in the bulk phase, is wide (Black *et al.*, 1991; Gunn *et al.*, 1972; He *et al.*, 1995a; He *et al.*, 1995b; Leeden *et al.*, 1995; Liu *et al.*, 1976). On the other hand, focus on the formation kinetics of BaSO₄ on surface was given the last two decades when it was clear that the precipitation mechanisms of the barium sulphate in the bulk phase and the ones that dominate the formation of barite on the surface are different (Carosso *et al.*, 1984;

Graham *et al.*, 2004; Quddus *et al.*, 2000). One appropriate and most advanced technique used to follow the development of the scale on the surface is the SXRD. *In-situ* application of the SXRD allows the assessment of the growth of the deposited scale on the chosen substrate.

X-Ray synchrotron diffraction measurements have so far proved to be very useful for *in-situ* scaling studies. The evolution of calcareous scale formed on different substrates was followed in real-time with SXRD at different temperatures and in the presence of conventional (DETPMP and PPCA) inhibitors (Chen *et al.*, 2009; Martinod *et al.*, 2008). The data provided from the SXRD tests, pointed out the effect of temperature on the growth of the different planes of CaCO₃ on different substrates. The role of inhibitors (e.g PPCA) in changing the polymorphic form of CaCO₃ was discussed and the effect of Mg^{2+} on the scale morphology was shown (Chen, 2005).

In terms of using the *in-situ* SXRD technology for barium sulphate studies, only a few efforts have been made so far. The X-Ray Synchrotron Diffraction measurements of barium sulphate precipitation made by other researchers, have lead to determination of the main faces of barite (Hartman *et al.*, 1989; Hennessy *et al.*, 2002; Jones *et al.*, 2008). Additionally SXRD was used in order to study the effect of inhibitors such as nitrilotriacetic acid (NTA) and nitrilomethylenephosphonic acid (NTMP) on barite precipitation at 80°C (Jones *et al.*, 2008).

CP1202, *Neutron and X-Ray Scattering in Advancing Materials Research: International Conference – 2009*
edited by A. Saat, H. A. Kassim, M. H. H. Jumali, J. M. Saleh, M. R. Othman, A. Ibrahim, F. M. Idris, and M. H. A.-R. M. Ahmad
© 2009 American Institute of Physics 978-0-7354-0739-8/09/$25.00

This paper presents the first *in-situ* SXRD results on the barite scale system in the presence of divalent cations. The *in-situ* SXRD measurements revealed (i) the formation kinetics and the crystallography of $BaSO_4$ at two different supersaturation indexes (ii) the effect of PPCA on the crystallographic nature of the deposited barium sulphate and (iii) the presence of the Sr within the barite crystals. Hence the application of the *in-situ* synchrotron X-ray diffraction on a challenging scaling system like the one examined here, makes the assessment of the formation kinetics and mechanisms real.

MATERIALS AND METHODS

SXRD procedures

The set up used for the *in-situ* measurements of the barium sulphate formation as presented in Figure 1 allows the flow of the solutions and their further mixing in the capillary cell. The design of the capillary cell allows the formation of the scale under non-ambient conditions with temperature limits up to 200°C and maximum line pressure 34 MPa. The two chosen temperatures for the barite formation experiments were 57 °C and 95 °C as they represent two basic temperature levels that characterize the different parts of the oil industry. As far as the absolute pressure, it was adjusted at 15 psi. In the experimental set up Figure 1 an oven is included, which is placed between the pumps and the mixing chamber. Further the control panel adjusted to the experimental set up, shows the indications of pressure and temperature during the *in-situ* measurements of barite precipitation and deposition. The flow rate of the Formation and Synthetic Seawater was 5ml/min respectively, resulting in a total flow rate equal to 10 ml/min.

The cell used as substrate during the formation of barium sulphate, it was stainless steel with a bore of 2 mm and length 10 mm. The roughness of the substrate was 0.014 μm. The stainless steel cell was chosen as material as it is able to resist corrosion phenomena yet is a more realistic industrial component than the Si used previously (Hennessy *et al.*, 2002).

For the *in-situ* SXRD measurements of barium sulphate formation, the station X17B1 at the NSLS (National Synchrotron Light Source) was used. The type of radiation used was Cu Ka X-rays (λ= 0.17712 Å) with energy equal to 70 keV. The detector was a siemens D500 model which offers two - dimensional data collection. The monochromator used in the set up was a Si (311) sagittal focusing with Laue crystals (Zhong *et al.*, 2001). The crystals are cut from 311 silicon wafer crystal and bend to the same sagittal bending radius. In this way crystals contribute to the sagittal focusing of the X-rays. This type of crystals is preferable at high energy levels due to the small area of the beam's footprint on them (Zhong *et al.*, 2001).

A sets of brines named A was prepared and mixed for the *in-situ* SXRD measurements to take place. The composition of the Formation and Synthetic Seawater for A is shown in Table 0.1 The brines after being prepared were both filtered (0.2 μm nitrate cellulose membranes). The pH of each brine was equal to 5.5 and in this way a neutral environment in the barite formation system is maintained. The supersaturation index of brine A is 3.89 and 3.65 at 57°C and 95ºC respectively. The calculations of the supersaturation index were made with the ScaleSoft Pitzer software. Brine A was highly supersaturated for these experiments, since in the past the formation of the $BaSO_4$ crystals in the SXRD cell has proved to be difficult (Hennessy *et al.*, 2002).

Table 0.1: Composition of A

Type of ions	Formation Water (ppm)	Synthetic Seawater (ppm)
K^+	1906	380
Ca^{2+}	2033	405
Mg^{2+}	547	1215
Ba^{2+}	3982	0
Sr^{2+}	417	0
SO_4^{2-}	0	2780
Na^+	26535	10900

The formation of barium sulphate was recorded in inhibited and uninhibited conditions. The chemical additive tested in this study for their inhibition effect on the barium sulphate formation is a commercial inhibitor already used in the oil and gas sector.

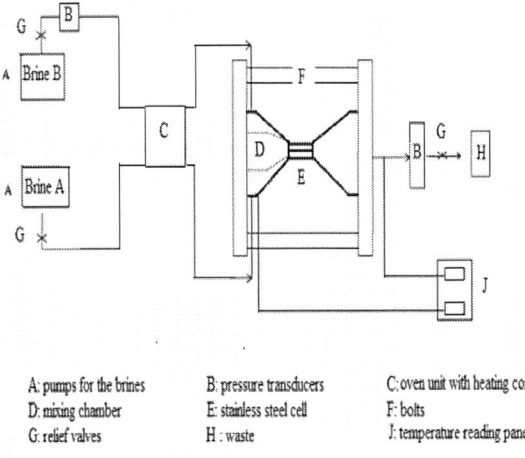

A: pumps for the brines   B: pressure transducers   C: oven unit with heating coils
D: mixing chamber   E: stainless steel cell   F: bolts
G: relief valves   H : waste   J: temperature reading panel

Figure 1: Diagram of the SXRD set up

Polyphosphinocarboxylic acid (PPCA) provided by Biolab has a molecular weight 3600 g/mole and activity 42%. For the inhibition tests the additive was added in the seawater at concentration 1, 4 and 10 ppm before the mixing. The duration of the experiments was set up for 60 minutes and 30 frames were collected during this time. The time gap between each frame was 2 minutes. For most of the experiments the duration was less than 1 hour as the mixing chamber was blocked by the high amount of the formed scale.

Data analysis

The data received from the *in-situ* measurements are integrated and converted to appropriate file formats by the beam manager. The data were calibrated with NIST standard Al_2O_3 powder. The package of data is next analyzed by the use of FullProf Suite program. The specific program consists the main tool for the basic analysis as through this program the raw data are converted to data reflecting the relationships of the intensity versus the θ angle or d space of the recorded diffraction peaks. Since every diffraction peak detected at a specific θ angle represents a different crystal plane of the examined material the determination of the crystallography can be accomplished. Next the correlation of the recorded diffraction peaks to the formed crystal faces of barite follows, based on the established crystallographic database for $BaSO_4$ (Pnma system) (Sawada H, 1990). Some of the experimental d space values recorded, are slightly shifted when compared to the theoretical ones. This was pointed out in the SXRD patterns received from the tests where the use of high supersaturation indexed brine A resulted in formation of big size $BaSO_4$ crystals.

SXRD PATTERNS

Scanning of the received *in-situ* SXRD patterns revealed the presence of three intense peaks recorded. Two of these peaks were present in the patterns due to the strong background of the stainless steel that was used as substrate. The two stainless steel peaks were detected at d space= 875 and 2.165 Å respectively. It is expected that a few peaks representing crystal faces of barite are hidden in the region where the stainless steel peaks are detected as well. For these missing barite planes no further determination can take place. Regarding the third peak which showed stable intensity and it always present in every SXRD pattern, was due to the formation of maghemite (Fe2 34O26.27) during the barite scaling tests (Cvejic *et al.*, 2006). The source of the Fe^{2+} in the system was the stainless steel substrate. The peak representing maghemite was marked at d space= 2.47 Å.

It is quite likely that in the same d-space where the stable peaks appear, other peaks due to the formed scale are hidden. Nevertheless the number of potentially hidden peaks is small allowing the analysis to be complete.

The data collected from the *in-situ* SXRD measurements are presented below as a qualitative

function of integrated intensity versus time for every crystal plane of barite present on the stainless steel cell. The integrated intensity is proportional to the crystal volume formed in the capillary cell (Nunes *et al.*, 2005). Moreover the determined crystallographic nature of the deposited scale can be clearly illustrated. For one case the SXRD pattern is also presented in order some details of the $BaSO_4$ crystallography can be highlighted.

RESULTS AND DISCUSSION

In-situ SXRD measurements in the absence of inhibitors

The SXRD data received during the deposition of the barium sulphate after precipitating from the mixture A, at 95°C is presented in Figure 2.

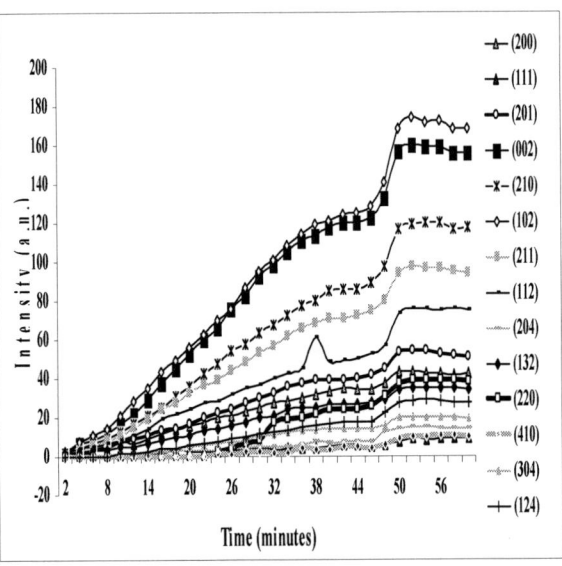

Figure 0.1: SXRD plot of uninhibited $BaSO_4$ formation at 95°C

During the *in-situ* X-ray diffraction measurement of the $BaSO_4$ formation on the cell 14 crystal planes of barite were present. Among the formed barite crystal faces the most important surfaces (210), (002), (200), (111) and (102) can be identified as published by Hartman and Strom *et al.*, 1989. The (002) and (210) crystal planes of barium sulphate are dominant since they have the lowest energy and they consist bounding faces for the equilibrium form of the barite (Allan *et al.*, 1993; Black *et al.*, 1991).

The formation kinetics of the different planes are shown in Figure 2. The recorded intensities reveal the slow growth of the barite planes until the 8th minute of the deposition and then an increase in the growth until the 50th minute. The last stage of the plots is characterized by a plateau for all the recorded surfaces of barium sulphate as the supersaturation index approached the value of

The barite surfaces that illustrate higher intensity values by the end of 1 hour precipitation are the (102), (002), (210), (211) and (112). The highest intensity values were received by the growth of the (102) and (002) lattice planes of barite with the (002) crystal face having a relative rough surface (Leeden et al., 1995). Yet the (111) which is also one of the most important planes of the barium sulphate due to its surfaces binding properties, by the end of the process was characterized by very low intensity.

The formation of the barite on the surface was further followed in-situ at 57°C and the SXRD data are presented in Figure 2. Clearly most of the barite planes recorded at 57°C were the same with the ones recorded at 95°C.

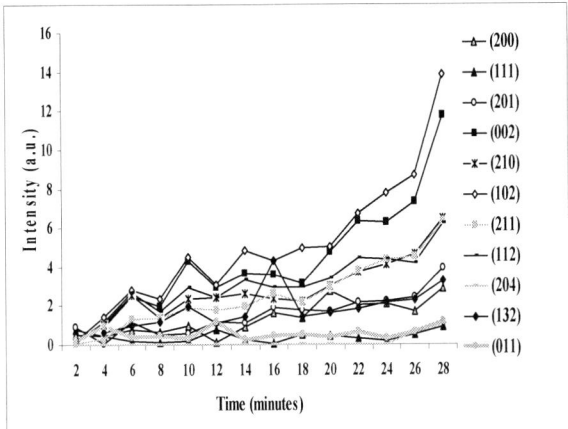

Figure 2: SXRD plot of uninhibited BaSO₄ formation at 57°C

However at 57°C, there are changes in the crystallography of the formed barite as the number of the formed surfaces is smaller. In addition the appearance of the (011) has been recorded. Generally the presence of the (011) crystal plane of $BaSO_4$ identifies the slower kinetics of the barium sulphate precipitation process.. As far as the kinetics of the crystal faces again they seem to follow a general increasing trend as time evolves, although the kinetics was recorded only until the 28th minute.. The intensity values of the dominant planes increase in time with the (102) and (002) again showing the maximum intensity recorded by the end of the test, however all the intensity values recorded for the tests at 57°C are lower than at 95°C. This sharp decrease in the intensity values of the barite crystal faces in combination with the decrease in the experimental duration suggests that the crystals are orientated in a preferred way within the capillary cell. A comparison of the Debye rings images formed during the in-situ measurements at the two different temperatures (Figure 3) is a good way to prove if preferred orientation of the developing material in the capillary cell occurs or not, under the specific examined conditions.

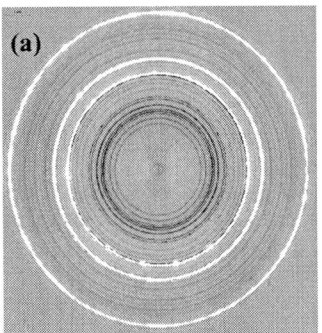

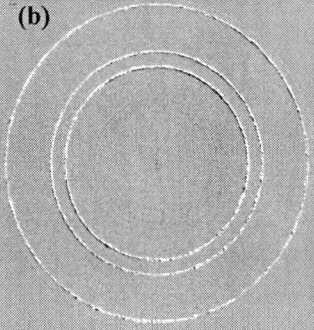

Figure 3: Debye rings of formed barite (a) 95°C and (b) 57°C

The white coloured rings as shown in both images of Figure 3 are due to the presence of the three stable peaks. In Figure 3 key point for observation is the Debye rings formed in the areas surrounded by the white coloured rings. It is clear that the intensity level of the Debye rings of $BaSO_4$ formed during the in-situ measurement at the two different temperatures changes. At 95°C the diffraction rings are quite dark coloured and they seem homogeneous Figure 3 (a). No variations in the intensity of the Debye rings are observed. On the other hand the diffraction rings formed during the in-situ test at 57°C are not intense at all and they can hardly be noticed. This difference in the intensity of the rings developed at both temperatures confirms that the barite deposited on the surface at 57°C is more textured compared to the one formed at 95°C (Jawad et al., 2007).

In-Situ SXRD Measurements In The Presence Of Ppca At 95°C

PPCA was tested in-situ for its performance on the formation of barium sulphate on the capillary cell. The formation kinetics and the crystal planes growing on the surface at 95°C after the addition of 1 ppm PPCA are presented in Figure 5.

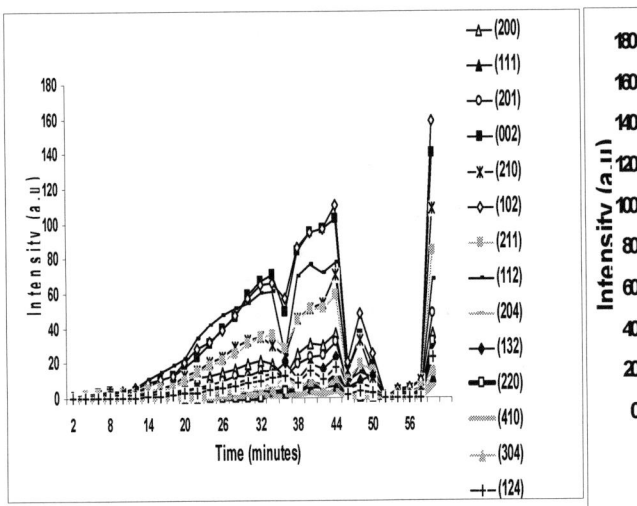

Figure 5: Effect of 1 ppm PPCA at 95 °C

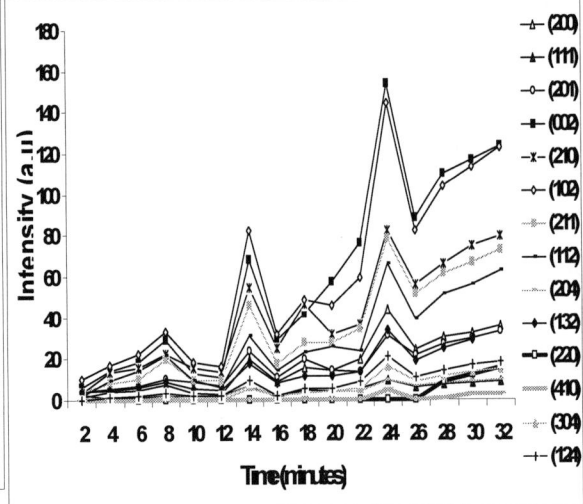

Figure 6: Effect of 4 ppm PPCA at 95 °C

According to the recorded barite crystal planes on the surface the crystallographic nature of the formed barium sulphate did not change when 1 ppm of PPCA was added in the mixing brine at 95°C. The only plane of barite not recorded during this *in-situ* test under inhibited conditions was the (220). As far as the surfaces of barite characterized by the highest intensity are the (102) and (002) just as recorded in the uninhibited systems. What is being more affected by the presence of the inhibitor in the supersaturated brine is the time that the planes of barite appear on the surface and the way they grow in time. All the crystal planes reveal a prolonged induction time on the surface which approaches the 16th minute after the mixing of the brines in the capillary cell. This lag in the barite formation is then followed by an increase in the intensities of all the crystal faces of barite which show a similar kinetics trend. An important feature in the recorded plots is the decrease in the intensity between the 44th and 56th minute.

The next concentration examined was 4 ppm PPCA. As shown in Figure 6 all the dominant planes of barite are present from the start of the precipitation process. The plots reveal a general increasing trend of the intensity. It is worth noting the two peaks recorded at the 14th and 24th minute of the formation of barium sulphate on the stainless steel surface, which illustrate the interactions of the inhibitor on the growth of the present planes of barite. Concerning the experimental duration of this test was 32 minutes. The capillary cell of 2 mm bore leads to show signs of blocking earlier and this could be due to the position of the crystals in the cell or due to interaction between the calcium ions and the inhibitor and will be added in future work. Potentially the increased concentration of the inhibitor (>4 ppm) results in formation of Ca-phosphonate complexes, which are able to block the cell quickly.

With further increase in the concentration of the inhibitor up to 10 ppm the formation kinetics of the barite surfaces seem to grow in a similar way on the surface. Although the barite planes emerge from the 2nd minute after mixing the brines with no obvious prolonged induction time on the surface, the intensity values at the end of the process were significantly low. The SXRD pattern of the *in-situ* measurement when 10 ppm of PPCA was added is shown in Figure 7.

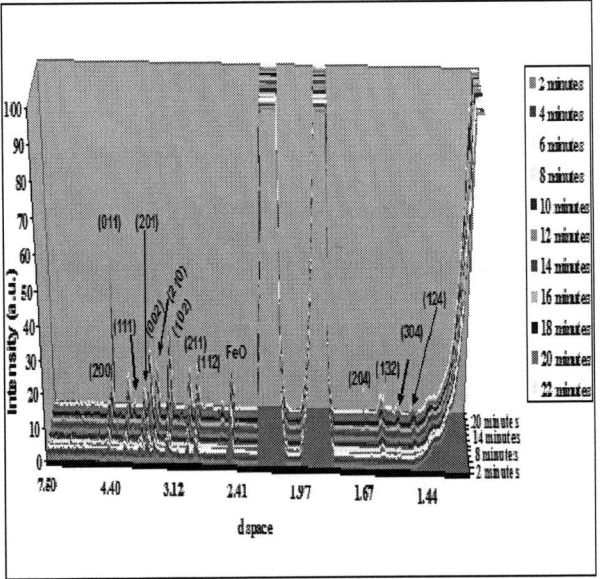

Figure 7: SXRD pattern of BaSO₄ after treatment with 10 ppm PPCA

The diffraction peaks as well the crystal faces of the barite are presented. Main feature of this pattern is the emergence of the (011) plane of barite at the 12th minute. The (011) is being recorded although this is difficult especially for tested systems like the one studied here, where the kinetics is really fast. Generally the formation of the (011) lattice plane of BaSO$_4$ is quite important, however it is predicted to grow too fast thus it is difficult to capture the appearance of the (011) in the system (Black et al., 1991). Yet the presence of the (011) at the 12th minute and its disappearance after that time point highlights the performance of PPCA. The inhibitor at the concentration of 10 ppm managed to reduce the growth rate of the specific plane thus the (011) is detected revealing a high intensity value. After the 12th minute the diffraction peak responding to the (011) barite plane cannot be identified as probably the (011) could not further being retarded by the presence of the inhibitor.

An overview of the effect of PPCA on the main barite planes at 95°C is presented in Figure 8. The intensity values in this graph response to the value recorded at the end of every experimental run.

extreme high supersaturation index conditions. The formation kinetics is shown to be the same for all the planes of barite recorded on the surface. Concerning the planes which dominate the growth of the barite on the surface are the same as in the previous tests. The total effect of PPCA on the BaSO$_4$ crystal planes present on the surface is illustrated in Figure 0.9. According to the graph the addition of 4 ppm PPCA in the system resulted in higher intensity values of the dominant barite planes. The enhancement of the growth of the barite surfaces is remarkable for the (002), (210) and (102). Concerning the effect of 10 ppm PPCA on the barite planes is more promising at this high supersaturation index conditions compared to the effect of 4 ppm PPCA. All the dominant planes have been inhibited by the presence of 10 ppm PPCA. Only the (111) and (132) plane are characterized by stable intensity under uninhibited and inhibited conditions.

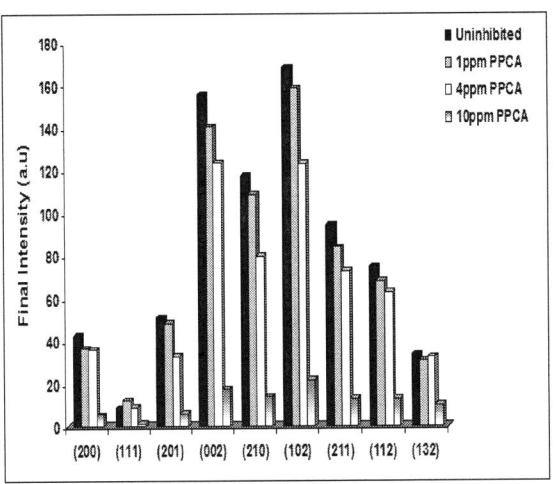

Figure 8: Overview of PPCA effect on dominant barite planes at 95 °C

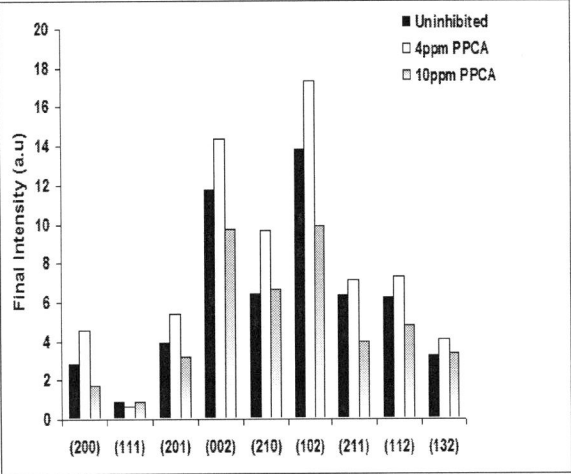

Figure 9: Overview of the PPCA effect on dominant barite planes at 57 °C

It is clear that the increase in the concentration of the applied inhibitor resulted in strong retardation of all the barite planes. The concentration of 10 ppm PPCA inhibits the formation of the dominant barite planes at high degree. More important is the inhibition of the (111) plane of barite when 10 ppm of PPCA is present. The growth of the (111) seems to be strongly blocked despite the known fast growth rate that characterizes the specific plane as it belongs to the kinked faces group (Hartman et al., 1989).

In-situ SXRD measurements in the presence of inhibitors at 57 °C

The effect of PPCA as inhibitor of barium sulphate was further examined at the temperature of 57°C. The decrease in the applied temperature leads to extended increase in the supersaturation index compared to the one at the temperature of 95°C. Thus it is interesting to probe any changes in the formation kinetics and in the crystal planes of the formed barite on the surface under

A summary of the crystal faces of barium sulphate grown on the surface at both temperatures in the presence and absence of PPCA is given in Table 0.1.

Effect of divalent cations

As the mixed brines contain other divalent cations besides barium, it is necessary to use all the respective databases in order to determine any diffraction peaks present due to other deposited scale type than barium sulphate. Thus the SXRD patterns were investigated for any potential crystal faces present due to SrSO$_4$ or CaSO$_4$ formation and/or any possible known mixtures containing Ba^{2+}, S, O, Sr^{2+}, Ca^{2+} and Mg^{2+} (Boeyens et al., 2002; Brigatti et al., 1997). None of the d space values for strontium sulphate or calcium sulphate were in agreement with the d space values received. This was expected as according to other researchers pure SrSO$_4$ is quite soluble at the tested temperatures (Risthaus et al., 2001) and also concerning the calcium ions they reveal a trend to precipitate within the barite lattice instead of depositing as CaSO$_4$

99

Table 0.1: Crystal faces of BaSO$_4$

Barite crystal faces formed at 95°C	Uninhibited	1 ppm PPCA	4 ppm PPCA	10 ppm PPCA	Barite crystal faces formed at 57°C	Uninhibited	4 ppm PPCA	10 ppm PPCA
(200)	√	√	√	√	(200)	√	√	√
(111)	√	√	√	√	(111)	√	√	√
(201)	√	√	√	√	(201)	√	√	√
(002)	√	√	√	√	(002)	√	√	√
(210)	√	√	√	√	(210)	√	√	√
(102)	√	√	√	√	(102)	√	√	√
(211)	√	√	√	√	(211)	√	√	√
(112)	√	√	√	√	(112)	√	√	√
(204)	√	√	√	√	(204)	√	X	X
(132)	√	√	√	√	(132)	√	√	√
(220)	√	√	√	X	(011)	√	X	X
(410)	√	√	√	X	(220)	X	X	X
(304)	√	√	√	√	(410)	X	X	X
(124)	√	√	√	√	(304)	X	√	√
(232)	X	√	X	X	(124)	X	√	√
(422)	X	√	X	X	(413)	√	√	X
					(223)	X	√	X

formations (Benton *et al.*, 1993). The only case where the recorded diffraction peaks were matching with non barite d space values was for the case of formed celestine barian on the substrate. This indicated that strontium ions incorporation into the barite lattice occurred on the stainless steel surface. The formation of celestine barian (Sr 0.87 Ba 0.13) (SO$_4$) was confirmed from the peaks detected at d space= 4.04 Å, 2.40 Å and 73 Å representing the (111), (212) and (031) crystal planes of celestine barian respectively (Brigatti, 1997). For the other peaks in the SXRD pattern which cannot be linked to the known databases, it is suggested that their presence is due to formation of non-well established mixtures of barite with the divalent cations and especially with Ca^{2+}

ACKNOWLEDGMENTS

The authors would like to thank the FAST II joint industrial project and the University of Leeds for the financial support of E.M. Also the beam manager of X17B1 station of Brookhaven National Laboratory, Dr. Zhong Zhong for his help and support.

REFERENCES

Allan N. L., Rohl A. L., Gay D. H., Catlow C. R. A., Davey R. J. and Mackrodt W. C., (1993), Calculated Bulk and Surface Properties of Sulfates, *Faraday Discussion* 95: 273-280

Benton W. J., Collins I. R., Grimsey I. M., Parkinson G. M. and Rodger S. A., (1993), Nucleation, growth and inhibition of barium sulfate-controlled modification with organic and inorganic additives, *Faraday Discuss* 95: 281-297

Black S. N., Bromley L. A., Cottler D., Davey R. J., Dobbs B. and Rout J. E., (1991), Interactions at the Organic/Inorganic interface: Binding motifs for phosphonates at the surface of barite crystals, *Chem. Soc. Faraday Trans.* 87: 3409-3414

Boeyens J. C. A. and Ichharam V. V. H., (2002), Redetermination of the crystal structure of calcium sulphate dihydrate, CaSO$_4$ x 2H$_2$O, *Zeitschrift für Kristallographie* 217: 9-10

Brigatti M. "Celestite, barian - (Sr0.87Ba0.13) (SO$_4$) - [PNMA]." 1997.

Brigatti M. F., Galli E. and Medici L., (1997), Ba -rich celestite: new data and crystal structure refinement, *Mineralogical Magazine* 61: 447-451

Carosso P. A. and Pelizzetti E., (1984), A stopped flow technique in fast precipitation kinetics - The case of barium sulphate, *Journal of crystal growth* 68: 532-536

Chen T., (2005), New insights into the mechanisms of calcium carbonate mineral scale formation and inhibition, Pages

Chen T., Neville A., Sorbie K. and Zhong Z., (2009), *In-situ* monitoring the inhibiting effect of polyphosphinocarboxylic acid on $CaCO_3$ scale formation by Synchrotron X-ray diffraction, *Chemical Engineering Science* 64: 912-918

Cvejic Z. and Rakic S. "Maghemite C - Fe2 34O26.67- [P4332]." 2006.

Graham A. L., Vieille E., Neville A., Boak L. S. and Sorbie K. S., (2004), Inhibition of $BaSO_4$ at a Hastelloy metal surface and in solution: The Consequences of Falling below the Minimum Inhibitor Concentration (MIC), *SPE 114049* 87444:

Gunn D. J. and Murthy M. S., (1972), Kinetics and mechanisms of precipitations, *Chem. Eng. Sci.* 27: 1293-1313

Hartman P. and Strom C. S., (1989), Structural morphology of crystals with the barite ($BaSO_4$) structure: A revision and extension, *Journal of Crystal Growth* 97: 502-512

He S., Oddo J. E. and Tomson M. B., (1995a), The nucleation kinetics of Barium Sulfate in NaCl solutions up to 6 m and 90°C, *Journal of Colloid and Interface Science* 174: 319-326

Hennessy A., Graham G., Hastings J., Siddons D. P. and Zhong Z., (2002), New pressure flow cell to monitor $BaSO_4$ precipitation using synchrotron in situ angle dispersive X - ray diffraction, *J. Synchrotron Rad.* 9: 323-324

Jawad M. A., Steuwer A., Kilcoyne S. H., Shore R. C., Cywinski R. and Wood D. J., (2007), 2D mapping of texture and lattice parameters of dental enamel, *Biomaterials* 28: 2908-2914

Jones F., Jones P., Marco R. D., Bobby Pejcic and Rohl A. L., (2008), Understanding barium sulfate precipitation onto stainless steel, *Applied surface science* 254: 3459-3468

Leeden M. C. V. D. and Rosmalen G. M. V., (1995), Adsorption behavior of polyelectrolytes on barium sulfate crystals, *Colloid and Interface Science* 171: 142-149

Liu S. T. and Nancollas G. H., (1976), Scanning electron microscopic and kinetic studies of the crystallization and dissolution of barium sulfate crystals, *Journal of Crystal growth* 33: 11- 20

Martinod A., Neville A., Sorbie K. and Zhong Z., (2008), Assessment of $CaCO_3$ inhibition by the use of SXRD on a metallic substrate, *NACE*

Nunes C., Mahendrasingam A. and Suryanarayanan R., (2005), Quantification of crystallinity in substantially amorphous materials by synchrotron X-ray powder diffractometry, *Pharmaceutical research* 22: 1942-1953

Quddus A. and Allam I. M., (2000), $BaSO_4$ scale deposition on stainless steel, *Desalination* 127: 217-224

Risthaus P., Bosbach D., Becker U. and Putnis A., (2001), Barite scale formation and dissolution at high ionic strength studied with atomic force microscopy, *Colloids and Surfaces A: Physicochem. Eng. Aspects* 191: 201-214

Sawada H T. Y. "Barite - $BaSO_4$ -[PNMA]." 1990.

Zhong Z., Kao C. C., Siddons D. P. and Hastings J. B., (2001), Sagittal focusing of high-energy synchrotron X-rays with assymentric Laue crystals II Experimental studies, *Journal of Applied Crystallography* 34: 646-653

X-Ray Diffraction Microstructural Analysis of Bimodal-Size-Distribution MgO Nanopowders

Suminar Pratapa and Budi Hartono

Department of Physics, Faculty of Mathematics and Sciences, Sukolilo Campus, Jl. Arief Rahman Hakim, Institute of Technology Sepuluh November (ITS), Surabaya, Indonesia
E-mail (corresponding author): suminar_pratapa@physics.its.ac.id

ABSTRACT

Investigation on the characteristics of x-ray diffraction data for MgO powdered mixture of nano and sub-nano particles has been carried out to reveal the crystallite-size-related microstructural information. The MgO powders were prepared by co-precipitation method followed by heat treatment at 500, 800 and 1200°C for 1 hour, being the difference in the temperature was to obtain two powders with distinct crystallite size and size-distribution. The powders were then carefully blended in air to give the presumably strain-free, bimodal-size-distribution MgO nanopowder. High-quality laboratory X-ray diffraction data for the powders were collected and then analysed using Rietveld-based MAUD software using the lognormal size distribution. Results show that the single-mode powders exhibit spherical crystallite size (D_v) of 29(1) nm, 36(1) and 185(0) nm for the 500, 800 and 1200°C data respectively with the nanometric powder displays slightly narrower crystallite size distribution character, indicated by lognormal dispersion parameter (σ) of 0.22 as compared to 0.18 for the sub-nanometric 1200°C powder. The mixture exhibits relatively more asymmetric peak broadening. By analysing the x-ray diffraction data of the latter specimen by using the single phase approach the results obtained was not according to experimental finding. Introducing two phase models for the 'double-phase' 500-1200 mixture to accommodate the bimodal-size-distribution characteristics give $D_v = 34(2)$ and $\sigma = 0.10$ for the 'nanometric phase' and $D_v = 363(0)$ and $\sigma = 1.38$ for the 'sub-nanometric phase'.

Keywords: bimodal-size-distribution, nanopowder, MgO, x-ray diffraction

INTRODUCTION

Size distribution is an important aspect in development of material science due to its effect on physical as well as chemical properties of materials, for example when bimodal size of zirconia powder was used for ink-jet printing [1]. The most powerful method to examine the distribution so far is probably by electron microscopy, particularly, with transmission electron microscopy (TEM). This method, however, is difficult especially in sample preparation.

On the other hand, diffraction method offers rather simple approach in extracting microstructural information from the broadening characters of diffraction peaks from their basic Bragg's law delta-function form. It is well-known that a measured diffraction profile h is a convolution between instrumental effects g and microstructural (or nanostructural if size is of concern) effects f, or,

$$h = g \otimes f$$

(1)

where $\otimes$ denotes convolution. Profile fitting to measured data using the expression probably is not quite acceptable from physical sense, but some studies showed good agreement with experiments particularly for isotropic (or, structurally cubic) materials [2, 3].

While physical approach through, for example, Fourier method should be used, despite its time consuming character [4], analysing diffraction data from materials exhibiting bimodal size distribution may lead to complex solution and give biased results. Fitting through (1) is therefore potential as a starting point to further accurate size analysis.

In this paper, we report our investigation on the characters of bimodal size distribution of mixed MgO nanopowders. Bimodal distribution effect on diffraction peak profiles has been investigated previously [5] by applying bimodal model using pseudo-Voigt function but no microstructural information were reported. The emphasize of this study is to provide immediate cautions to ones who have concerns on nanosized particle analysis by inspecting and analyzing the broadening characters of the diffraction peaks.

CP1202, *Neutron and X-Ray Scattering in Advancing Materials Research: International Conference – 2009*
edited by A. Saat, H. A. Kassim, M. H. H. Jumali, J. M. Saleh, M. R. Othman, A. Ibrahim, F. M. Idris, and M. H. A.-R. M. Ahmad
© 2009 American Institute of Physics 978-0-7354-0739-8/09/$25.00

METHOD

MgO nanopowders were produced by co-precipitation method where Mg powder was dissolved in HNO_3. NH_4OH was repeatedly added at low amount to produce initial precipitate, which was then washed with distilled water to give acid-free precipitate. The precipitate was heated at several desired temperatures, after knowing from DTA measurement that MgO formed at around 370°C, i.e. 500, 800, and 1200 °C to give MgO nanopowders with different size characters. The process is presumed to yield strain-free powders. These powders were labelled as MgO500, MgO800 and MgO1200 respectively. Mixtures between two powders were carefully made by hand mixing; labelled as MgO500-800, MgO500-1200, and MgO800-1200.

X-ray diffraction (XRD) data collection of these single and mixture powders was done using Philips X'Pert MPD machine at 40 kV and 30 mA applied to Cu target with 2θ range 5-120° and step size 0.02°. Collecting time per step was adjusted to provide acceptable counting statistics.

the refinement and fixed.

RESULTS

Figure 1 shows the diffraction patterns of all MgO samples. The samples obviously contain pure MgO. Figure 2 shows the patterns of the associated 420 reflection of the only phase. Using quick inspection, it is clear from the figure that (1) peak breadth decreases with temperature and (2) peak intensity increases with temperature for the original powders, while those characters are in between for the associated mixtures. Assuming that the powders do not contain residual strain, the decrease in peak breadth indicates changes in crystallite size characters. There is no simple way, however, to quickly recognize whether peak broadening of any pattern is caused by one or more microstructural effects. The following analysis is to provide detail of the size characters revealed using MAUD, a Rietveld-based fitting software [6].

The output of the refinement using the software were

Tabel 1. Rietveld-based refinement output for single and bimodal-mixture MgO nanopowders using MAUD. Numbers in parantheses are standard deviations to the associated values in the same digits. The mixtures are assumed with a single-phase model

	MgO500	MgO800	MgO1200	MgO500-800	MgO500-1200	MgO800-1200
R_{wp}	14.56	14.37	16.37	14.64	17.89	16.29
GoF	1.34	1.31	1.48	1.33	1.62	1.48
D_V (nm)	29(1)	36(1)	185(0)	36(1)	81(10)	85(9)
Variance	0.22(3)	0.20(4)	0.18(6)	0.18(4)	0.54(21)	0.36(16)

Analyses of the diffraction data were done using MAUD [6], with correction to the instrument contribution [g profile in (1)] was embedded in the refinement. In constructing the instrument correction profile, an yttria powder heated at 1100°C for hour was selected after showing comparable broadening characters to MgO standard ceramic [7] and LaB_6 NIST Standard [8].

All patterns from six samples were analysed using single MgO phase model, while the patterns of the mixtures were also analysed using what so called 'double phase model', where lattice parameters from the single phase model of the single powder were keyed in

directly collected and used for microstructural analysis since the instrument effects have been included. Table 1 lists the figures-of-merit (FoM) of the refinement and shows slight increases of both R_{wp} and GoF values when single phase was assumed. These FoMs indicated more acceptable fitting when double phase model was used, as shown in Table 2. Table 1 and 2 also show the volume-weighted size (D_v) and the size distribution parameter (σ), which denotes the degree of dispersion, for the single and mixture powders, analysed with one phase model (Table 1) and two phase model (2). Obviously, crystallite size increases with temperature and dispersion degree becomes broader with temperature. The tables also show that larger differences in D_v of the original

Tabel 2. Rietveld-based refinement output for bimodal-mixture MgO nanopowders using MAUD with double-phase model

	MgO500-800		MgO500-1200		MgO800-1200	
R_{wp}	13.79		13.94		13.83	
GoF	1.25		1.26		1.26	
	Phase 1 (MgO500)	Phase 2 (MgO800)	Phase 1 (MgO500)	Phase 2 (MgO1200)	Phase 1 (MgO800)	Phase 2 (MgO1200)
D_V (nm)	22(1)	49(3)	34(2)	363(0)	60(3)	153(4)
Variance	0.04(4)	0.12(8)	0.10(8)	1.38(40)	0.02(61)	0.52(45)

powders in the associated mixture cause slightly more adverse fitting, indicated by larger FoMs.

Example graphical presentations for the MgO500, MgO1200 and MgO500-1200 size characters are shown in Figure 3. Analyzing the data for the mixture using single phase model gives information on that it exhibits slightly lower D_v with broader region. By contrast, when the same data was analysed using double phase model, two distributions were resulted with fair agreement. The agreement can be, intuitively, improved by retaining the original lattice parameters as done by others [5].

In terms of unknown tested samples, separating such bimodal size distribution characters is probably a difficult task. Some aspects maybe recommended to recognize the presence of such characters, i.e. by inspecting the degree of asymmetry of the diffraction peaks and the presence of relatively high standard deviations on D_v and σ, when unimodal model is used. Provided no other source of asymmetry presents, it is believed that the former aspect can be easily performed. It should be noted, however, that FoMs, D_v and σ values cannot be used to provide indication on the presence of bimodal-size-distribution behaviour.

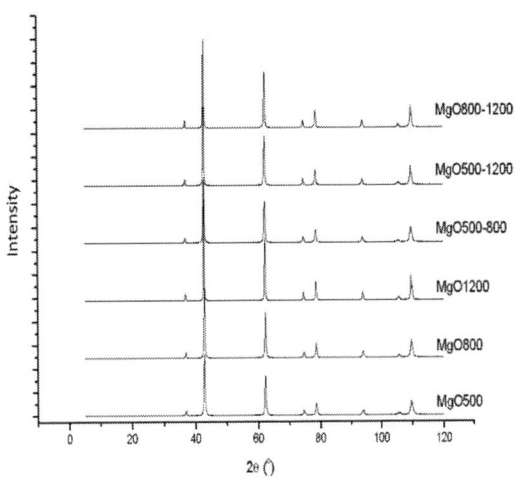

Figure 1. XRD (CuKα radiation) patterns for MgO single and mixture MgO nanopowders.

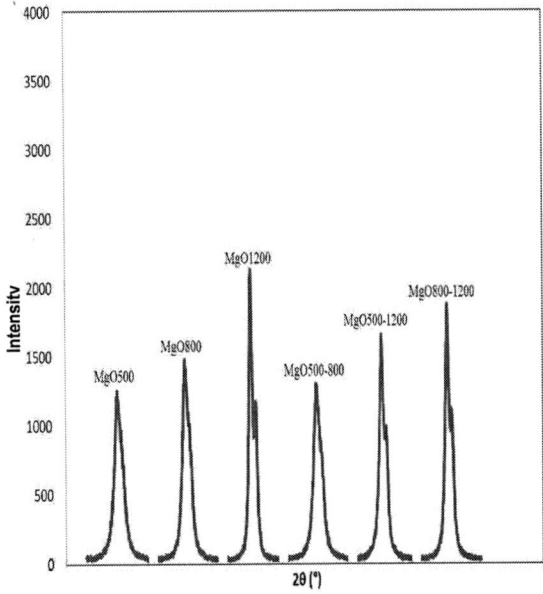

Figure 2. Comparison of the 420 reflection of single and mixture MgO nanopowders.

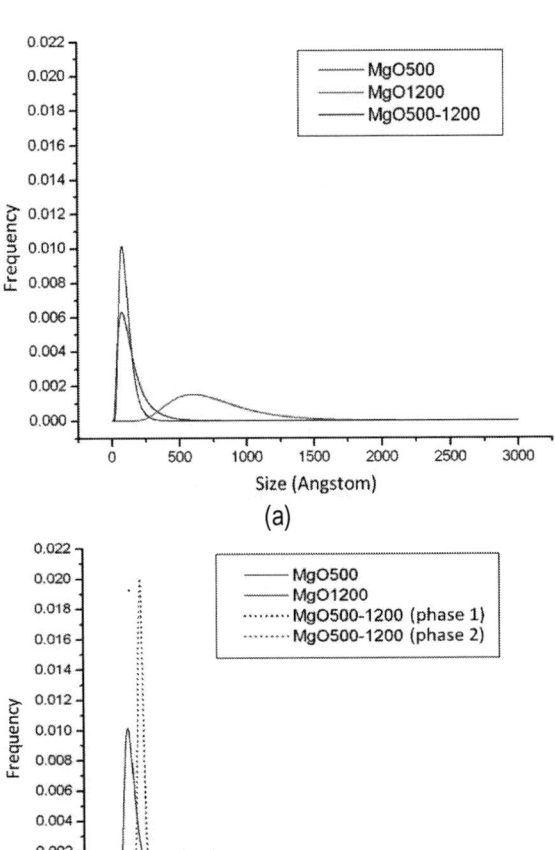

(a)

(b)

Figure 3. Lognormal size distribution profile of the MgO500, MgO1200 and mixture of them resulted from the x-ray diffraction data analysis using MAUD. Size distribution for the mixture is after (a) unimodal and (b) bimodal assumptions.

CONCLUSION

Mixing two MgO nanopowders produced by different heat treatments causes new diffraction broadening behaviour arising from the bimodal-size-distribution characters. Careful measurement and analyses should be a primary concern to gain the appropriate original size and distribution parameter values, for example using the bimodal model rather than unimodal model. Qualitative inspection such as peaks with strong asymmetry and relatively large errors on size and distribution parameter values can be useful prior to quantitative analysis.

ACKNOWLEDGEMENT

We wish to thank the Directorate of Higher Education (DIKTI), Ministry of National Education, Republic of Indonesia, for providing research funding through Hibah Penelitian Pascasarjana, 2008.

REFERENCES:

1. Blazdell, P. and S. Kuroda, *Bimodal Ceramic Ink for Continuous Ink-Jet Printer Plasma Spraying.* Journal of American Ceramic Society, 2001. **84**(6): p. 257-259.

2. Langford, J.I., D. Louër, and P. Scardi, Effect of a crystallite size distribution on x-ray diffraction line profiles and whole-powder-pattern fitting. Journal Applied Crystallography, 2000. **33**: p. 964-974.

3. Balzar, D., *Voigt function model in diffraction-line broadening analysis*, in *Defect and Microstructure Analysis by Diffraction*, R. Snyder, J. Fiala, and H.J. Bunge, Editors. 1999, IUCr/Oxford University Press: Oxford. p. 94-126.

4. Scardi, P. and M. Leoni, *Whole powder pattern modelling.* Acta Crystallographica, 2001. **A58**: p. 190-200.

5. Young, R.A. and A. Sakhtivel, *Bimodal distributions of profile-broadening effects in Rietveld refinement.* Journal of Applied Crystallography, 1988. **21**: p. 416-425.

6. Lutteroti, L. *MAUD: Materials Analysis Using Diffraction.* 2006 [cited 2008; Available from: http://www.ing.unitn.it/~maud.

7. Pratapa, S. and B.H. O'Connor, Development of MgO ceramics standards for x-ray and neutron line broadening assessments. Advances in X-ray Analysis, 2002. **45**: p. 41-47.

8. Freiman, S.W. and N.M. Trahey, Standard Reference Material 660a. Lanthanum Hexaboride Powder - Line Position and Line Shape Standard for Powder Diffraction. 2000, National Institute of Standards & Technology (NIST): Gaithersburg, MD. USA

X-Ray Diffraction Studies of Cross Linked Chitosan With Different Cross Linking Agents For Waste Water Treatment Application

Nurhidayatullaili Muhd Julkapli, Zulkifli Ahmad and *Hazizan Md Akil

School of Materials and Mineral Resources Engineering, Universiti Sains Malaysia, 14300 Seberang Perai Selatan, Pulau Pinang, Malaysia

**Corresponding author:* hazizan@eng.usm.my

ABSTRACT

Chitosan is a polysaccharide derived from N-deacetylation of chitin and receiving increased attention as metal ion absorbent in wastewater treatment application. To improve the performance of chitosan as an absorbent, the cross linking approach was applied. Introduction of cross-linking agent would break the crystal zone in chitosan system, making it less crystal and consequently enhanced the absorption area. Therefore, in this study, cross-linked chitosan were prepared using different of cross-linking agents. The chitosan powder was weighed, dissolved in acetic acid (0.1 M), and dropped slowly into absolute N-methyl pyyrolidone solvent containing cross-linking agent. The cross linking reaction was carried out in N_2 environment at 150°C for 6 hours. X-ray diffraction (XRD) analysis was applied to characterize the crystallinity of native and cross linked chitosan. Generally, the XRD patterns of all types of chitosan show two crystalline peaks approximately at 10° and 20° (2θ). However, the cross linked chitosan with longer length of cross linking agents show lower and broader crystalline peaks as compare to those with shorter length. Similarly, the calculated crystalline index (Cr I) also showed this decreasing tendency.

Keywords: Cross linked chitosan; X-ray Diffraction; Relative crystallinity

INTRODUCTION

In recent years, much attention has been focused on bio-absorbents because of their biocompatibility, biological function, non-toxic, cost effective and biodegradable. Chitosan absorbent is one of the most promising bio-absorbent. Chitosan is the second most abundant polysaccharide found on earth next to cellulose. It composed of two different copolymer which are glucosamine and N-acetyl glucosamine linked in β (1-4) linkage (Figure 1) [1-2].

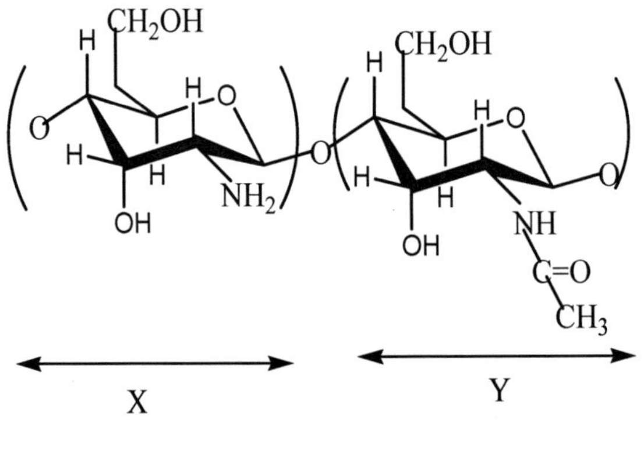

X > 50 % = Chitosan polymer
Y > 50% = Chitin polymer

Figure 1: Chemical structure of chitin and chitosan with glucosamine (X) and acetylglucosamine (Y)

Chitosan has received considerable interest as bio-absorbent for metal ions absorptions due to its excellent metal binding capacities by the presenting of hydroxyl and primary amine functional groups on its backbone [3].

Many methods have been used in order to improve its absorption capacities, including physical modifications (grinding, breads transformation, and others) [4-5] and chemical modifications [6] (cross linking [7], radiation [8] grafting [9], functional groups substituent [10-11] and others). Cross-linking was reported as the most effective way to modify chitosan structure [12]. Cross-linking occurs when a reagent (namely as cross-linking agent) introduces intermolecular bridge and form crosslinks between polymer structures. As generally known, chitosan is semi crystalline polymer; whereas cross-linked structure will produce materials that are more amorphous. Therefore, the main idea of cross-linking process is to reduce the amount of the crystalline

CP1202, *Neutron and X-Ray Scattering in Advancing Materials Research: International Conference – 2009*
edited by A. Saat, H. A. Kassim, M. H. H. Jumali, J. M. Saleh, M. R. Othman, A. Ibrahim, F. M. Idris, and M. H. A.-R. M. Ahmad

domains and change the crystalline nature of chitosan. This parameter will significantly influence the sorption properties of chitosan. Cross linked chitosan remark high swelling capacity in water, and consequently its network is sufficiently expanded as to allow a fast diffusion process [13]. Besides, cross linked chitosan was reported to poses faster and better absorption capacity rather than native chitosan due to its porous structure [14].

Cross-linked chitosan have been prepared and investigated by many researchers. Generally, chitosan is chemically cross-linked using glutaraldehyde, epiclorohydrine and ethylene glycol diglycidyl ether [15-17]. However, these cross linking agents are toxic and must be washed properly to remove any free cross linking agents before consumption. Therefore, in the presence study, we prepared cross linked chitosan with carboxylic acid based cross linking agents. Carboxylic acid based cross linking agents was selected because they differ in length of the spacer in the kind and number of functional groups on the spacer. Besides, it also water-soluble and this factor will make our cross linked chitosan is environmental friendly bio-absorbent. It is believe that, the longer of length of cross-linking agent will obtain more porous structure of cross linked chitosan.

The specific objective of this study is to produce the crosslinked chitosan with different length of cross-linking agents. Consequently, the crystallinity properties of native and cross-linked chitosan are investigated using X-ray diffraction (XRD) analysis.

MATERIALS AND METHODOLOGY

Chitosan powder was obtained from Hunza

Infrared (FTIR) method, which gave values of 95 % respectively. Two types of carboxylic acid based cross linking agents were purchased from Sigma Aldrich. The descriptions of the cross linking agents are listed in Table 1.

Preparation of cross linked chitosan

Cross linking agents powder was dissolved in freshly distilled NMP solvent in 3-neck round bottom flask under N_2 atmosphere. Chitosan powder in acetic acid (0.1 M) was dropped into the solution through dropping funnel for ½ hour with stirring under reflux at 150°C. The reaction continues for another 6 hours until the clear solution was obtained. The clear solution was dissolved in distilled water at ratio of 1:20 for 1 day followed by filtering the precipitate obtained inside the distilled water. Then the residue of filtering was collected and dried in vacuum oven for 24 hours at 35°C, Then the solution was extract with chloroform for 24 hours at 100°C.

The reflux product was washed with methanol 4 times followed by drying at room temperature before dissolved in acetic acid (0.1 M) solution. Then, the solution was filtered and dried at room temperature, before it was neutralizing with NaOH (0.1 M) solution. Lastly, the cross linked chitosan powder was washed with methanol and dry under vacuum at room temperature overnight.

Details of the composition of the produced cross linked chitosan with the respect of different cross linking agents are presented in Table 2.

Table 1: Descriptions of carboxylic acid based cross linking agents used in the study

Cross linking agents	Type of cross linked chitosan
  1,2,4,5- benzenetetracarboxylic dianhydride, 97 % $C_{10}H_2O_6$	MC 1
  3,3',4,4'- biphenyltetracarboxylic dianhydride,97 % $C_{16}H_6O_6$	MC 2

Pharmaceutical Sdn Bhd (Malaysia). The degree of deacetylation was determined by Fourier Transform

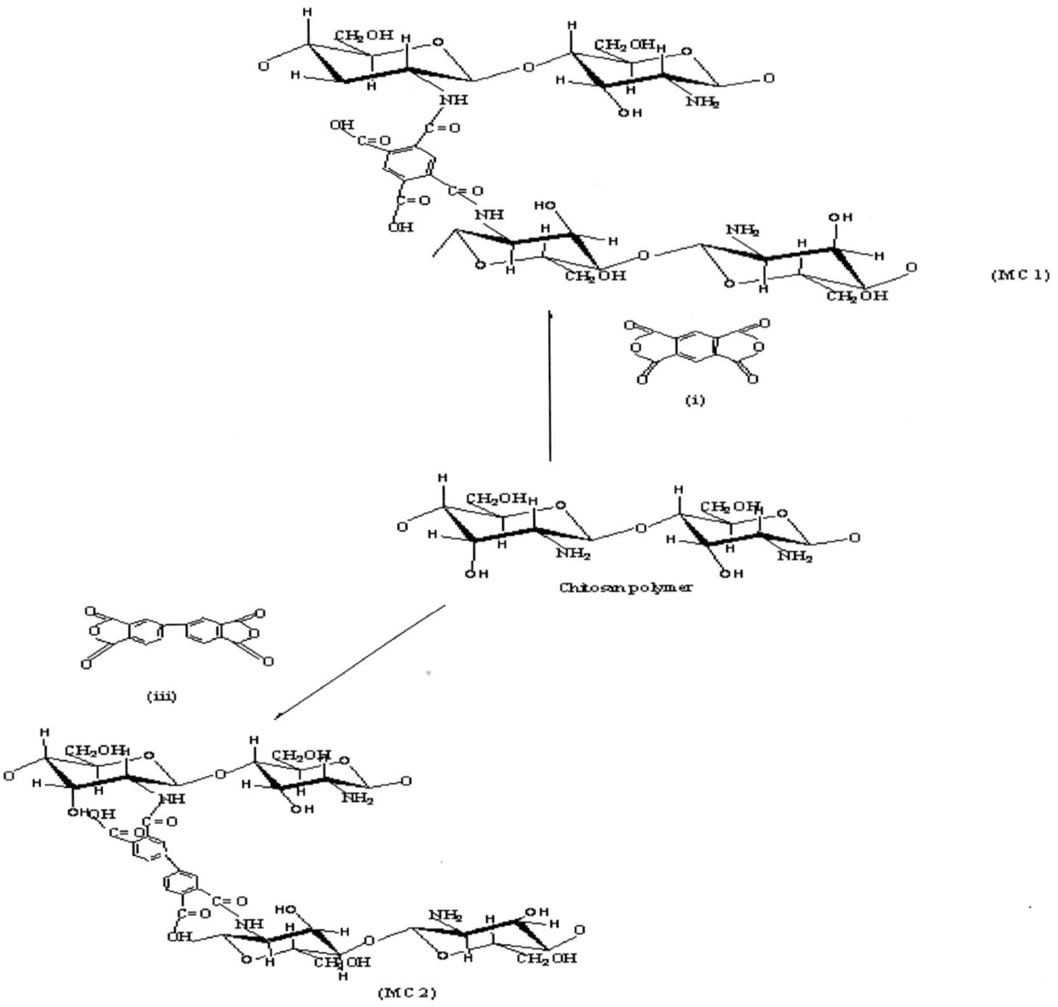

Figure 2: Chemical structure of cross linked chitosan derived from native chitosan

(i) MC 1 = 1, 2,4,5-benzentetracarboxylic dianhydride-chitosan

(ii) MC2 = 3,3',4,4'-biphenyltetra dianhydride chitosa

Table 2: Composition of cross linked chitosan (MC 1 and MC2)

Type of chitosan	Weight of chitosan powder (g)	Weight of cross linked powder (g)	Volume of 0.1 M acetic acid (ml)	Volume of NMP solvent (ml)
MC 1	2	1.26	100	40
MC 2	2	1.70	100	40

X-Ray Diffraction (XRD) Analysis

X-ray diffractograms of powdered samples were obtained using a Bruker AXS D 8 diffract meter under the following conditions: 40 kV and mA with Cu Kα radiation at 1.5418 Å and acceptance slot of 0.1 mm. About 20 mg of sample was spread on a sample stager and the relative intensity was recorded in the scattering range (2θ) of 5° to 40°.

RESULTS AND DISCUSSION

Formation of cross linked chitosan

Figure 2 illustrates the modification reaction of cross linked chitosan. Chitosan polymer chains were cross linked by carboxylic acid based cross linking agent covalently at primary amine (NH_2) groups of chitosan polymer. Consequently, imide linkage (N-H and C=O) was performed. The covalent cross linking leads to chitosan systems with a permanent network structure because of the formation of irreversible chemical links. On the other hand, the presence of covalent bond could change the crystalline nature of chitosan.

108

X-ray Diffraction (XRD) analysis

Figure 3 illustrates the XRD patterns of native and cross linked chitosan (MC1 and MC2). From the figure, there were two obvious crystalline peaks detected at $2\theta=12°$ (I_{010}) and $2\theta=20°$ (I_{020})

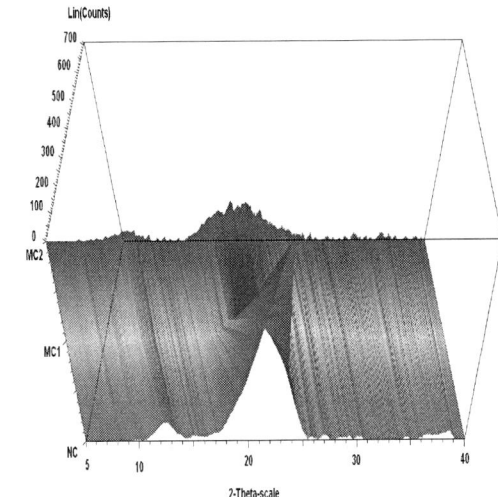

Figure 3: X-ray diffractogram of native and cross linked chitosan at $2\theta=5-40°$

According to the reports [18, 19], the peaks around $2\theta=12°$ (Figure 4) and $2\theta=20°$ (Figure 5) are related to crystal I and crystal II in chitosan structure, respectively. The unit cell of crystal I is characterized by a=7.76A°, b=10.91A°, c=10.30A° and β= 90°. Meanwhile, the cell unit of crystal II is characterized at a=4.4A°, b=10.0A°, c=10.30A° and β=90°. Crystal I was reported to has larger crystal size as compared to crystal II. Therefore, the characteristic peak at $2\theta=20°$ is attributed to the allomorphic tendon of chitosan structure. Moreover, both of characteristics peaks in NC diffractogram indicated a high degree of crystallinity of chitosan in agreement with previous reports [20].

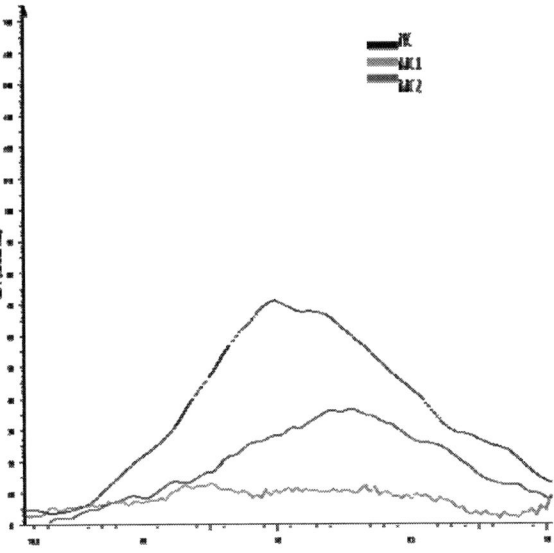

Figure 4: X-ray diffractogram of native and cross linked chitosan at $2\theta=12°$

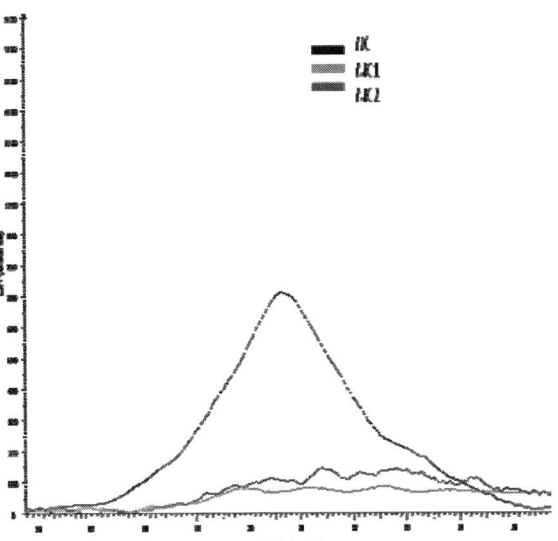

Figure 5: X-ray diffractogram of native and cross linked chitosan at $2\theta= 20°$

The characteristics peaks also indicate that, the crystalline state exist in chitosan polymer. However, since it involves molecules instead of just atoms or ions, therefore it can conclude that, the crystallinity of chitosan polymer depends strongly on the atomic arrangement and be more complexes. Generally, the crystalline structure of chitosan polymer as the packing of molecular chains which linked by inter molecular hydrogen bond [1,2]. The structure would produce an ordered atomic array (Figure 6).

For cross linked chitosan (MC 1 and MC 2), it can be seen that the characteristic peaks at $2\theta=12°$ was gradually absent and only one broad peak at $2\theta=20°$ was present.

To gain a better understanding of the crystallinity of the samples, the relative crystallinity (Cr I (%)) and increment of amorphous region were calculated. Cr I was calculated at the respective crystalline peaks using Equation (1) and (2) [20]. The relevant data is shown in Figure 7.

$$Crl_{100} = \left[\frac{I_{020} - I_{am}}{I_{020}} x100 \right] \quad (1)$$

$$Crl_{100} = \left[\frac{I_{110} - I_{am}}{I_{110}} x100 \right] \quad (2)$$

Where:

I_{020} = maximum intensity at $\approx 12°$
I_{am} = intensity of amorphous diffraction at $16°$
I_{110} = maximum intensity at $\approx 20°$

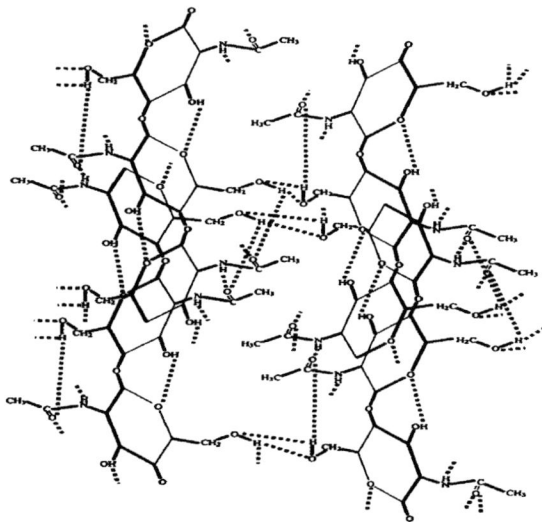

Figure 6: Inter and intra hydrogen bonding of chitosan polymer

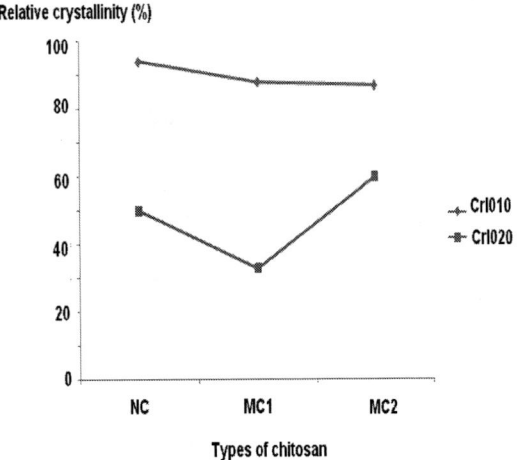

Figure 7: Relative crystallinity (Cr I (%)) of samples

The increments of amorphous region for sample were calculated by using Equation (3):

$$Increment\ of\ amorphous\ region\ (\%) = \frac{A_{MC} - A_{NC}}{A_{NC}} X\ 100$$

(3)

where:

A_{MC} = area under the X-ray diffractogram of MC

A_{NC} = area under the X-ray diffractogram of NC

From Figure 7, it can be concluded that cross-linking reduce the relative crystallinity at both characteristics peaks. It is highly anticipated that the cross linking will have a significant influence on the amorphous properties of chitosan. The amorphous region of chitosan increase almost of 40 to 41 % as cross linking agents were introduced into chitosan system (Figure 8). The interaction of chitosan with carboxylic acid based cross linking agents of various length results in interesting changes in the structure of the chitosan. The reduction is significantly observed at MC 1 rather than MC 2.

Initially, the relative crystallinity of native chitosan is relatively high (94 % at CrI $_{020}$ and 50% at CrI$_{010}$). The structure of chitosan is sensitive with others components such as cross linking agents, salt, acids, water and others due to the presence of hydroxyl and primary amine groups. The structural sensitivity may be implied by the results of modifying the chemical structure of chitosan. Therefore, as the cross linking agents were introduced; the relative crystallinity becomes lower significantly. It illustrates that intra and inters molecular hydrogen bonds of chitosan network would break apart.

The phenomenon eliminates the hydrated crystalline structure of native chitosan. In the other words, the lower crystallinity of cross linked chitosan was ascribed to the presence of cross linking agents in chitosan system, which might hindered the formation of hydrogen bonds. The results also indicated that, there was connection between relative crystallinity of chitosan and length of cross linking agents used. Increasing of cross linking agent length will reduce the crystallinity of cross linked chitosan.

Besides, the relative crystallinity of NC, MC1 and MC2 depends on the accessibility of solvent (NMP and acetic acid) during solidification process as well as on the chain configuration. During solidification, the polymer chains are highly random and entangled in the viscous liquid. As NC is a linear polymer, the crystallization process is easily accomplished because there are virtually no restrictions to prevent chain alignment.

However, the three-dimension structure of crosslinked chitosan (MC1 and MC2) gives some difficulty of solvent evaporation. The branched structure of MC1 and MC2 would interfere with crystallization and prevent any crystallization whatsoever. As MC1 has bulkier side bonded groups of atoms, therefore it has fewer tendencies for crystallization. Meanwhile, more flexible structure of MC2 facilitates the process of fitting together adjacent chains. Consequently, make its crystallization process become easier rather than MC1.

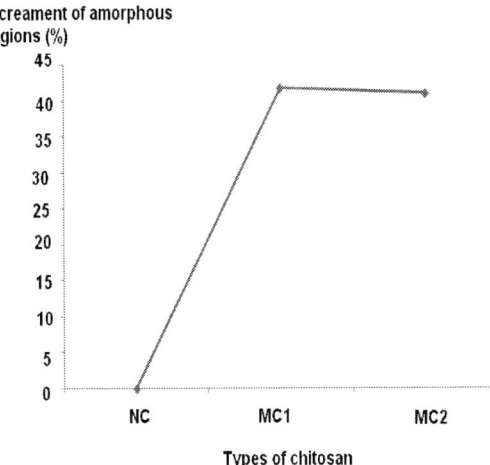

Figure 8: Increment of amorphous region of samples

CONCLUSION

Crystallinity of native and cross-linked chitosan was investigated using X-ray diffractogram. The results show that cross-linking process breaks the crystal zone in chitosan making it less ordered structure. The length of cross-linking agents had significantly influence on crystallinity of cross linked chitosan. The reduction on crystallinity will produce absorbent that are more porous and consequently enhance the absorption area of the absorbent.

ACKNOWLEDGEMENT

Appreciations are given to Fundamental Research Grant Scheme (FRGS) under Ministry of Science, Technology, and Innovative (MOSTI) Malaysia, for funding this research work. The first author acknowledges the National Science Fellowship (NSF) under Ministry of Science, Technology, and Innovative (MOSTI) Malaysia for the scholarship.

REFERENCES

Nurhidayatullaili Muhd Julkapli and Hazizan Md Akil, Degradability properties of kenaf filled chitosan bio-composites, Materials Science and Engineering C 28 (2008) 1100

Nurhidayatullaili Muhd Julkapli, Zulkifli Ahmad and Hazizan Md Akil, Preparation and properties of kenaf dust filled chitosan bio-composites, Composites Interfaces, 7-9 (2008) 851

Sandhye Babel, Tonni Agustiono Kurniawan, Low cost adsorbents for heavy metals uptake from contaminated water; A review, Journal of Hazardous Materials B 97 (2003) 219

Choong Jeon, Walfgang H. Holl, Chemical modification of chitosan and equilibrium study for mercury ion removal, Water Research, 37 (2003) 4770

W.S.Wan Ngah, S. Fatinathan, Chitosan flakes and chitosan-GLA beads for adsorption of p-nitophenol in aqueous solution, Colloids and Surfaces A; Physicochem Eng. Aspects 277 (2006) 214

Hitoshi Sashiwa, Sci-ichi Aiba, Chemically modified chitin and chitosan as biomaterials, Progress in Polymer Science 29 (2004) 887

C.G.T.Neto, J.A.Giacometti, A.E. Job, F.C. Ferreira. J.L.C. Fonseca, M.R. Pereira, Thermal Analysis of Chitosan Based Networks, Carbohydrate Polymers 62 (2005) 97

Fei Yang, Xunhu Li, Mingyu Cheng, Yandao Gong, Nanming Zhao, Xiufang Zhang and Yinye Yang, Performance modification of chitosan membranes induced by Gamma Irradiation, Journal of Biomaterials Applications 16 (2002) 215

R. Jayakumar, M Prabaharan, R.L. Reis, J.F. Mano, Graft copolymerized chitosan present status and applications, Carbohydrate Polymers 62 (2005) 142

Z. Zong, Y.Kimura, M. Takahashi, H. Yamane, Characterization of chemical and solid state structures of acylated chitosans, Polymer 41 (2000) 899

X.Qu, A.Wirsen. A.C. Alberstsson, Effect of lactic/glycolic acid side chains on the thermal degradation kinetics of chitosan derivatives, Polymer 41 (2000) 4841

W.S. Wan Ngah, S. Ab Ghani, A.Kamari, Adsorption behavior of Fe (II) and Fe (III) ions in aqueous solution on chitosan and crosslinked chitosan beads, Bioresource Technology 96 (2005) 443

J. Berger, M. Reist, J.M. Mayer, O. Felt, N.A. Peppas, R. Gurny. Structure and interactions in covalently and ionic ally crosslinked chitosan hydrogels for biomedical applications, European Journal of Pharmaceutics and Biophamaceutics 57 (2004) 19

Iyabo Adekogbe, Amyl Ghanem, Fabrication and characterization of DTBP-crosslinked chitosan scaffolds for skin tissue engineering, Biomaterials 26 (2006) 7241

Robert Y.M. Huang, Rajinder Pal, Go Young Moon, Crosslinked chitosan composite membrane for the pervaporation dehydration of alcohol mixture and enhancement of structural stability of chitosan/polysulfone composite membranes, journal of membrane Science 160 (1999) 17

V. K. Mourya, Nazma N. Inamdar, Chitosan-modifications and applications opportunities galore, Reactive and Functional polymers 68 (2008) 1013

Kamlesh Kumari, K.K. Raina, P. Kundu, Studies on the cure kinetics of chitosan-glutamic acid using glutaraldehyde as crosslinker through differential scanning calorimeter, Journal of Applied Polymer Science 108 (2008) 681

Ebru Gunister, Dilay Pestroli, Cuney H. Unlu, Oya Atici, Nurfer Gungor, Synthesis and characterization of chtiosan-MMT biocomposite systems, Carbohydrate Polymers 67 (2007) 358

Marguerite Rinaudo, Chitin and chitosan : Properties and applications. Progress in Polymer Science, 31 (2006) 603

Nurhidayatullaili Muhd Julkapli and Hazizan Md Akil, X-ray Powder Diffraction (XRD) studies on kenaf dust filled chitosan bio-composites, AIP Conference Proceedings, 989 (2007) 111

X-Ray Structural Analysis Of Some Indian Coals

Binoy K Saikia [1]*, Rajani K Boruah [2]

[1]Department of Chemical Sciences, Tezpur University, Tezpur-784028, India

[2] North-East Institute of Science & Technology (CSIR), Jorhat-785006, India

* e-mail: binoyrrl@yahoo.com

Abstract

Coal is one of the most abundant energy resources and has the capability to meet future energy needs with high reliability. The use of coal as an energy source and as a source of organic chemicals feedstock may become more important in the future. It is physically and chemically a heterogeneous and carbonaceous rock which consists of organic and inorganic materials. Assam coal has been, and continuous to be, a valuable energy source, especially for the various industry in India and for liquefactions of coal. The basic chemical structure of coal that has been widely accepted today was built up from the synthesis of results obtained from X-ray diffraction data. The present paper reports a comparative investigation of coals from different collieries/areas of Makum coalfield, Assam viz. Ledo, Tikak, Baragolai, Tipong and Tirap collieries Makum coalfield, Assam with the help of X-ray diffraction (XRD). The X-ray diffraction patterns indicate that the coals are amorphous in nature. The present XRD method includes the evaluation of Function of Radial Distribution of Atoms (FRDA) and structural interpretations of the coals from their Radial Distribution Function (RDF) plots after proper corrections for air scatter, absorption by sample and polarization. The curve intensity profiles in FRDA clearly show quite regular molecular packets for these coals. The first maxima in the FRDA curves was obtained at r= 0.4 A° for Ledo, Baragolai and Tipong coals whereas for Tikak coal it was observed at r= 0.5 A°. The first maximum in the pair distribution function plots, G (r) of Ledo, Tikak, and Tipong coals was obtained at r=0.15 nm whereas for Baragolai and Tirap coals it was observed at r=0.14 nm and r=0.12 nm respectively, which relates to the C-C (aliphatic/aromatic) bonds in coal matrix. The Assam coal samples from Ledo, Tikak, Baragolai, Tipong and Tirap collieries of Makum coalfield have almost the same RDF inter-atomic distances except slight differences. This study reveals the absence of graphite like structures in Assam (India) coals.

Keywords: Indian coal; X-ray diffraction; Radial Distribution Function

INTRODUCTION

Coal is a sedimentary rock composed mostly of carbon and its composition varies with the locations. It is one of the most abundant energy resources and has the capability to meet future energy needs with high reliability. The use of coal as a source of organic chemicals feedstock also may become more important in the future. The structure of coal also varies with its matrix. Coal has a physical microstructure that is discernibly derived from plant and a chemical structure containing a wide variety of polymeric organic compounds and crystalline minerals. Detailed structural characterization has been found to be extremely difficult and therefore, research on coal structure is still a challenging task and continues to be pursued intensively (Hanel, 1982).

The basic chemical structure of coal that has been widely accepted today was built up from the synthesis of results obtained from X-ray diffraction studies. The role of X-ray diffraction study in coal science is enormous. It can provide information about the structure of the molecular core of coal. X-ray diffraction data for coal and coal-derived solids are more appropriately considered in terms of their trends, and can provide valuable information about structural variations. For amorphous substances, such as coal produce X-ray diffraction spectra of continuous scatter curves on which only more or less grossly broadened maximum appear.

Assam coal has been, and continuous to be, a valuable energy source, especially for the various industry in India and for liquefactions process. The coal of the region is mostly of sub-bituminous variety. The Makum coalfield of Assam are by far the most important as they practically constitute the major share to the total tertiary coal output of the country. It lies between latitudes 27°13'-27°23' N and longitudes 95°35'-96°00' E in the Tinisukia district of Assam, India. At present

CP1202, Neutron and X-Ray Scattering in Advancing Materials Research: International Conference – 2009
edited by A. Saat, H. A. Kassim, M. H. H. Jumali, J. M. Saleh, M. R. Othman, A. Ibrahim, F. M. Idris, and M. H. A.-R. M. Ahmad
© 2009 American Institute of Physics 978-0-7354-0739-8/09/$25.00

there are five working collieries in Makum coalfield, which are located at Ledo, Baragolai (Namdung), Tipong, Tirap and Tikak of Tinisukia district.

The present paper reports a comparison of some structural aspects of some representative high sulphur coal samples from various collieries of Makum coalfield, Assam with the help of X-ray diffraction (XRD) which includes the evaluation of Function of Radial Distribution of Atoms (FRDA) and structural interpretations of the coals from their Radial Distribution Function (RDF) plots.

EXPERIMENTAL

Freshly mined (ROM) coal samples were collected from Makum coalfield (latitudes 27°13'-27°23' N and longitudes 95°35'-96°00' E) Assam (India) by adopting standard sampling methods. The samples were ground to –200 BS fineness sizes and preserved in polyethylene airtight containers.

The coal samples were washed with distilled water for several times. Effect of low mineral matter in coal is neglected and effect of heteroatom is insignificant (Iwashita and Inagaki, 1993). A very slow step scan (0.03°) X-ray diffraction data were collected in a computer controlled X-ray Diffractometer Type XPERT PRO (PHILIPS) using silicon as the internal standard. Start angle: 3.015, Target: Cu (Fe-filtered), Stop angle: 100.0, Measuring time per step: 0.4s, step angle: 0.03 and Data processing condition: Smoothing points, Goniometer radius (R): 240 mm, Equatorial angle subtended at the specimen by the detector slit (β): 1°.

The observed diffraction profiles have been corrected for absorption, polarization and atomic scattering factors (these depend on diffraction and measurement conditions) using the methods given elsewhere (Iwashita and Inagaki, 1993; Klug and Alexander, 1974).

RESULTS DISCUSSION

A careful observation of the diffused x-ray diffraction profiles (Figure 1) of the coals indicates that there is a significant amount of amorphous carbons. The diffuseness has been attributed to structures in which the arrangement of carbon atoms is that of a graphite crystal, but with extremely small size of the elemental crystallites (Warren, 1941).

The obtained Radial Distribution Function (RDF) and atom pair distribution function (PDF), G (r) plots for the coal are depicted in Figures 2 and 3. The radial distribution function (RDF), which oscillates around $4\pi r^2 \rho_0$ calculated for Tikak coal. The RDF is the weighted probability of finding two atoms separated by a distance r and r+Δr. The inter-atomic distances in the coal were calculated from the atom-pair correlation function curve, obtained after Fourier transformation of intensity data. The C-C atom-pair distances in this coal are reflected by maxima in the RDF curve and the magnitude of each maxima provides a measure of the number of the C-C atom pairs. Structural determination

based on RDF fall into two main categories. In the first, the RDF yields the mean distribution of interatomic distances of carbons and the second category is the evaluation of the structure of coal. In either category, deductions about structure are based on the peak position and their areas in the RDF plot. The RDF provides meaningful information about the average number of nearest neighbours and their mean distances of approaches, and in certain cases intra-molecular configuration may be deduced. Although in the later case configuration of structure is the attained by agreement between observed intensity and that calculated on the basis of the some model (Kruh, 1962)

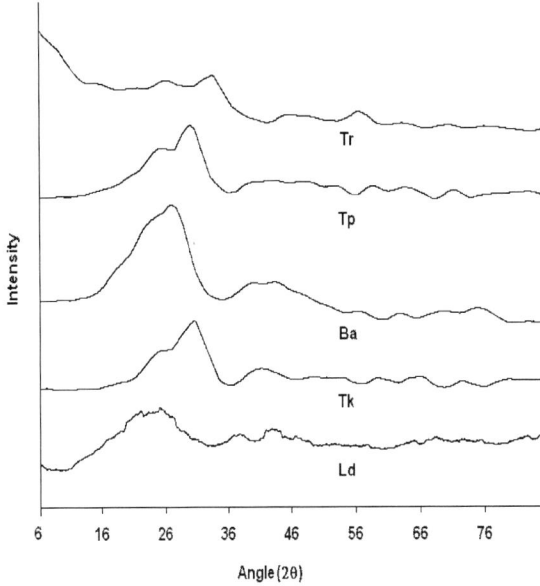

Figure 1: X-ray diffraction patterns of the Coals

Table 1 shows the inter-atomic distances of carbon atoms in a single graphene layer of coal obtained from the RDF curves. The compositional differences like elemental composition, volume of vitrinite, mineral matters etc. between these coals lead to difference in the FRDA. The curve intensity profiles clearly show quite regular molecular packets for these coals. The first maxima positions in the FRDA plots were observed to be similar for Ledo, Baragolai and Tipong coals at r= 0.4 Å except for Tikak coal which shows r= 0.5 Å. Grigoriew (Grigoriew, 1990) in his study of coal vitrinite with about the same chemical composition found no inter atomic distance of 0.50 A°. This indicates that the 0.50 A° distance comes from non-vitrinite portion of the coal.

From the table it can be concluded that Assam coals do not have the graphite like structures. Baragolai coal may be considered to show a very little similarity with the graphite layer structure. It has 3 nearest neighbours at a distance at 1.4 Å, which is similar with the nearest neighbour of graphite structures at a distance of 1.42 Å. It may be mentioned that particular arrangements of bond lengths, bond angles, and torsion angles within a given molecule may force pairs of non-bonded atom into closer contact than that given by the appropriate non-bonded radii (Dunitz, 1995). Thus the 1,3 distance between carbon atoms is about 2.55 Å in paraffin chain, about 2.40 Å in benzene, and about 2.20 Å in a cyclobutane ring. It is to be noted that in such cases the intermolecular distance are much shorter than typical non-bonded contact distances between carbon atoms (3.4-3.6 Å), but they are much longer than typical bonded distances (1.2-1.5 Å).

Table 1: Inter-atomic distances (Å) in a single layer of Assam coals and graphite structure

Tikak		Ledo		Baragolai	
Distances (Å)	No of neighbours	Distances (Å)	No of neighbours	Distances (Å)	No of neighbours
0.5	--	0.4	--	0.4	--
1.5	7	1.5	6	1.4	3
2.6	11	2.5	12	2.7	10
3.7	26	3.6	21	3.8	21
4.6	33	4.6	33	4.6	30
5.7	49	5.6	47	5.6	45
6.7	69	6.6	64	--	--
7.6	84	7.6	84	--	--

Table 1: Inter-atomic distances (Å) in a single layer of Assam coals and graphite structure (cont.)

Tipong		Tirap		Graphite	
Distances (Å)	No of neighbours	Distances (Å)	No of neighbours	Distance (Å)	No of neighbours
0.4	--	--	--	--	--
1.5	4	1.3	3	1.42	3
2.6	10	2.7	10	2.46	6
3.6	19	3.8	21	2.86	3
4.6	30	4.8	32	3.75	6
5.4	41	--	--	4.25	6
--	--	--	--	4.92	6
--	--	--	--	5.11	6

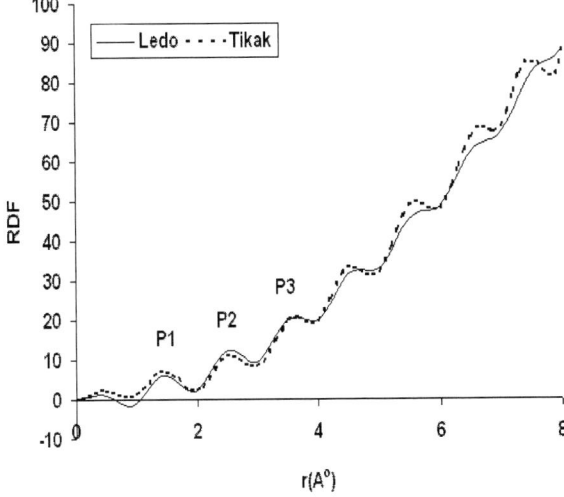

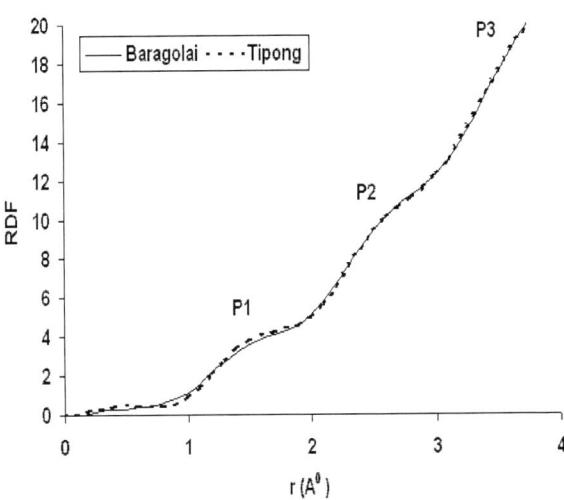

Figure 2: X-ray RDF plots of the coals

The layers present in this coal are thus not graphite like in their internal co-ordination system, which can be also attributed from the average C-C-C valence angle. This angle may be estimated from the RDF by, $\phi = \cos^{-1}[p_2^2 - 2p_1^2]/2p_1^2$, where ϕ is found greater than 180° which is very unlikely. Therefore, it may be concluded that the aromatic layers in Tikak coal are not as like as in graphite. It can be compared with the lignite type coal of different localities already reported (Elliot, 1981).

Figure 3 shows the G(r) curves of five numbers of coal samples from different collieries of Assam. The first maximum in the plots of Ledo, Tikak, and Tipong coals was at r=0.15 nm, whereas, r=0.14 nm for Baragolai coal which relates to the C-C aliphatic bonds (type C-C=C-C) in coals. The second maximum at r=0.26 nm for Tikak coal, r=0.25 nm for Ledo, r=0.27 nm for Baragolai, r=0.26 nm for Tipong and r=2.7 nm for Tirap coal are made to see its variation. These relates to the distance between carbon atoms of aliphatic chains that are located across one carbon atom. Further, the third maximum for Tikak coal r=0.37 nm, for Ledo coal r=0.36 nm, for Baragolai coal r=0.38 nm, for Tipong coal r=0.36 nm and for Tirap coal r=3.8 nm which is more intense in the case of Tirap coal can probably be explained by the

availability of molecular packets (similar to graphite structure) in humic coal. All the five coals have maxima near at r=0.46 nm. This value is similar to the distance between parallel aliphatic chains of oriented polyethylene (Dunitz, 1995). From the results it can be attributed that the five Assam coals have almost the same RDF inter-atomic distances with slight differences. The slight differences in the inter-actomic distances for coals are attributed to their difference in contents of clay minerals like illite, quartz and chlorite. Lignite and brown coals contain 10-40 % of humic acid (Compbell, 1964). The FRDA results show the presence of humic acid in Baragolai, Ledo, Tipong, Tikak and Tirap coals. This can be explained probably due to location and predominance of other part including mineral matter association in these coals.

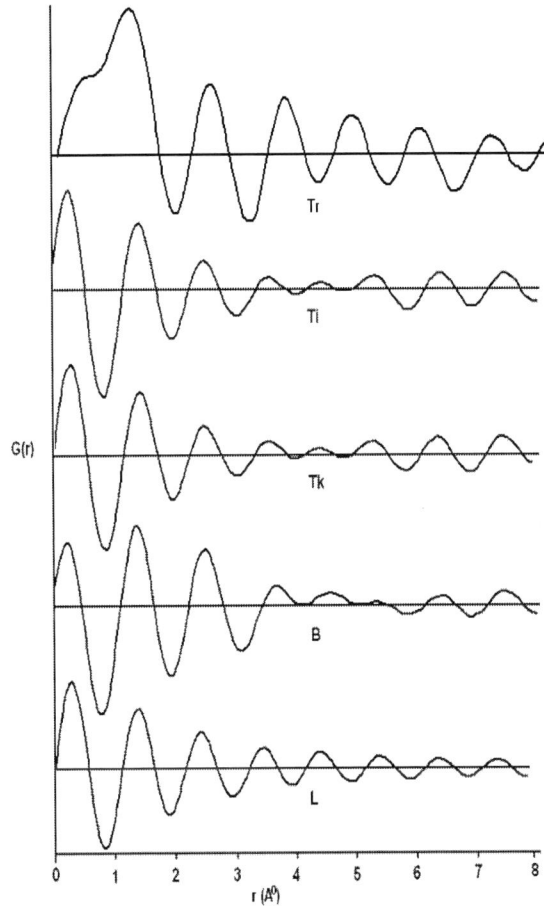

Figure 3: Pair distribution, G (r) plots of the coals

Bodoev et al. (Bodoev et al., 1996) investigated the structure of Taimylyr, Balknashite, Matagan coals by RDF study and reported the first, second and third maximum in G(r) plotes to be at r-=0.15 nm, r=0.25 nm and r=0.39 nm respectively. The slight differences in the inter-atomic distances for different coals may be attributed to their difference in mineral contents viz. clay minerals like illite, quartz and chlorite.

CONCLUSION

Assam (India) coals are lignite in type and consist of mainly amorphous carbon with little crystalline form. Absence of graphitic arrangement is observed in the graphene layers of these coals. The results obtained from this investigation will help in gaining an insight in to the differences in structural properties of these coals for better understanding the potential problems for their industrial utilizations.

ACKNOWLEDGEMENT

Author is thankful to Dr Robin K Dutta, Tezpur University, India for his support in carrying the research works. Author (BKS) is thankful to International Union of Crystallography (IUCr) for providing the Young Scientist Travel Grant award to attend the ICNX2009

REFERENCES

Bodoev W. V., Guet J. M., Grber, R., Dolgopelov, N. J., Wiilhelm, J. C., Bazarrova, O., Fuel, 1996, 75, 837-842

Compbell, N. (Edt) A Text book of organic chemistry, Oliver & Boyd London, 1964, 875

Dunitz, J. D. X-ray analysis and the structure of organic molecules, Verlag Helvetica Chimica Acta, Baset, 1995, p. 338-339

Elliot Martin A (edt.) Chemistry of coal utilization, 2nd Supplementary Volume, New York: John Wiley & Sons 1981, p. 89

Grigoriew G. Fuel, 1990, 69, 840

Hanel, M. W. Fuel, 1992, 71, 1211-1223

Iwashita, N., Inagaki, M. Carbon, 1993, 31, 1107-1113

Klug H. P., Alexander L. E., X-ray diffraction procedures, Wiley-Interscience Publication, London, 1974

Kruh, R. F., Chem Rev, 1962, 62, 318-345

Warren B. E., Diffraction in random layer lattice, Phy Rev, 1941, 59, 693-698

Effect Of Milling Time On Microstructure Of Mechanically Alloyed Al-Ti Powders

Adolf Asih Supriyanto and Abdul Razak Daud

School of Applied Physics, Faculty Science and Technology. University Kebangsaan Malaysia, 43600 Bangi, Selangor, Malaysia
E-mail address: adolfsusi@yahoo.com

ABSTRACT

Mechanical alloying of Al-Ti was performed by high-energy ball milling at ambient temperature under argon atmosphere. A Fritsch Pulverisette-5 planetary type ball mill was used for the mechanical alloying for 2 up to 30 hours with the ball milling size of 15 mm and the rotational speed of about 360 rpm. The balls-to-powder ratio was 20 : 1 (in weight percent). The final products have been characterized by SEM and XRD. The SEM results showed that the average particle size of mechanically alloyed Al-Ti powders decreased with increasing milling time and the particle size was smaller than 1.63 μm after 30 hours milling. The XRD results showed that the crystallite size of mechanically alloyed Al-Ti powders decreased with increasing milling time and the steady-state crystallite size was approximately 17.6 nm. It also found that the peaks of Ti begin disappear with the increasing of milling time, which indicates the forming alloying of Ti atoms in the Al matrix. The SEM results confirmed that the nanocrystalline produced were binary alloy of Al-Ti and no other impurities detected in the particles. The optimum milling time for the mechanical alloying of nanocrystalline Al-Ti was 30 hours.

Keywords: **Al-10wt.%Ti alloy, nanocrystalline, mechanical alloying, stearic acid**

INTRODUCTION

Microcrystalline materials are polycrystals with a grain size of a few micrometers and nanocrystalline materials are defined as polycrystalline solids having crystallite sizes usually less than 100 nm. Because their reduced grain size, nanocrystalline materials have superior mechanical properties compared to their microcrystalline counterparts. Some successful production of nanocrystalline materials in particular Al alloys have been reported (Kyoung Il Moon et al., 1999: Marek Krasnowski et al., 2000: C. E. Wen et al., 2000). In particular Al-Ti Alloys have been one of the materials used for aerospace structural and engine applications, because they have high melting points and lower densities together with high strength (E. Swewczak et al., 1997: K.I. Moon et al., 1998: K. B. Gerasimov et al., 1996: Xiaoying Zhu et al., 2006: C. Suryanarayana et al., 1997: G. J. Fan et al, 1995).

High-energy ball milling has been used to improve particle distribution throughout the matrix. This technique, first developed by John Benjamin, is known as mechanical alloying or mechanical milling. The process in which mixtures of powders are milled together is denominated mechanical alloying. It involves material transfer to obtain a homogeneous alloy by repeated welding, fracturing, and rewelding mechanisms. Nanocrystalline materials can be synthesized by mechanical alloying. The advantages of using mechanical alloying for synthesis of nanocrystalline materials are because of its simplicity, cost effectiveness, room temperature synthesis process, and capability of production on industrial scale (A. Belyakov et al., 2003: I. Chicinas et al., 2004: J. Bonastre, 2007: Marjoni Imamora Ali Oemar et al., 2004: W. Barona Mercado et al., 2006).

One of the effective methods to avoid excessive cold welding is the application of surface-active substances, generally known as process-control agents (PCAs). Many researchers have used PCAs to prevent excessive cold welding of Al particles during milling. Mostly organic compounds, such as stearic acid, oxalic acid, hexane, ethanol, methanol are used as PCAs (Marjoni Imamora Ali Oemar et al., 2004: W. Barona Mercado et al., 2006). Among them, stearic acid is one of the most effective and very often used as a PCA. This paper reports the use of mechanical alloying process and stearic acid as process control agent to investigate the

CP1202, *Neutron and X-Ray Scattering in Advancing Materials Research: International Conference – 2009*
edited by A. Saat, H. A. Kassim, M. H. H. Jumali, J. M. Saleh, M. R. Othman, A. Ibrahim, F. M. Idris, and M. H. A.-R. M. Ahmad
© 2009 American Institute of Physics 978-0-7354-0739-8/09/$25.00

effect of milling time on microstructure and lattice structure of nanocrystalline powder of Al-Ti alloy.

EXPERIMENTAL PROCEDURE

The starting materials were fine powders of aluminium and titanium. They were mixed to prepared a sample of Al-10wt%Ti. The aluminium powder used had a purity of 99.7 % with a mean particle size of about 71 μm, while the titanium powder used had a mean particle size of about 60 μm. Mechanical alloying of the mixture was conducted in a Fritsch Pulverisette-5 planetary type ball mill. Milling was done at room temperature in a cylindrical stainless steel container with stainless steel balls under an argon atmosphere. The container volume and the balls sizes were 250 ml and 15 mm respectively. The milling speed and ball to powder weight ratio were 360 rpm and 20 : 1 respectively. Stearic acid (CH_3-$(CH_2)_{16}$-COOH) of about 4 wt.% of the sample mixture was used as a process control agent to prevent the excessive welding in a cylindrical stainless steel container. Several milling times were used ranging 2 up to 30 hours, interrupted for 30 minutes every 2.5 hours in order to minimize excessive temperature rise and to limit adherence of the powder to container walls.

Samples before and after milling were characterized by X-ray diffraction (XRD) method (in a Siemen D5000 X-ray diffactometer) with Cu K_α radiation. A scanning microscope electron (SEM) Model Phllips XL30 was used to investigate the microstrusture of the samples.

RESULTS AND DISCUSSION

The morphologies and the change of particle size of the Al-10wt%Ti powders are shown in Figure 1. At the beginning of the mechanical alloying process (Figure 1.a and 1.b), the particles were flattened by the plastic deformation caused by the compressive forces induced by the contacts between ball and ball or ball and wall. After 10 hours, the particle become smaller as shown in Figure 1.c and after 20 hours of milling the particles were still same as before (Figure 1.d). The mean particle size were about 16.3 μm.

In order to evaluate the crystallite size of the Al-10wt%Ti powders, the powders were further examined by XRD. Figure 2 shows the XRD patterns of the Al-10wt%Ti powders with 4wt% stearic acid after mechanical alloying for various periods time. From the beginning, the peaks from both aluminium and titanium are clearly seen in Figure 2. The progressive mixing of the sample due to the high-energy ball milling process can be observed with the increase in milling time. The titanium peaks vanish gradually and disappear completely for the milling time longer than 20 hours. These phenomena can be showed that Ti diffuses into the bcc structures and form alloying of Ti atoms in the Al matrix (F.G.Cuevas *et al.*, 2006).

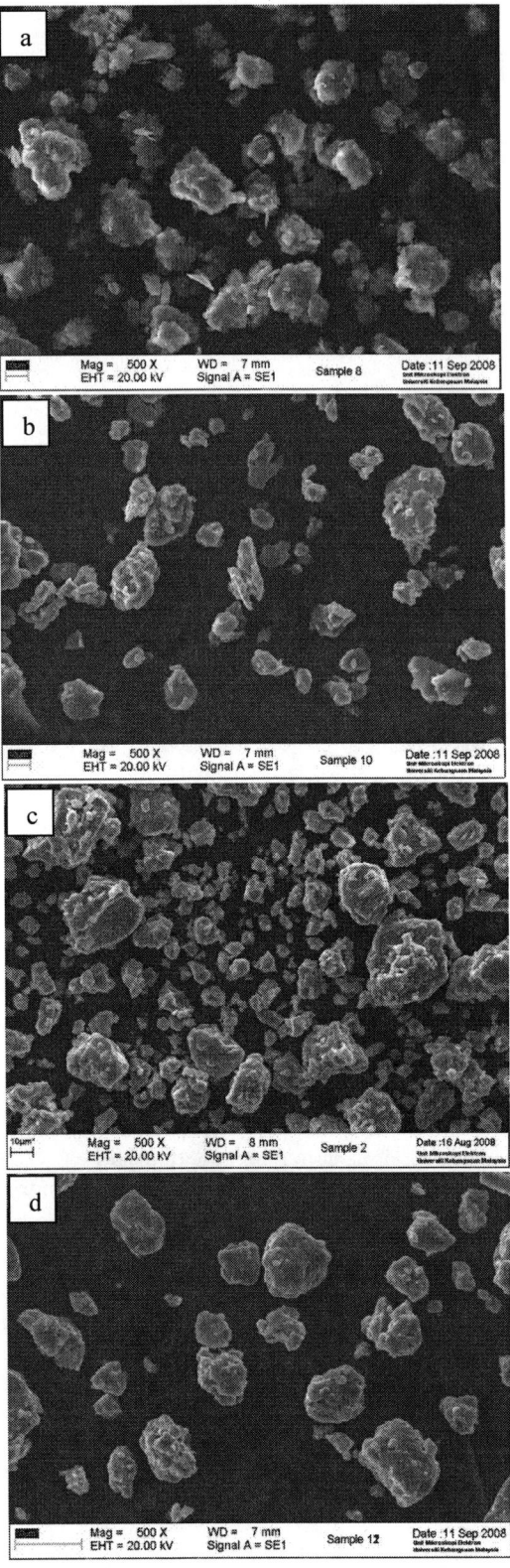

Figure 1. The SEM micrographs of the powders after ball milling for various times: a. 2 h, b. 5 h, c. 10 h and d. 20 h.

The broadening of the aluminium peaks as shown in Fig. 2 is commonly attributed to the structural change in the samples, such as the formation of amorphous phases, large lattice strains in grains, or the embedding of very small crystals in an amorphous matrix (P. B. Joshi et al., 2005: C. Suryanarayana, 2001). The crystallite size and lattice strain of the Al-10wt%Ti powders after mechanical alloying for various periods time re shown in Figure 3.

The crystallite size of the powders decreases from the initial sizes to about 17.6 nm after 30 hours milling. The lattice strain in Al-10wt%Ti powders as observed in Figure 3 reaches a maximum of about 0.58% after milling for 30 hours.

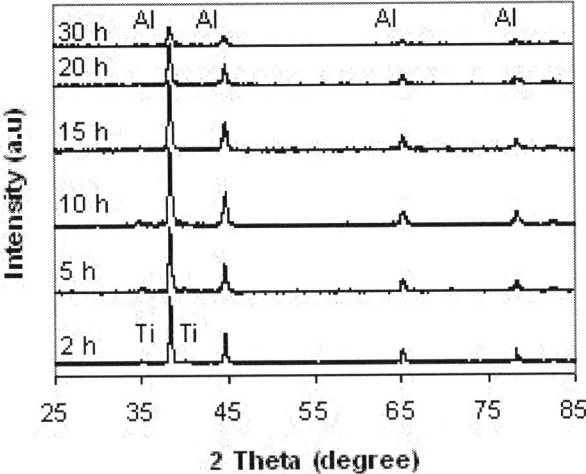

Figure 2. XRD patterns of the Al-10wt%Ti powders at different milling time.

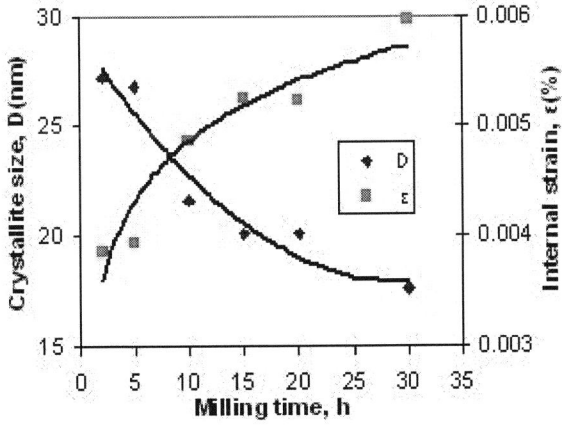

Figure 3. The crystallite size and lattice strain of the Al-10wt%Ti powders as a function of milling time.

The aluminium peaks in Figure 2 are observed to shift towards the higher diffraction angles, indicating the dissolution due to interdiffusion between the two elements during milling. The shifting of Al(111) peak positions is shown in Figure 4. At the beginning of mechanical alloying at 2 hours of milling 2θ is $38.4986°$. After 10 hours of milling 2θ become 38.5026 and after 30 hours of milling 2θ is 38.5626.

The shifting of Al(111) peak positions can be also related to a reduction of the Al lattice parameter, due to Ti dissolution in Al (Araki H et al., 1995). The Al lattice parameter, as calculated from the XRD pattern, is shown in Figure 4. At the beginning of mechanical alloying, the aluminium lattice parameter is 0.40505 nm which is close to the theoretical aluminium lattice parameter. After milling for 15 hours, the aluminium lattice parameter did not show apparent changes. The decrease of the aluminium lattice parameter indicates that atoms of titanium occupy some positions in the bcc lattice of aluminium during the process of mechanical alloying, causing the aluminium lattice to shrink. Further milling to 30 hours results in a significant decrease to 0.4044 nm. Beyond 30 hours milling, the aluminium lattice parameter likely achieve 0.4044 nm. This suggests that the alloying between aluminium and titanium mainly takes place at 30 hours milling and the formation of Al(Ti) solid solution has completed at about 30 hours. Suryanarayana also reported that the supersaturated solid solution was formed at the early stage of milling process (C. Suryanarayana et al., 1998). Therefore, 30 hours was selected as the optimum milling time of the mill under the experimental conditions.

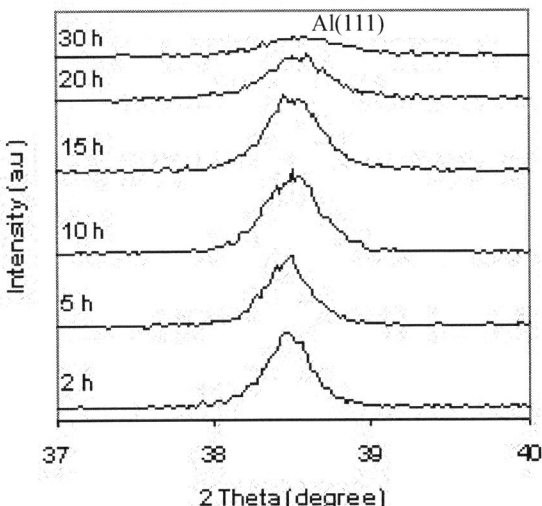

Figure 4. Enlarged fragments of the diffraction patterns in the range from 37 to 40°, for the 2, 5, 10, 15, 20 and 30h milled powder.

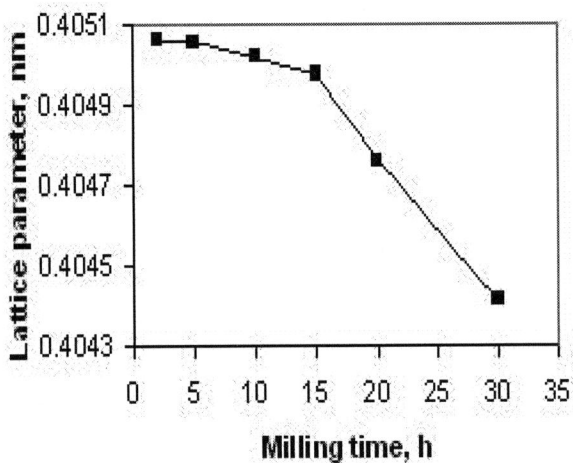

Figure 5. Lattice parameter as a function of milling time for MA Al$_{90}$Ti$_{10}$.

Figure 6 shows the energy dispersive X-ray spectrum of mechanically alloyed particles for 10 and 20 hours milling time. Aluminium and titanium were detected in the particle that concludes mechanical alloying has taken place between aluminium and titanium particles to form particles of binary alloy of Al-Ti. From Figure 6, the existence of C and O elements in the sample are believed to be from the original materials such as residual stearic acid (CH$_3$(CH$_2$)$_{16}$CO$_2$H which was used as a processing control agent (L. Lu *et al.*, 1998). There are no other impurities detected in the particles. This conforms that the ball milling procedure does not introduce any spurious impurity (J.L. Guimaraes *et al.*, 2003).

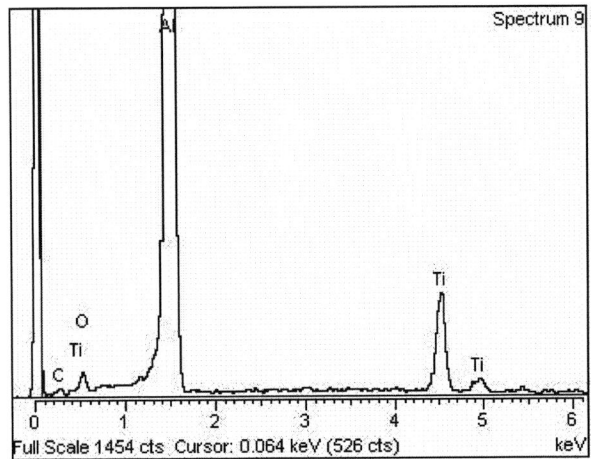

Figure 6. The powder energy dispersive X-ray spectrum of mechanically alloyed particles for 20 hours milling time.

CONCLUSIONS

From the above results the following conclusions were derived. The mechanically alloyed Al-Ti powder was mainly composed of nanocrystalline structure. With an increase in milling time of up to 30 hours, the crystallite size of mechanically alloyed Al-Ti decreased. The steady-state phase exhibited a crystallite size of 17.6 nm. A significant increase in solid solubility of titanium in aluminium was achieved by mechanical alloying. XRD and SEM/EDX results showed that the optimum milling time for the synthesis of nanocrystalline Al-Ti was 30 hours.

ACKNOWLEDGEMENTS

We would like to thank to the School of Applied Physics, FST, UKM for the providing laboratory facility and to En. Aswani and En. Razali for their technical assistance.

REFERENCES

Belyakov, Y. Sakai, T. Hara, Y. Kimura, and K. Tsuzaki, Metallugical and Materials Transactions, 34A (2003) 131-138.

Araki H, Saji S, Okabe T, Minamino Y, Yamane T, Miyamoto Y, Mater Trans JIM 36(3) (1995) 465.

C. E. Wen et al., Journal of Materials Science 35 (2000) 2099 – 2105.

C. Suryanarayana and M. Grant Norton, Xray diffraction A Practical Approach, Plenum Press, New York, (1998).

C. Suryanarayana, "Mechanical alloying and milling", Prog. Mater. Sci., Vol. 46, pp. 1-184, (2001).

C. Suryanarayana, G.E. Korth and F.H. Froes, Metallurgical and Materials Transactions 28A (1997) 293-302.

E. Swewczak et al., Materials Science and Engineering, A226-228(1997) 115-118.

F.G.Cuevas, J.Cintas, J.M Montes and J. M. Gallardo, J Master Sci, 41 (2006) 8339–8346.

G. J. Fan, M. X. Quan, Z. Q. Hu, Journal of Materials Science 30 (1995) 4847-4851.

I. Chicinas, V. Pop and O. Isnard, Journal of Materials Science 39 (2004) 5305 – 5309.

J. Bonastre, Journal of Thermal Analysis and Calorimetry, Vol. 88 (2007) 1, 83–86.

J.L. Guimaraes, M. Abbatte, S.B. Betim and M.C.M. Alves, J. Alloys Comp. 1 (2003) 20.

K. B. Gerasimov, S. V. Pavlov, Journal of Alloys and Compounds, 242 (1996) 136-142.

K.I. Moon, K. S. Lee, Journal of Alloys and Compounds, 264 (1998) 258-266.

Kyoung Il Moon and Kyung Sub Lee, Journal of Alloys and Compounds 264 (1998) 258–266.

Kyoung Il Moon, Kyung Sub Lee, Journal of Alloys and Compounds 291 (1999) 312–321.

L. Lu and M.O. Lai, Mechanical Alloying. Kluwer Academic Publisher, London (1998) 29.

Marek Krasnowski, Anna Antolak and Tadeusz Kulik, Rev. Adv. Mater.Sci. 10(2005) 417-421.

Marjoni Imamora Ali Oemar, Abdul Razak Daud, Shahidan Radiman and Noor Baa'yah Ibrahim, Physics Journal of The Indonesian Physical Society, A7 (2004) 0109.

P. B. Joshi, G. R. Marathe, Arun Pratap and Vinod Kurup, Hyperfine Interactions 160 (2005) 173–180.

W. Barona Mercado et al., Hyperfine Interact (2006) 168:943–949.

Xiaoying Zhu et al., Oxidation of Metals, Vol. 65, Nos. 5/6, June 2006 357-376.

La-Doped CaCu$_3$Ti$_4$O$_{12}$ Prepared By Conventional And Microwave Processing

Nur Shafiza A. Sharif, Zainal A. Ahmad, Sabar D. Hutagalung*

School of Materials and Mineral Resources Engineering, USM Engineering Campus, 14300 Nibong Tebal, Penang, Malaysia

Corresponding author: mrsabar@eng.usm.my

ABSTRACT

Two processing techniques were used to prepare separate samples of undoped and La-doped CaCu$_3$Ti$_4$O$_{12}$: the conventional furnace and microwave processing. Stoichiometric composition of undoped CaCu$_3$Ti$_4$O$_{12}$ was produced by mixing starting materials of Ca(OH)$_2$, CuO and TiO$_2$ powder. The mixed powder was milled and then calcined, compacted and sintered using either a furnace (conventional) or a domestic microwave oven (microwave processing). The La$_2$O$_3$ was added to undoped CaCu$_3$Ti$_4$O$_{12}$ in order to prepare the La-doped CaCu$_3$Ti$_4$O$_{12}$ with different doping concentrations. The conventional furnace heating technique requires a calcination temperature of 900°C for 12 hours before the mixture is sintered at 1000°C for 12 hours. However, a single phase CaCu$_3$Ti$_4$O$_{12}$ compound was successfully synthesized using a microwave oven for a calcination time of 30 minutes. Longer microwave sintering time tends to produce denser CaCu$_3$Ti$_4$O$_{12}$ pellets.

Keywords: CaCu$_3$Ti$_4$O$_{12}$; La-doped; calcination; furnace heating; microwave processing
PACS: 61.05.cp; 68.37.Hk; 81.05.Je; 81.20.Ev

INTRODUCTION

The CaCu$_3$Ti$_4$O$_{12}$ is obtained through the reaction of three raw materials: Ca(OH)$_2$, CuO and TiO$_2$. Recent studies show that CaCu$_3$Ti$_4$O$_{12}$ is generally prepared using the traditional solid-state method (Lunkenheimer et al., 2004). The ceramic CaCu$_3$Ti$_4$O$_{12}$ in particular, at 1 kHz, shows a dielectric constant of about 12,000 that is nearly constant from room temperature to 300 °C. The cubic structure of these materials is related to that of perovskite (CaTiO$_3$), but the TiO$_6$ octahedral is tilted to produce a square planar environment for Cu^{2+}. The structure remains cubic and centric. Most compositions of the type $A_{2/3}$Cu$_3$Ti$_4$O$_{12}$ (A = trivalent rare earth or Bi) show dielectric constants above 1000 (Subramanian et al., 2000). This good characteristic makes this material potentially applicable in microelectronics such as in capacitors and memory devices (Jin et al., 2009; Mandal et al., 2009).

The studies have shown that the dielectric constant of pure CaCu$_3$Ti$_4$O$_{12}$ ceramics ($\sim 10^4$) is much higher than La-doped CaCu$_3$Ti$_4$O$_{12}$ at room temperature (Capsoni et al., 2004). Although the pure CCTO gives a higher value in dielectric constant, but when La doping is used, the value for dielectric loss is, as expected, much lower than pure CCTO. As is known, the electrical behavior will be enhanced with lower dielectric loss (William Jr., 2000). Moreover, an interesting part of the investigation of La doping is that the radius of ions La^{3+} (1.15 A) and Ca^{2+} (1.05 A) are very close in value. This is another reason for the researcher's preference of La^{3+}, compared to other ions, as a substitute to Ca^{2+} (Buscaglia et al., 2002).

A conventional method for calcination and sintering processes of CaCu$_3$Ti$_4$O$_{12}$ (CCTO) are carried out by using furnace (Aygun, 2005; Adams et al., 2006; Shao et al., 2006). The mechanism of furnace heating is heat conduction from the material's surface inward, thus generate temperature gradient between outer and inner surface of the heated material. Therefore, furnace heating is slow and tend to overcure the surface while undercuring the interior. On the other hand, microwave heating is a quick and efficient method of heating materials because of the type of heating which is volumetric heating. Volumetric heating produces a better quality of CCTO and increases production rates (Hutagalung et al., 2008a; Hutagalung et al., 2008b). Microwave is an electromagnetic spectrum with a frequency between 300 MHz to 300 GHz. Microwave heating share several similar properties with visible light: both may be reflected or absorbed by the material, may be transmitted through materials without any absorption and when traveling from one material to another, the microwaves may change direction. Materials absorb microwave energy through a relaxation mechanism like dipolar, ion jump and ohmic loses. Two important parameters for microwave processing are power absorbed (P) and depth of microwave penetration (D). Unlike conventional heating, these parameters are highly dependent on the dielectric properties of the

CP1202, *Neutron and X-Ray Scattering in Advancing Materials Research: International Conference – 2009*
edited by A. Saat, H. A. Kassim, M. H. H. Jumali, J. M. Saleh, M. R. Othman, A. Ibrahim, F. M. Idris, and M. H. A.-R. M. Ahmad
© 2009 American Institute of Physics 978-0-7354-0739-8/09/$25.00

material and, in practice, can provide another degree of process flexibility. The factor to consider in microwave heating is the susceptor. Microwave processing provides rapid and uniform heating, decreased sintering temperatures and reduced processing time.

In this study, CCTO was prepared by microwave processing and compared with samples prepared through the conventional route. A detailed investigation on the phase structure, microstructure and crystallography properties of La-doped CCTO of different molar quantities, $Ca_{1-x}La_xCu_3Ti_4O_{12}$ (x = 0.01, 0.02, 0.03, 0.05 and 0.10), by the conventional and microwave routes will be presented in this paper.

EXPERIMENTAL

The conventional processing method used an electric furnace for the calcination and sintering processes in preparing undoped and La-doped $CaCu_3Ti_4O_{12}$. Undoped $CaCu_3Ti_4O_{12}$ (CCTO) was prepared from a stoichiometric composition of raw materials of $Ca(OH)_2$ (Merck, Germany), CuO (Merck, Germany), and TiO_2 (Fluka). La-doped $CaCu_3Ti_4O_{12}$ was prepared by adding La_2O_3 (Merck, Germany) to CCTO in 0.01, 0.02, 0.03 and 0.05 mole fractions. The raw materials were mixed via wet milling for 2 hours using zirconia balls as a medium. As reported, wet ball milling uses less time than dry milling in producing the homogenous samples (Moulson and Herbert, 1990). Subsequently, the powder mixture was dried in an oven. The dried powder was then calcined in an alumina crucible at 900 °C using an electrical muffle furnace for 12 hours with a 5°/min heating rate. This temperature and soaking time were chosen by referring to a previous study (Mohamed et al., 2007). Next, the powders were ground and pressed using hydraulic pressing under 1 Ton, 4.7 Tons and 7 Tons of pressure to form pellets with diameters of 5mm, 13mm and 18mm and a thickness of ≤ 3mm. Finally, the brown pellets were sintered at 1000 °C with a soaking time of 10 hours as referred to the previous study.

Instead of using a furnace to calcine the milled powder, the microwave processing method used a domestic microwave oven (Panasonic, Model NN-S554WF/MF). The milled powder was calcined using the microwave oven for 30 to 90 minutes, and then natural cooled in the oven after the irradiation process. This microwave oven is able to generate a maximum microwave power of 1100 Watt at an operation frequency of 2450 MHz. In this study, a silicon carbide (SiC) crucible was used as a susceptor to absorb microwave energy in the calcination process of the $Ca(OH)_2$-CuO-TiO_2 mixtures to form CCTO compound. The sample was placed in a crucible and then positioned in the centre of the microwave cavity chamber. The oven was operated at full power to ensure that the susceptor had enough energy to heat the samples. A shielded R-type thermocouple was attached to the sample for temperature sensing

The crystalline structures of the prepared samples were analyzed using X-ray powder diffraction (XRD) employing Cu Kα radiation at the wavelength of 1.5406 Å in a scan range (2θ°) starting at 10 to 90°. The

microstructures of the samples were examined using a scanning electron microscopy (SEM).

Table 1. The lattice parameter calculated from main peak (220) XRD pattern of CCTOs.

Sample	Lattice parameter
Undoped CCTO	a = b = c = 7.370 A
1 mol% La doped CCTO	a = b = c = 7.263 A
2 mol% La doped CCTO	a = b = c = 7.242 A
3 mol% La doped CCTO	a = b = c = 7.228 A
5 mol% La doped CCTO	a = b = c = 7.195 A

RESULTS AND DISCUSSION

Figure 1 shows the XRD profiles of powder calcined at 900 °C, 12 hours using the furnace for undoped $CaCu_3Ti_4O_{12}$ and $Ca_{1-x}La_xCu_3Ti_4O_{12}$ (x = 0.01, 0.02, 0.03, 0.05). XRD patterns of CLCTO ceramic samples are quite similar to undoped CCTO, which confirmed the formation of single phase. The crystallite size of the calcined powder can be calculated from the X-ray diffraction peak broadening using the well known Scherer's formula (Jin et al., 2007):

$$L = k.\lambda / \beta \cos \theta \qquad (1)$$

in which k is the shape coefficient, λ is the X-ray wavelength, β is the full width at half maximum (FWHM) of each phase and θ is the diffraction angle. It is known that the phase of pure CCTO and CLCTO is a cubic perovskite related structure with a lattice constant of a = 7.3930 Å and a space group $Im3$ (Almeida et al., 2004).

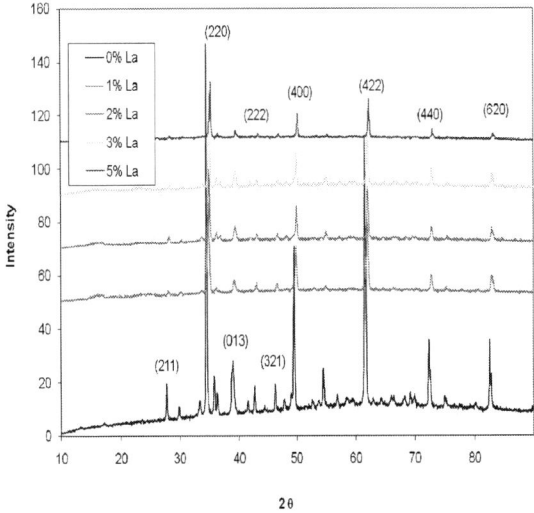

Figure 1. XRD curves of pure $CaCu_3Ti_4O_{12}$ and $Ca_{1-x}La_xCu_3Ti_4O_{12}$ (x = 0.01, 0.02, 0.03 and 0.05) powders calcinations at 900 °C, 12 hours using furnace.

Close observation of the XRD main peak (220 peak) found that the peak positions shifted to the right, which indicates that La^{3+} influences the crystalline structure (Figure 2). The calculated lattice constant values decrease when La doping concentrations increase as shown in Table 1.

Microstructure is reported to play a vital role in determining the properties of electroceramics (Ramirez *et al.*, 2000). Figure 3 shows the SEM images of undoped $CaCu_3Ti_4O_{12}$ and La-doped $CaCu_3Ti_4O_{12}$ ceramics sintered at 1000 °C for 10 hours using a furnace. It can be seen that all La-doped $CaCu_3Ti_4O_{12}$ samples exhibit dense and homogenous microstructures with grain size of about ~ 2μm. The dense microstructure is expected to be advantageous in the improvement of the electrical properties of La-doped CCTO. SEM photographs also show that the average grain size decreased with the increasing La_2O_3 amount because some La ions were inserted at grain boundaries and inhibited grain growth. Higher La_2O_3 amounts in the sample means more La ions will be deposited at the grain boundaries thus inhibiting grain growth, which at the same time reduces the average grain size.

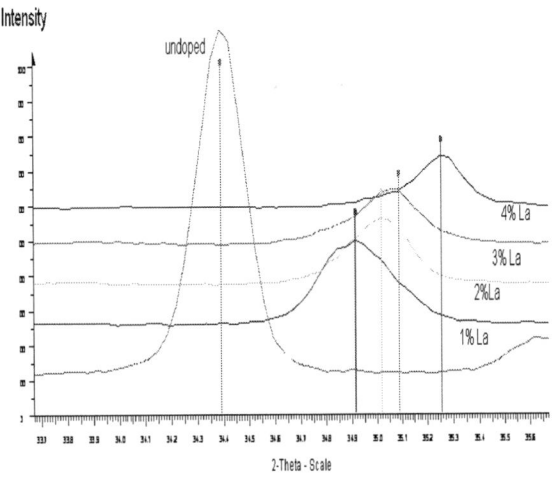

Figure 2. Right shifted behavior on XRD main peaks (220) of La-doped CCTO.

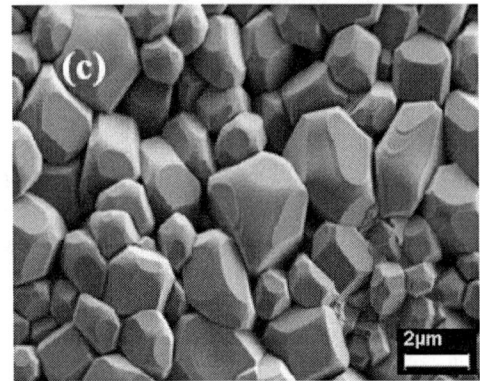

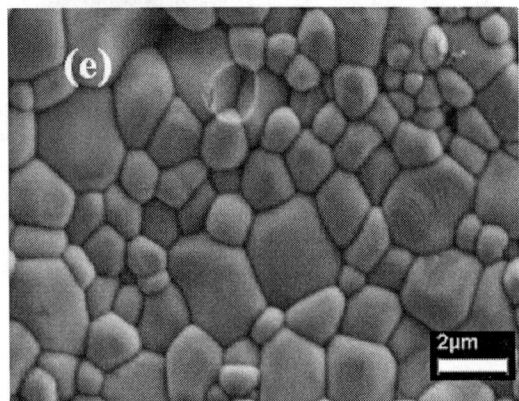

Figure 3. The SEM micrographs of (a) undoped $CaCu_3Ti_4O_{12}$ and $Ca_{1-x}La_xCu_3Ti_4O_{12}$ with x = 0.01 (b), 0.02 (c), 0.03 (d) and 0.05 (e) of pellets sintered at 1000 °C for 10 hours using furnace.

Figure 4 shows the XRD patterns of La-doped CCTO ($Ca_{0.99}La_{0.01}Cu_3Ti_4O_{12}$) prepared by calcination process in SiC crucible using a domestic microwave oven operated at high power for 30 to 90 minutes. Results show that all samples exhibit the formation of single phase CCTO compound. This means that microwave irradiation for 30 minutes is enough to form CCTO compound via microwave processing.

At the first 10 minutes of operation, the temperature inside the crucible as recorded by a shielded R-type thermocouple, reached 1100°C which. This data indicates that the microwave irradiation heating rate using a SiC crucible is much higher (about 105°C/min) than using an alumina crucible. Microwave heating using an alumina crucible can reach a maximum temperature of only about 700°C (Hutagalung *et al.*, 2008b). Therefore, the microwave processing time is much shorter compared to conventional processing using a furnace which requires at least 12 hours.

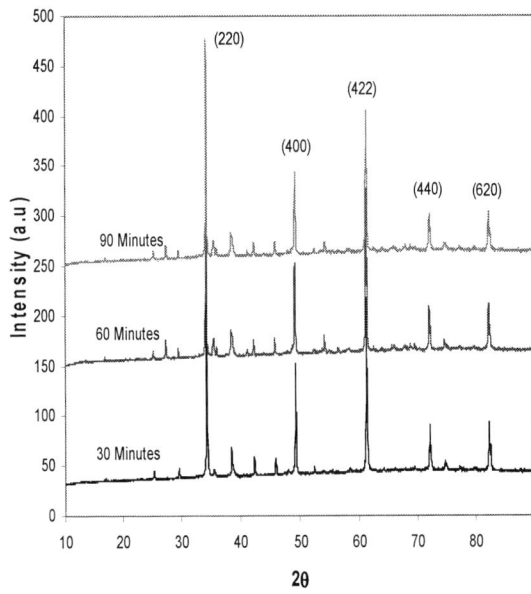

Figure 4. XRD pattern of of $Ca_{0.99}La_{0.01}Cu_3Ti_4O_{12}$ powder prepared by microwave calcination at high power for 30 to 90 minutes using a domestic microwave oven.

CONCLUSION

The nanocrystalline of undoped $CaCu_3Ti_4O_{12}$ and La-doped $CaCu_3Ti_4O_{12}$ were prepared by conventional and microwave processing and were confirmed by x-ray powder diffraction (XRD) and scanning electron microscopy (SEM). Although the conventional route is able to produce single phase CCTO compound, the processing time is very long which is at least 12 hours. However, using the microwave processing route, undoped and La-doped CCTO can be produced in a very short processing time of only 30 minutes.

ACKNOWLEDGEMENTS

This work was supported by the Science Fund, Ministry of Science, Technology and Innovation (MOSTI), Malaysia under project no. 03-01-05-SF0432.

REFERENCES

Adams, T.B., Sinclair, D. C. and West, A. R. (2006), Decomposition reactions in $CaCu_3Ti_4O_{12}$ ceramics, *J. Amer. Ceram. Soc.* 89: 2833-2838.

Almeida, A. F. L., Fechine, P. B. A., Goes, J. C., Valente, M. A., Miranda, M. A. R. and Sombra, A. S. B., (2004), Dielectric properties of $BaTiO_3$ (BTO)–$CaCu_3Ti_4O_{12}$ (CCTO) composite screen-printed thick films for high dielectric constant devices in the medium frequency (MF) range, *Mater. Sci. Eng. B.* 76: 118 –123.

Aygun, S., Tan, X., Maria, J.P., Cann, D., (2005), Effects of processing conditions on the dielectric properties of $CaCu_3Ti_4O_{12}$, *J. Electroceramics* 15: 203-208.

Buscaglia, M. T., Viviani, M., Buscaglia, V. and Bottino, C., (2002), Incorporation of Er^{3+} in $BaTiO_3$, *J. Am. Ceram. Soc.* 85:. 1569–1575.

Capsoni, D., Binia, M., Massarottia, V., Chiodellib, G., Mozzatica, M. C. and Azzonic, C. B., (2004), Role of doping and CuO segregation in improving the giant permittivity of $CaCu_3Ti_4O_{12}$, *J. Sol. State Chem.* 177: 4494–4500.

Hutagalung, S. D., Ibrahim, M. I. M. and Ahmad, Z. A., (2008a), Microwave assisted sintering of $CaCu_3Ti_4O_{12}$, *Ceram. International* 34: 939-942.

Hutagalung, S. D., Ibrahim, M. I. M. and Ahmad, Z. A., (2008b), The role of tin oxide addition on the properties of microwave treated $CaCu_3Ti_4O_{12}$, *Mater. Chem. Phys.* 112: 83-87.

Jin, S., Xia, H. and Zhang, Y., (2009), Effect of La-doping on the properties of $CaCu_3Ti_4O_{12}$ dielectric ceramics, *Ceram. International* 35: 309-313.

Jin, S., Xia, H., Zhang, Y., Guo, J. and Xu, J., (2007), Synthesis of $CaCu_3Ti_4O_{12}$ ceramic via a sol-gel method, *Mater. Lett.* 61: 1404-1407.

Lunkenheimer P., Fichtl, R., Ebbinghaus, S. G. and Loidl, A., (2004), Nonintrinsic origin of the colossal dielectric constants in $CaCu_3Ti_4O_{12}$, *Phys.l Rev. B* 70: 172102.

Mandal, K. D., Rai, A. K., Kumar, D. and Parkash, O., (2009), Dielectric properties of the $Ca_{1-x}La_xCu_3Ti_{4-x}Co_xO_{12}$ system (x = 0.10, 0.20 and 0.30) synthesized by semi-wet route, *J. Alloys Comp.* 478:.771-776.

Mohamed, J. J., Hutagalung, S.D., Ain, M. F., Deraman, K. and Ahmad, Z.A, (2007), Microstructure and dielectric properties of $CaCu_3Ti_4O_{12}$ ceramic, *Mater. Lett.* 61: 1835-1838.

Moulson, A. J. and Herbert, J. M., (1990), *Electroceramics Materials, Properties and Application*, Chapman & Hall, London.

Ramirez, A. P., Subramanian, M. A., Gardel, M., Blumberg, G., Li, D., Vogt, T. and Shapiro, S. M., (2000), Giant dielectric constant response in a copper-titanate, *Sol. State Commun.* 115: 217 - 220.

Shao, S.F., Zhang, Z.L., Zheng, P., Zhong, W.L., Wang, C.L., (2006), Microstructure and electrical properties of $CaCu_3Ti_4O_{12}$ ceramics, *J. Appl. Phys.* 99: 084106.

Subramanian, M. A., Li, D., Duan, N., Reisner, B. A. and Sleight, A. W., (2000), High dielectric constant in $ACu_3Ti_4O_{12}$ and $ACu_3Ti_3FeO_{12}$ phases, *J. Solid State Chem.* 151: 323-325.

William Jr., D. C., (2000), *Material Science and Engineering an Introduction*, John Wiley & Sons, Inc., 5th Edition, New York.

Three-Dimensional Imaging Using Microcomputed Tomography For Studying Gaharu Morphology

Khair'iah Yazid [1], Bert Masschaele [2], Mat Rasol Bin Awang [1], Mohd. Zaid Abdullah [3], Junita Mohamad Saleh [3], Abdul Aziz Mohamed [1], Mohd Ashhar Bin Hj Khalid [1]

[1] *Malaysian Nuclear Agency (Nuclear Malaysia), Bangi, 43000 KAJANG, MALAYSIA*
[2] *UGCT, Ghent Universit ,Belgium*

[3] *School of Electrical and Electronic Engineering,Universiti Sains Malaysia, Nibong Tebal, Seberang Perai Selatan, Pulau Pinang, Malaysia*

Abstract

To demonstrates the potential application of the high resolution X-ray micro-CT technique in the analysis of internal structure in Gaharu wood. Gaharu or internationally, Agar wood, is known for its fragrant resinous wood. The hardware device used in this study was an X-ray micro-CT scanner at Center of Tomography (UGCT), CT facility in Ghent University, Belgium. This technique allows the 3D investigation of the internal structure of the wood in a non-destructive way. Most of the data analysis was done with the software VG Studio Max and MATLAB. Here we present some preliminary results from three-dimensional images from a piece of high grade Gaharu. Micro-CT images of the specimens were obtained at 7 μm resolution. Besides a clear distinction between pores and material, some bright white areas occur in the reconstruction images. Not only the volume visualization is helpful, morphological parameters of open-pores and dark resins are calculated from these 3D data set. The micro-CT technique is a valid support for evaluating the pores structure and resin distribution in Gaharu.

Keywords: X-ray micro-computed tomography, 3D visualization, wood, resin, gaharu grade

INTRODUCTION

Internationally, Agarwood, otherwise known as Eeaglewood, Aloeswood or Gaharu (The Indonesian and Malay name), is the resinous wood from the Aquilara tree, an archaic tropical evergreen tree native to northern India, Laos, Cambodia, Malaysia, Indonesia, Southern China and Vietnam. Its scientific name is Aquilara Malaccensis or Aquilaria agallocha. The Aquilaria tree grows up to 40 meters high and 60 centimeters in diameter. It bears sweetly-scented, snow-white flowers. The trees frequently become infected with a parasite fungus or mold and begin to produce an aromatic resin in response to this attack. The fungus and decomposition process continue to generate a very rich and dark resin forming within the heartwood. Thus, Agar wood develops very, very slowly over time, typically several hundred years.

Agarwood becomes one of the earths most valuable forest products compared to others because of its quality that makes the wood is high in demand. Agarwood is the most rare and precious wood on the planet, prized for its rich, wonderful and healing fragrance. Agarwood has three principal uses: medicine, perfume and incense.

Agar wood has been used for medicinal purposes for thousands of years, and continues to be used in Ayurvedic, Tibetan and traditional East Asian medicine. The resin is also used in perfumery. Recently reported *Yves Saint Laurent* uses Agar wood as a base in perfumes product [1].

The purpose of this project is to advance understanding internal structure of agar wood. For this a non-destructive radiation imaging method is needed. Since the visualization of the products inside wood is based on the same principles as for the visualization of products inside solid material, the micro-CT technique was selected for this research.

X-ray CT is a powerful tool which enables the non-destructive visualization of the internal structure of objects. The technique of X-ray CT is based on the interaction of X-rays with matter. When X-rays pass through an object they will be attenuated in a way depending on the density and atomic number of the object under investigation and of the used X-ray energies. By using projection images obtained from different angles a reconstruction can be made of a virtual slice through the object. When different consecutive slices are reconstructed a 3D visualization can be obtained.

CP1202, *Neutron and X-Ray Scattering in Advancing Materials Research: International Conference – 2009*
edited by A. Saat, H. A. Kassim, M. H. H. Jumali, J. M. Saleh, M. R. Othman, A. Ibrahim, F. M. Idris, and M. H. A.-R. M. Ahmad
© 2009 American Institute of Physics 978-0-7354-0739-8/09/$25.00

The aim of the present experiments is to evaluate the performance and the potentials of the high resolution micro-CT technique on visualizing micro structure inside Agarwood.

EXPERIMENTAL AND METHOD

Micro-CT system

The CT scans were conducted in December 2007 using micro-CT scanner at Center of Tomography (UGCT), CT facility in Ghent University, Belgium. The open type micro-focus x-ray source manufactured by Fein focus with can generate x-rays with medium energy level up to 160 kV. It incorporates an amorphous silicon (a-Si) based flat panel detector (Varian PAXSAN 2520V) with pixel matrix 1880 x 1496 elements with pixel size 127 μm. This indirect detection array was coupled to a commercial scintillating screen Cesium Iodide. The flat-panel detector technology eliminates the need for scan time consuming. It utilizes the concept of "volume tomography" where the entire specimen area is projected onto the CT detector and the x-rays attenuation is collected simultaneously.

The sample was placed on a sample holder between the detector and X-ray source. As a consequence of the cone beam of the source the distance of the sample to the source determines the magnification of the system. This magnification will be set so that the sample stays within the field of view of the detector for the full rotation cycle. In other words for doing micro-CT experiments the diameter of the sample is the most important for reaching a good resolution.

Parameter of measurement

For this study, high grade Agarwood was chosen. From a bigger size, sample was cut to a size 3x3x3mm, suitable for the micro-CT measurements, and used for the basic characterization of the wood. The investigations were performed with voltage 80 kV and a current of 50μA, small angle interval 0.3 degree. The distances from the x-ray source to the sample and from the sample to the flat panel detector were 24mm and 806mm respectively. The beam hardening was suppressed by pre-filtering the radiation with 0.5 Al. The sample was placed on a sample holder between the X-ray source and detector. As a consequence of the cone beam of the source the distance of the sample to the source determines the magnification of the system. This magnification will be set so that the sample stays within the field of view of the detector for the full rotation cycle. By using detector binning, i.e. 2 by 2 pixels taken together giving 940 pixels on a detector row instead of 1880.The number of projections taken was 1200 over full rotation. The integration time per projection was 0.5 sec. In order to increase signal-to-noise ratio, radiographic projections have been acquired out by using averaging 4 frames. Total scanning time is about 2 hours with rotation motor speed 0.3 /sec.

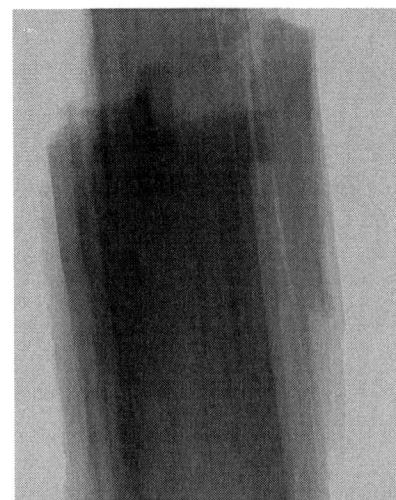

Figure 1 Projection of Agar wood sample

Reconstruction Parameter

Parameters for cone beam reconstruction are summarized in Table 1.

Table 1 Parameters for cone beam reconstruction

Parameters for FDK	
Pixel Size	0.254[mm]
Detector to source distance	830[mm]
Source to object distance	24[mm]
Total projections	1200
Magnification	34.5
Voxel size	0.007[μm]

After image acquisition, projections (2D radiography) of the wood sample were reconstructed with cone beam reconstruction software. This software is based on the FDK algorithm. The raw projection data were performed normalization procedures to correct the influence of the dark current and correction of center of rotation, no further image processing was applied. Finally the 2D cross-sectional images of the sample were obtained in consecutive slices throughout the stone.

Image Analysis

Most of the data analysis was done with the software VG Studio Max [2] and MATLAB.VG Studio software views the CT images one by one, or it can build a volume by stacking the cross sectional slices and view the whole volume at once, with different color and resolution options. Start point for the image analysis of CT data and the separation of the main two components, white spot(resin) and black spot(pores). A local surface calibration is used to extract an iso-surface of the border between materials and pores. An example of histogram is shown in Figure 2.

Table 2 Calculated Pores and resin total amount in sample			
	Σ *Voxel*	Σ *Volume* [mm³]	*Volume* [%]
Pores	4347933	1.64	3.45
Resin	2047677	0.77	1.62

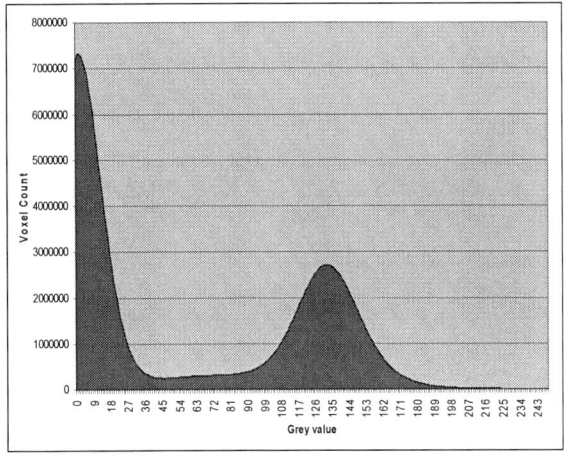

Figure 2 Histogram of sample

RESULTS AND DISCUSSION

X-ray highlights the internal structure in the wood. There are three main regions with different range gray value in the images as shown in Figure 3. The first region which contains composite olio-resins spots, correspond to totally white spot as signified by the highest X-ray attenuation. The second region of drier and contained pores which there are no medium to attenuate X-ray, should map as low X-ray absorbed (dark spot). Other than these regions are the wood itself.

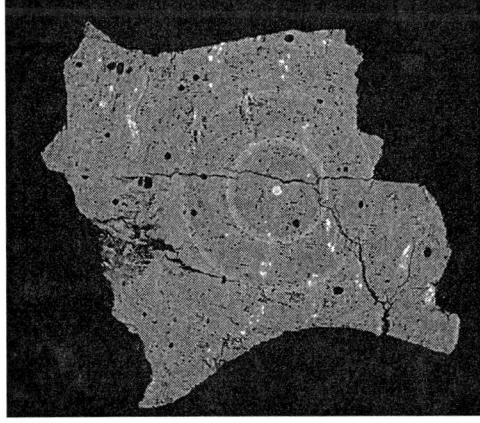

Figure 3 Reconstructed top view cross section of Agar wood scanned at pixel size 0.254mm

The result of different parameters are calculated from the 3D data set can be found in Table 2. Full 3D visualization samples allow for numerical measurements of the pore size distribution as shown in Figure 4. Figure 5 shows the volume renderings of the sample.

For the visualization of pores in inside wood material, the minimum resolution is crucial. Micro-CT with a voxel resolution of $(0.007\mu m)^3$ can distinguish pores from material as shown in Figure 3 . Besides a clear distinction between pores and material, some bright white areas occur in the reconstruction images. The identification of these bright areas in reconstructed images, caused by a denser medium component in the wood which attenuates more radiation .These areas has higher X-ray attenuation. The quality of the overall images was good, except there were several ring artifacts cause by individual pixels in the flat panel detector. Moreover, prior knowledge about the multi component material in wood and the density of each component are essentially important in order to analysis wood.

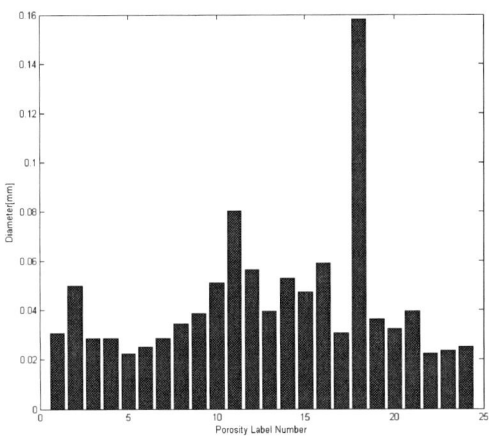

Figure 4 Pore size distributions

Figure 5 CT volume renderings of the sample

CONCLUSION

As a conclusion of this preliminary research, the micro-CT technique is a valid support for evaluating internal structure in Agarwood. Undoubtedly, further studies will be necessary, in particular, with different grade.

REFERENCES

[1] http://en.wikipedia.org/wiki/Agarwood

[2] VG Studio MAX, Version 1.2.1, Volume Graphics

Design and Simulation of FPGA-Based Readout Control of A High-Energy Electron-Proton Collision Calorimeter

Faridah Mohamad Idris[ab], Wan Ahmad Tajuddin Wan Abdullah[b], Zainol Abidin Ibrahim[b], and Burhanuddin Kamaluddin[b]

[a] Agensi Nuklear Malaysian, Bangi, 43000, Kajang, Selangor, Malaysia
[b] Jabatan Fizik, Universiti Malaya, 50603 Kuala Lumpur, Malaysia

Abstract

The readout control of the former ZEUS detector comprises of five analogue modules, i.e. table control, pipeline control, buffer control, format control and generator module. A new readout control in form of Programmable Gate Array (FPGA) module, was designed by integration the five analogue modules into a single module. This paper discusses the design of the readout FPGA-based readout control and its simulation on Quartus II.

Keywords: readout control, integration, simulation.

PACS: 10

INTRODUCTION

In the high energy physics experiment at HERA (Hadron-Electron Ring Accelerator), the proton beam with 920 GeV energy collided with an electron/positron beam (at 30GeV). As a result of the collision, quarks interactions within the accelerated protons and incoming electron/positron were observed and recorded by ZEUS detector, synchronized by the HERA clock at 96 ns or 10MHz . The read-out system controlling the data-taking of the calorimeter part of the detector consists of five analogue modules i.e. table, pipeline, buffer, format and generator modules, with more than 140 input and output signals interconnected to each other.

In this project the analogue circuit diagrams, as well as the block diagram of the table, pipeline, buffer and format modules were used as bass to form the building blocks of a Field-Programmable Gate Array (FPGA) - based readout control, using Verilog as the hardware description language.

THE READOUT CONTROLLING MODULES AND ITS FUNCTIONS

The readout electronics were 'data driven', i.e. the operation of the components was completely determined by the context provided by the data themselves [1]. The table module gave preset controlling data to the readout system; the pipeline selected which particular cell out of 96 samples [2], to trigger; the buffer keep interim data storage from the pipeline; and the format set the timing for the digitization of the output.

FPGA PROGRAMMING

The table module accepts 8-bit serial data from the universal computer interface card i.e. from table, format, pipeline, generator for its RAM (random access memory) data. The controller bits in the table module were synchronized with the 10MHz serial clock. Here, the readout control system was first isolated by giving flag 0, before each subsequent byte pushed the prior byte onto the next register in the chain [1]. Once set, the readout control was put on-line again, where the signals from GFLT (Global First Level Trigger) would determine the controlling sequence of the readout control. In the table module, an FPGA 16-bit shift register and 8-bit shift register were design with Verilog to accept serial data and serial clock and compare them with the GFLT signals.

In the pipeline module, the signals from table control would determine which data in the 96 samples of the physics events for accept (ACT) or abort (ABT). While the ACT was true, the buffer controller would continue taking the physics data event and forward them to format controller for digital outputs. On receiving the ABT signal form the table module, the pipeline controller would notify the buffer to reject the current data taking.

Figure 1 gives the sequence of the readout control development; the table, pipeline, buffer and format modules were integrated into one FPGA-based module with Verilog.

CODING WITH VERILOG

The converting of the modules were carried out based on the logic block diagram of the readout modules. Connecting inputs and outputs from/into each of the

CP1202, Neutron and X-Ray Scattering in Advancing Materials Research: International Conference – 2009
edited by A. Saat, H. A. Kassim, M. H. H. Jumali, J. M. Saleh, M. R. Othman, A. Ibrahim, F. M. Idris, and M. H. A.-R. M. Ahmad
© 2009 American Institute of Physics 978-0-7354-0739-8/09/$25.00

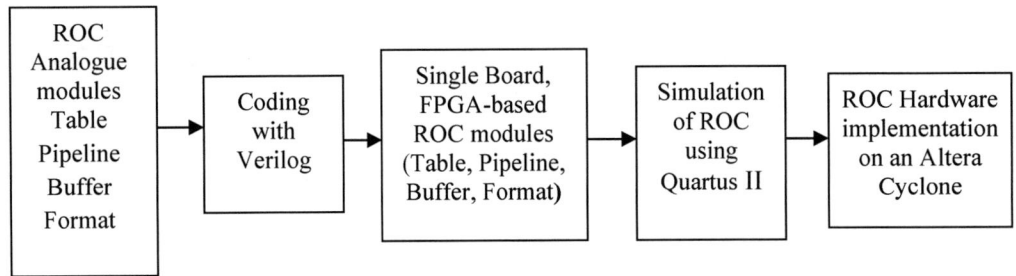

Figure 1A. The analogue modules of readout control (ROC) of the ZEUS detector were coded into single board, FPGA-based using Verilog before being simulated on Quartus II.

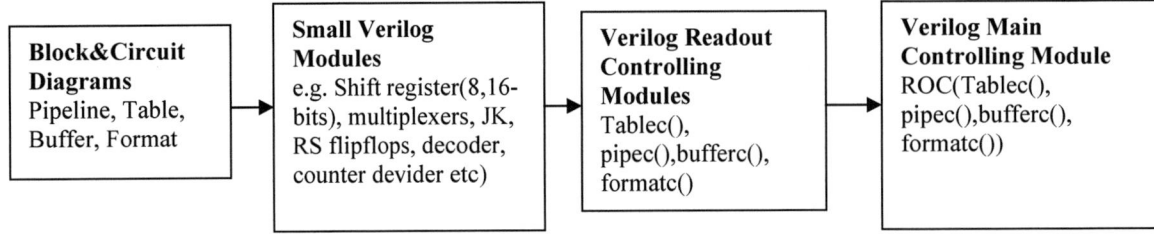

Figure 1B. Coding sequence of the controlling analog read-out modules using Verilog. Coding were carried our starting with basic blocks, later combined to become the main controlling block

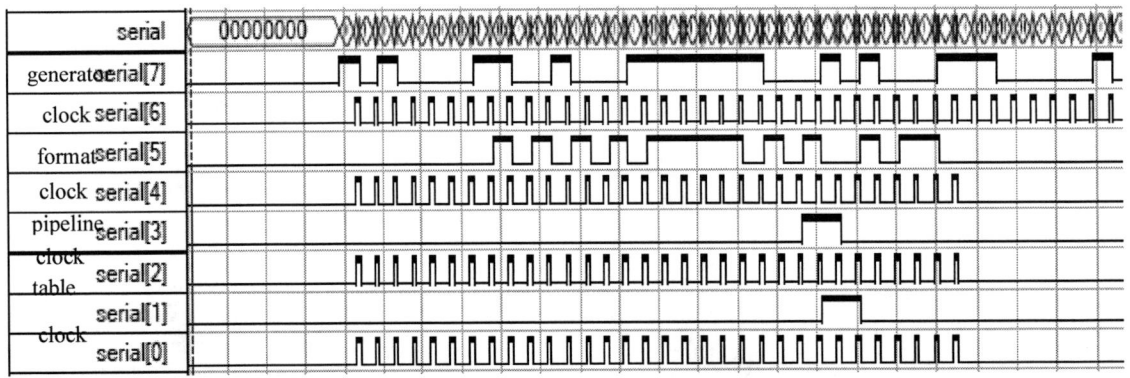

Figure 2. Serial data input to the FPGA-based readout control (serial[0] for table control, serial[3] for pipeline, serial[5] for format control, serial[7] for generator control; while serial[0],[2],[4].[6] were serial clock 10MHz)

modules were indentified. Smaller sub-modules i.e. shift register(8,16-bits), multiplexers, JK, RS flipflops, decoder, counter divider etc. were created and combined to form the four controlling i.e. modules table, pipeline, buffer and format, which were later combined to form the main controlling module. Figure 1.B gives the coding sequence used in converting the analog table, pipeline, buffer and format modules into the FGGA-based read-out controlling module.

FPGA SIMULATION AND RESULTS

The integrated FPGA-based modules in Verilog were simulated on Quartus II using the device settings of Altera Cylcone I. **Figure 2** shows the part of the vector waveform used in Quartus II for the simulation for serial data input. Each serial data consist of 4 bytes control data given to RAM in table module.

Each bit of the serial data is only counted on negative change of the clock edge, where subsequent byte pushes the a prior byte up the 16-bit shift register.

In **Figure 3A**, when the pipeline accept data PACT is

CONCLUSION

In this project, we have demonstrated that an FPGA-based readout control for a high energy physics electron-

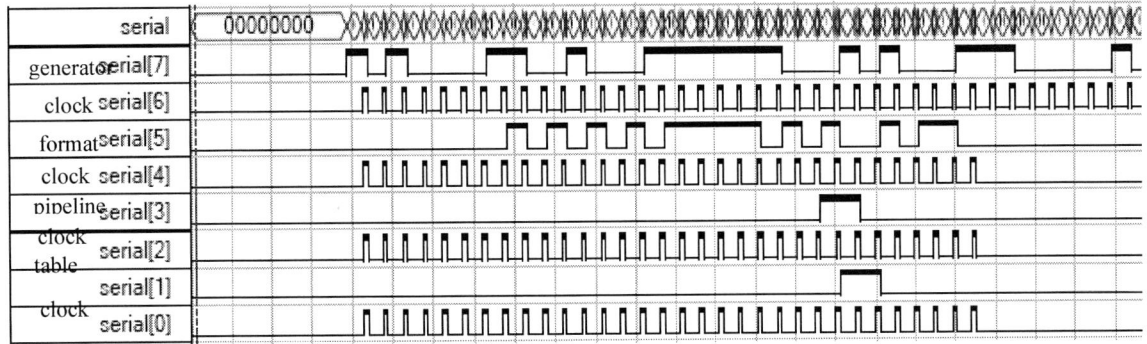

FIGURE 3,A. Serial data input to the FPGA-based readout control (serial[0] for table control, serial[3] for pipeline, serial[5] for format control, serial[7] for generator control; while serial[0],[2],[4].[6] were serial clock 10MHz)

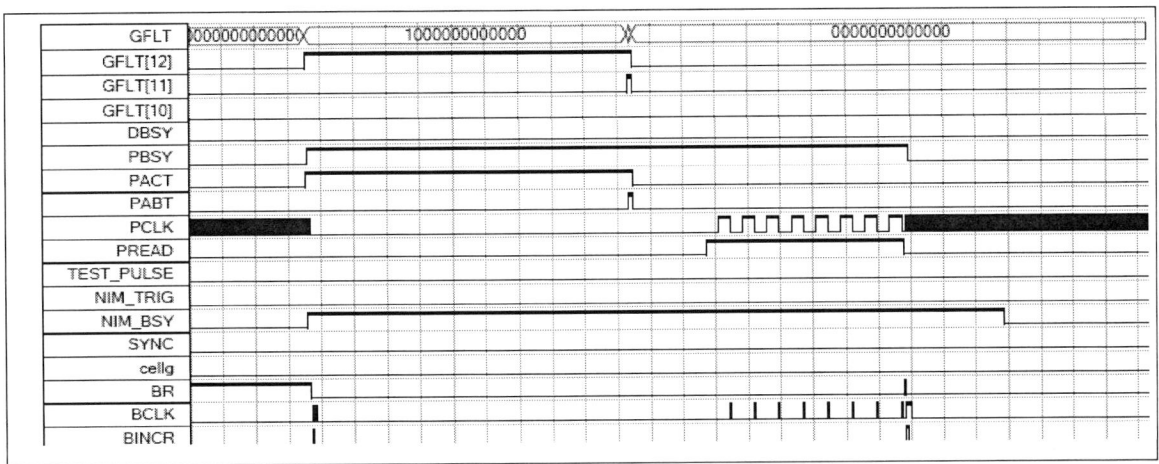

FIGURE 3B. A close-up of the FPGA-based readout control showing the abort ABT signal from the pipeline control

triggered, the pipeline busy PBSY and the pipeline read PREAD are triggered with PCLK temporarily disabled. Here, the buffer read is flagged 0 and the BCLK is temporarily disabled. **Figure 3B** gives a closer look at the sequence of pipeline and buffer triggers upon abort ABT signal by GFLT. On a negative edge of ABT from GFLT, the pipeline abort PABT triggers and pipeline accept PACT is flagged down to 0. During the abort trigger, the buffer is still flagged 0 and only changes to 1 about 0.07 ms later, resulting in unused data being taken by the buffer instead of rejecting it.

proton collision calorimeter is feasible. The integrated FPGA-based module on a single chip besides compact, is easier to modify in future. More work would have to be carried out to overcome glitches of the timing sequence of the present readout control.

TABLE . Some of the output label from FPGA-based readout control (ROC) as shown in Figure 3 and its status,

Ouput label	Output Status	Flag
GFLT_busy	GFLT busy	1
Offline	System is off-line	1
STRB	Strobe signal	0
TYP	Type of event	000
TSTEN	Test mode enable	0
DBSY	Data busy	0
PBSY	Pipeline busy	1
PACT	Pipeline accept data	1
PABT	Pipeline abort data	1
PCLK	Pipeline clock	counter
PREAD	Pipeline read	1
cellg	Number of bunch crossing	1
BR	Buffer read	1

ACKNOWLEDGMENTS

We would like to acknowledge and extend our gratitude the ZEUS Collaboration DESY, Jabatan Fizik Universiti Malaya and Agensi Nuklear Malaysia MOSTI in making this project a success.

REFERENCES

1. A. Caldwell and S. Ritz, User Interface to the CAL Electronics Readout, ZEUS-Note 92-046

2. B. Schmidke, The CAL Readout; Columbia University/ZEUS Collaboration

3. F. Mohamad Idris, W.A.T. Wan Abdullah, Z.A. Ibrahim, M.A.F. Mat Jusoh, Y. Yamazaki, P. Goettlicher, W. Schmidke; FPGA Implementation of the Readout Control; CAL-HES Meeting ZEUS Collaboration February 2006n

Contribution To Degradation Study, Behavior Of Unsaturated Polyester Resin Under Neutron Irradiation

D.Abellache, A.Lounis, and K.Taïbi

Laboratory of Sciences and Material Engineering. Faculty of Mechanical Engineering and Chemical Engineering. U.S.T.H.B, BP32 El Alia. Algiers. Algeria.

ABSTRACT

Applications of unsaturated polyester thermosetting resins are numerous in construction sector, in transport, electric spare parts manufactures, consumer goods, and anticorrosive materials. This survey reports the effect of thermosetting polymer degradation (unsaturated polyester): degradation by neutrons irradiation. In order to evaluate the deterioration of our material, some comparative characterizations have been done between standard samples and damaged ones. Scanning electron microscopy (SEM), ultrasonic scanning, hardness test (Shore D) are the techniques which have been used. The exposure to a neutrons flux is carried out in the column of the nuclear research reactor of Draria (Algiers-Algeria) .The energetic profile of the incidental fluxes is constituted of fast neutrons ($\Phi_R = 3.10^{12}$ n.cm^{-2}.s^{-1} ,E = 2Mev) of thermal neutrons ($\Phi_{TH} = 10^{13}$ n.cm^{-2}.s^{-1} ; E= 0.025ev) and epithermal neutrons ($\Phi_{epi} = 7.10^{11}$ n.cm^{-2}.s^{-1} ; E > 4,9ev). The received dose flow is 0,4Kgy. We notice only a few scientific investigations can be found in this field. In comparison with the standard sample (no exposed) it is shown that the damage degree is an increasing process with the exposure. Concerning the description of irradiation effects on polymers, we can advance that several reactions are in competition : reticulation, chain break, and oxidation by radical mechanism. In our case the incidental particle of high energy fast neutrons whose energy is greater or equal to 2 Mev, is braked by the target with a nuclear shock during which the incidental particle transmits a part of its energy to an atom. If the energy transfer is sufficient, the nuclear shock permits to drive out an atom of its site the latter will return positioning interstitially, the energy that we used oversteps probably the energy threshold (displacement energy). This fast neutrons collision with target cores proceeds to an indirect ionization by the preliminary creation of excited secondary species that will generate ionization. Scanning electron microscopy (SEM) performed with an acceleration tension of 0,7 kV shows clearly the caused damage. This observation seems to indicate the presence of major chain breaks for the sample bombarded during 90 minutes. Let us note that the presence of benzenic cores improves behavior toward radiations indeed the chemical function recognized as the most stable to radiations is the aromatic ring. In order to value the rigidity of our material we have determined the Young's modulus . The values are 7.17, 7.60, 8.39 and 8.96 Gpa respectively for blank samples, 30, 60 and 90 minutes exposure ones. Thus, we remark an increase of Young's modulus that can be interpreted in terms of reticulation, provided to use the level of irradiation dose.

Key words: Unsaturated polyester, Chemical, Neutrons radiation, Degradation.

INTRODUCTION

Development and research activities on systems design processes of polymeric composites are being carried out to optimize its performances, aiming at exceptional features. This objective is attained by a better control of matrix properties. In the nuclear field, polymers resistance to radiations is barely known, few works mention polymers radiation interaction, Devanne [1] studied peroxide network radiochemical ageing. Three identical samples constituted by an unsaturated polyester resin were exposed to a neutron flux 10^{12} n/cm^2.s (nuclear reactor) during 0.5h, 1.0h and 1.5 h.

The purpose of the present work is assessment of the deterioration and damage of an unsaturated polyester matrix. In order to value the behaviour of these materials, comparative characterizations were performed between samples (standard and damaged) by optical microscope and Scanning electron microscopy (SEM) observations as well as ultrasonic scanning.

EXPERIMENTAL

During the experiments, the same samples compositions have been maintained *e.g.*, 100 g of unsaturated polyester resin + 2 % of (peroxide) hardener + 1 % of accelerator (cobalt acetate). The dimensions of

CP1202, *Neutron and X-Ray Scattering in Advancing Materials Research: International Conference – 2009*
edited by A. Saat, H. A. Kassim, M. H. H. Jumali, J. M. Saleh, M. R. Othman, A. Ibrahim, F. M. Idris, and M. H. A.-R. M. Ahmad

the samples are. : diameter 10 mm; and thickness 4 mm. The resin is composed of a solution polymer within styrene. Polymer is resulting from polycondensation reaction of propylene glycol, maleic anhydrideacid and phthalic anhydride acid. The polymeric solution analysis has been led with an UV-visible spectrophotometer of type Lambda, UV/visible. It should be noted that prior to this experiment the purity of present product by precipitation test was confirmed. In order to evaluate the material rigidity the Young's modulus has been determined. The measurer 26MG6XT uses the ultrasonic scanning principle [2]. The exposure to a neutrons flux is carried out in the column of the nuclear research reactor of Draria (Algiers). The energetic profile of the incidental fluxes is constituted of fast neutrons (Φ_R = 3.10^{12} n cm^{-2} s^{-1}, E = 2 Mev) of thermal neutrons (Φ_{TH} = 10^{13} n cm^{-2} s^{-1} n cm^{-2} s^{-1}; E =0.025 ev) and epithermal neutrons (Φ_{epi} = 7.10^{11} n .cm^{-2} s^{-1}; E > 4.9 ev). The received dose flow is 0.4 KGy.

The spectra of standard samples and damaged ones have been compared. micrographies obtained by optical microscopy (Nikon 500) and SEM (Jeol JSM6360) permit to display and confirm samples surface deterioration..

RESULTS AND DISCUSSION

From the absorption spectrum (Fig. 1), it is assessed the absorbance value with the wavelength of λ = 262 nm that corresponds to molar absorbance coefficient of styrene (ε = 2.19 cm^{-1} mol^{-1}). The optical path, thickness of the used tank is 1 cm. 10 mg of resin mass is dissolved in 10 mL of THF, which permits to calculate styrene proportion (41 %).

Concerning the irradiation test, the incidental flux energetic profile is constituted of fast, thermal and epithermal neutrons. In comparison with the standard sample (non-exposed) it is shown that the damage degree is an increasing process with the exposure. From literature data and bibliographic material concerning the description of irradiation effects on polymers, one can advance several reactions, which are in competition such as reticulation, chain break and oxidation by radical mechanism [3][4]. This oxidation modifies the resistance of polymers in a notable way starting from 5Mrd. Figure 4 gives the level of irradiation proportions to leave whose the mechanical properties are affected [4].

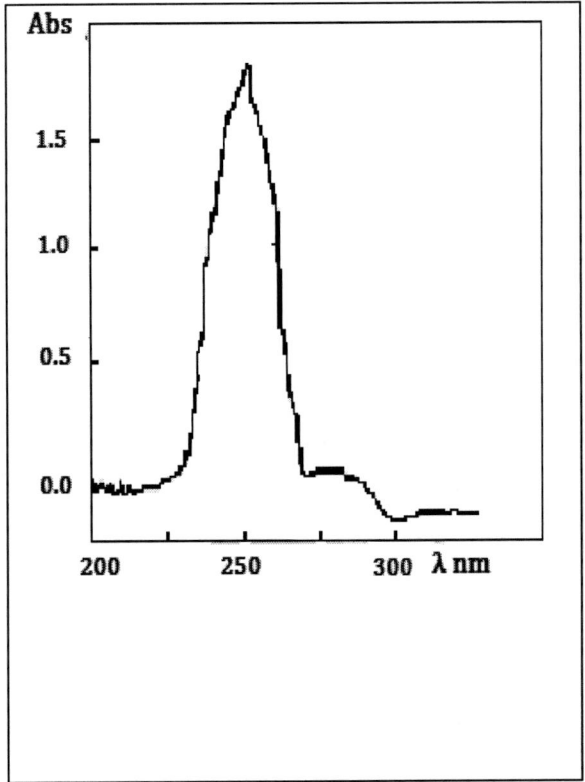

Fig. 1. Absorption spectrum of standard resin sample.

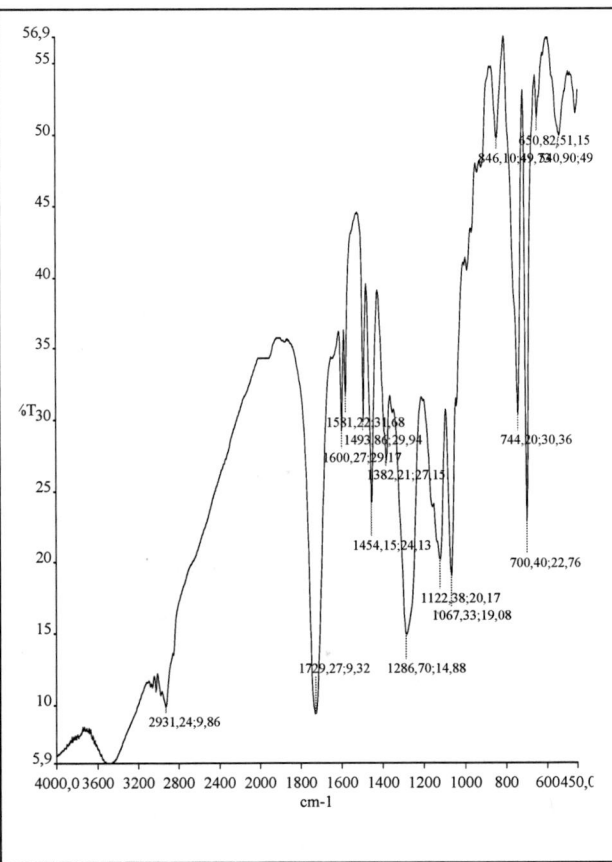

Figure 2 : The FT-IR spectrum of standard resin sample (non damaged)

136

It is now established that the chains tendency to reticulate or to break depends strongly of the bond energy between monomers units. In present case the incidental particle of high energy fast neutrons whose energy is greater or equal to 2 Mev, is braked by the target with a nuclear shock during which the incidental particle transmits a part of its energy to an atom. If the energy transfer is sufficient, the nuclear shock permits to drive out an atom of its site the latter will return positioning interstitially, the energy used oversteps probably the energy threshold (displacement energy). This fast neutrons collision with target cores proceeds to an indirect ionization by the preliminary creation of excited secondary species that will generate ionization. Scanning electron microscopy (SEM) performed with an acceleration tension of 0.7 kV (Fig. 5) shows clearly the caused damage. This observation seems to indicate the presence of major chain breaks for the sample bombarded during 1.5 h. Let us note that the presence of benzenic cores improves behaviour toward radiations [5] indeed the chemical function recognized as the most stable to radiations is the aromatic ring. This ring after excitation returns to the fundamental state with a very weak rate of bond break. This property is assigned to the relocation of π electrons that distribute absorbed energy on a significant number of bonds [6]. In order to value the rigidity of these material, the Young's modulus (ultrasonicscanning principle) was determined. The values are 7.17GPa, 7.60GPa, 8.39GPa and 8.96 GPa, respectively for blank samples (non-exposed), 0.5h, 1.0h and 1.5 h exposure ones.Thus, an increase of Young's modulus can be interpreted in terms of reticulation,

provided to use the level of irradiation dose this result correspondings with Wilski's works [7].

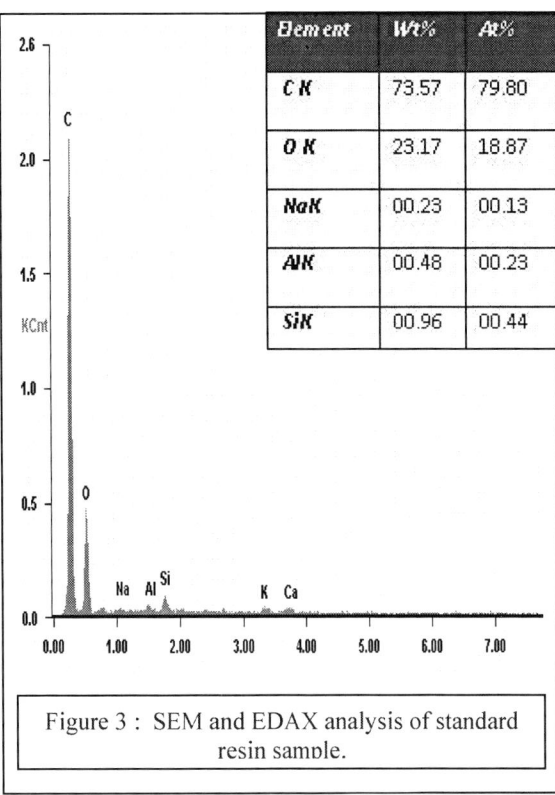

Element	Wt%	At%
C K	73.57	79.80
O K	23.17	18.87
NaK	00.23	00.13
AlK	00.48	00.23
SiK	00.96	00.44

Figure 3 : SEM and EDAX analysis of standard resin sample.

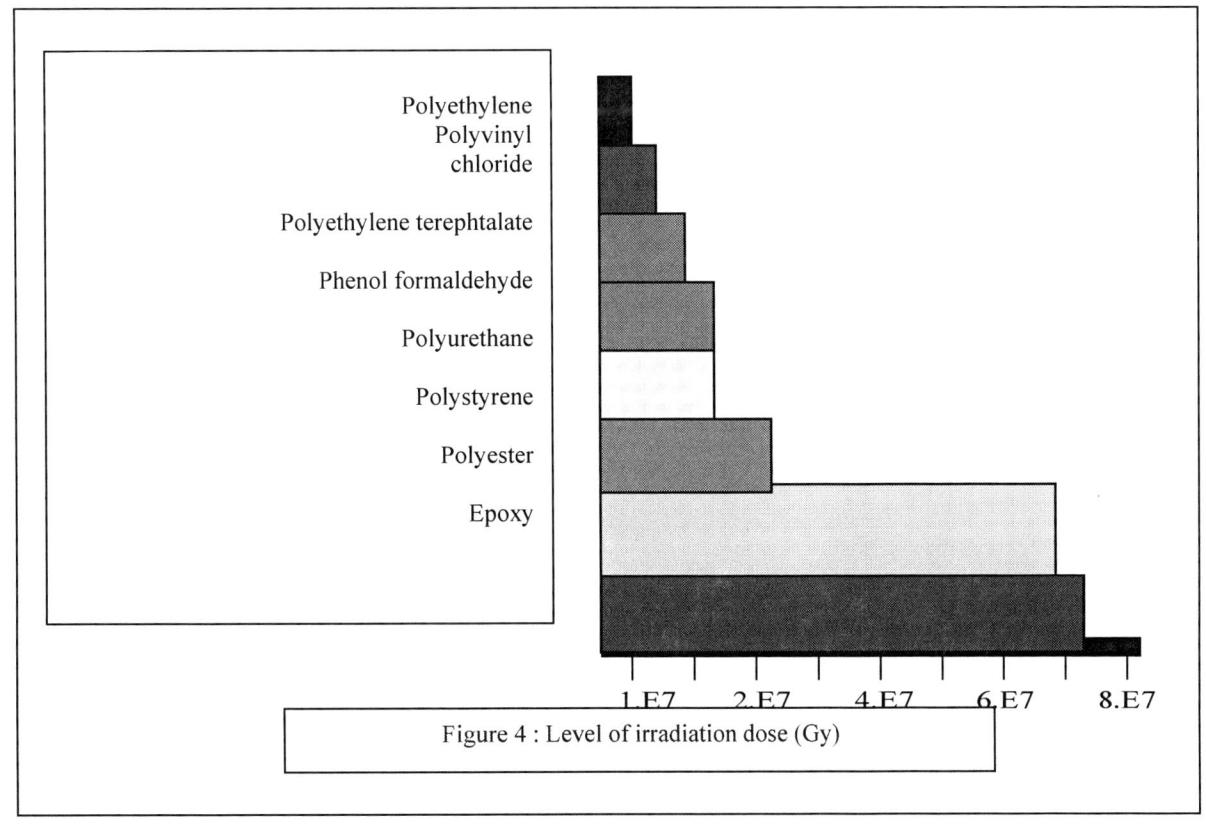

Figure 4 : Level of irradiation dose (Gy)

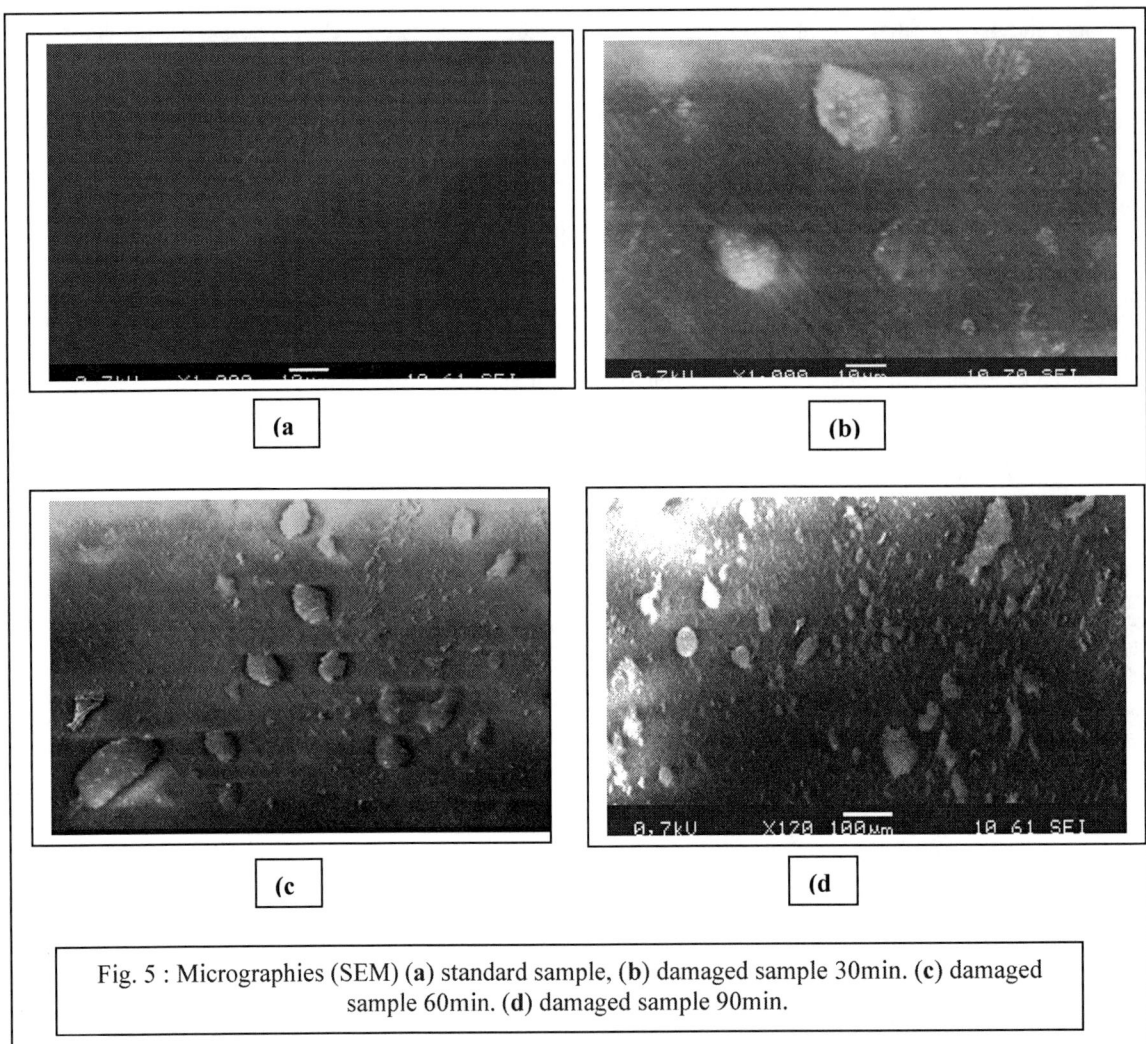

Fig. 5 : Micrographies (SEM) (a) standard sample, (b) damaged sample 30min. (c) damaged sample 60min. (d) damaged sample 90min.

Polymer properties evolution depends on operative conditions of the atmosphere in which irradiations are applied. This reticulation that provokes resin hardening strengthens the sample. It is noted as well that the released γ-photons have insufficient energy to provoke nuclear reactions in radiated material but they rather break bonds between atoms and generate free radicals in the material. The modulus value varies from 6 to 25 %, this suggests an approach for quite important point related to mechanical behaviour *i.e.*, it is ruled by structural state in macromolecular scale that is to say chain length, reticulation density or rather by the structural state on molecular scale *i.e.*, carbonyl, acids, hydroperoxides, *etc*.

CONCLUSION

We were interested in a matrix constituted by an unsaturated polyester resin. In order to evaluate the deterioration and the damage of unsaturated polyester resin material, a comparative characteristics between the standard and deteriored samples are carried out. The exposures to a neutron flux led to an increase in the Young modulus which can be interpreted in term of reticulation considering the amount of irradiation. The relations between structure and mechanical properties has been developed.

REFERENCES

[1]. T. Devanne, Vieillissement Radiochimique d'un réseau époxyde". Thèse de Doctorat. École

Nationale Supérieure d'Arts et Métiers Centre de Paris. France (2003).

[2]. J.H. Bungey, Testing of Concrete in Structures, Surrey University Press, New York, edn. 2 (1989).

[3]. J.W. Chin, J. Martin, T. Nguyen J.W. Chin, T. Nguyen and K. Aouadi, *J. Compos. Technol. Res.*,**19**, 205 (1997).

[4]. D.W. Clegg and A.A. Collyer, Irradiation Effects on Polymers, Elsevier Applied Science, NewYork, p. 110 (1991)

[5]. F.A. Bovey, The Effects of Ionizing Radiation on Natural and Synthetic High Polymers,

Interscience Publisher, NewYork, p. 17 (1958).

[6]. F.A. Makhlis, Radiation Physics and Chemistry of Polymers, John Wiley and Sons, New York, Ch. 3 (1975).

[7]. H. Wilski, *Radiat. Phys. Chem.*, **39**, 407 (1992).

Application of Image And X-Ray Microtomography Technique To Quantify Filler Distribution In Thermoplastic-Natural Rubber Blend Composites

Sahrim Ahmad [a], A. Aziz Mohamed [b], Jaafar Abdullah [b], Hafizal Yazid [a, *], M. Dahlan [b], Rozaidi Rasid [a], W. Saffiey W. Abdullah [b], Mahathir Mohamad [b], Rafhayudi Jamro [b], M. Hamzah Harun [b], Mouad A. T. [a]

[a] *Faculty of Applied Science and Technology, Universiti Kebangsaan Malaysia (UKM), Bandar Baru Bangi, 43000 Kajang, Malaysia.*
[b] *Malaysian Nuclear Agency, Bangi, 43000 Kajang, Malaysia.*
Corresponding Author: Hafizal Yazid, Faculty of Applied Science and Technology, Universiti Kebangsaan Malaysia (UKM), Bandar Baru Bangi, 43000 Kajang, Malaysia.
E-mail: hafizal@mint.gov.my

Abstract

X-ray microtomography and ImageJ 1.39u is used as a tool to quantify volume percentage of B_4C as fillers in thermoplastic-natural rubber blend composites. The use of percentage of area occupied by fillers as obtain from ImageJ from the microtomography sliced images enables the proposed technique to easily obtain the amount volume percentage of B_4C in the composite non-destructively. Comparison with other technique such as density measurement and chemical analysis proves the proposed technique as one of the promising approach.

Keywords: Non-destructive testing, X-ray Microtomography, Thermoplastic-natural rubber blend, Boron Carbide, Composites

INTRODUCTION

Filler distribution in composite is one of the critical issues once the composite is fabricated. The information is very important since the resulted properties of the composite are affected by the spatial distribution of the filler or particles [1-3]. This is true especially in the case of particulate composite. Therefore a systematic approach is used to characterize the filler distribution in polymer matrix composite. The approach is to detect the present of the filler or particles and to quantify the particle distribution through the use of X-ray microtomography and other techniques. Similar approach has been developed and used to quantify the particle distribution in metal matrix composite with promising results [4]. Another approach to characterize the filler distribution in thermoplastic is through the use of X-ray phase contrast imaging, which also gives a promising result [5].

Stringent requirement of product specification leads to the improvement in product quality control and testing. Thus brings the proposed technique as an attractive method as it is non-destructive in nature. The non-intrusive nature of the measurement makes it possible for repetition and as a volume is mapped there is no need for statistical approximation to the final population size or distribution. It is apparent that tomography is useful in many fields and much work has been done and reported in materials applications [6-10].

Conventional method in obtaining some quantitative information on the particle distribution in composites is achieved through micrograph. The micrographs are captured using microscope coupled with image analysis software. The captured micrographs are taken from slices of prepared sample. The obtained information is only in 2D and extended to 3D by extrapolation and assumption. One typical assumption is to treat the particle as point form or having a circular particle shape. These assumptions may lead to inaccuracy or totally wrong information. Furthermore, sample preparation to turn it into sliced form is necessary and this may damage the sample. The damage is possibly brought by torn-out particle or smeared boundaries that result in misrepresentation of size or population [4]. Therefore the non-intrusive method is required as the obtained data is more accurate and reliable.

X-ray microtomography is a non-destructive technique that produces accurate images of 3D volumes by reconstruction from multiple X-ray projections, allowing the direct characterization of the 3D microstructure of samples [11]. The X-ray images are taken around a single axis of rotation and currently able to resolve details as small as few microns in size, even

CP1202, *Neutron and X-Ray Scattering in Advancing Materials Research: International Conference – 2009*
edited by A. Saat, H. A. Kassim, M. H. H. Jumali, J. M. Saleh, M. R. Othman, A. Ibrahim, F. M. Idris, and M. H. A.-R. M. Ahmad
© 2009 American Institute of Physics 978-0-7354-0739-8/09/$25.00

when imaging objects are made of high density materials. As long as the density different between reinforcement and the matrix is comparatively significant, the resulted image would yield a good image contrast. This is the common criterion that has to be met in the use of any X-ray imaging technique.

The purpose of the development of thermoplastic-natural rubber (TPNR) composites is to produce a boronated thermoplastic-rubber used as neutron shielding material in nuclear research reactor. The excellent " in between property " of elastic and rigid of thermoplastic natural rubber make it an attractive materials to be used as the matrix [12] and it could provide wide area coverage that require radiation shielding. High thermal neutron cross-section material such as B_4C is used as filler in order to provide shielding effect against thermal neutrons [13-15]. The aim of the current work is ultimately to quantify the distribution of boron carbide in the thermoplastic-natural rubber with the use of X-ray microtomography and other techniques such as density measurement and thermogravimetry analysis. All the techniques are compared and used to acquire information pertaining to the filler distribution and concentration within the sample.

METHODOLOGY

Thermoplastic-natural rubber used in this study was incorporated with various percentage of boron carbide with average size of 13.5μm. Boron carbide with natural rubber, thermoplastic and other additives were compounded in Haake twin blade internal mixer for a predetermined optimum time and mixing scheme. Finally the composite compound was heated press to form a slab sample with 1mm in thickness. The sample is depicted in Fig. 1 as observed from the top view of the slab.

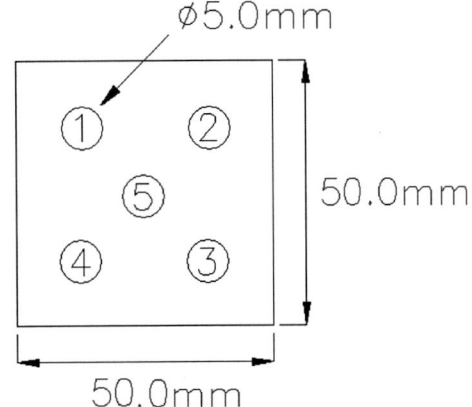

Figure 1: Sample in the form of slab as observed from top view

Slab with different formulation is given in Table 1.

For each slab, five samples or volume of interest are cut. They are denoted as volume of interest 1 or VOI 1, VOI 2, VOI 3, VOI 4 and VOI 5. The size of each VOI is 5mm in diameter and 1mm in thickness. The location

of VOI on the slab are shown in Figure 1 and numbered accordingly. Each of the VOI is the test sample and subjected to testing and analysis.

Table 1: Slab samples with different formulation

Slab	Percentage by weight of B_4C as filler in TPNR
A	0
B	10
C	20
D	30
E	40
F	50

A systematic approach is used to obtain information on reinforcement within the sample. The samples were subjected to density measurement, X-ray microtomography scanning and chemical analysis. Archimedes principle is used to obtain the density of each small cut section that contain VOI from which the volume content of B_4C could be calculated, as the value of standard blank sample is known. VOI samples were subjected to Thermogravimetry analysis (TGA) to quantitatively obtain the weight percent of boron carbide in the test samples. X-ray microtomography measurements were carried out on VOI using a Sky Scan 1172 desktop x-ray microscanning. Data were collected at 60kV and 167μA. The image was detected on a high resolution (1280 x 1024 pixels) CCD camera as the sample was rotated and the output from the CCD was fed through a reconstruction algorithm [9]. Beam hardening was also computationally corrected, to produce a binary grey-scale bitmap image of the sample. The datasets collected for the tomography were 360° angular range with a step size of 0.7° between images. The sliced binary images are thresholded to obtain the areas occupied by B_4C. These values are used with results from chemical analysis to establish a parameter named correlation factor of B_4C in the composites.

RESULTS AND DISCUSSIONS

Chemical analysis gives an accurate value of percentage of filler within the tested volume of interest. In this case, TGA is used to obtain the amount of filler in the composite sample. The drawback of this technique is that it is destructive in nature. This technique gives the overall vol% of B_4C but the location of the particle within the VOI is not obtained. Result of TGA is given in Figure 2.

TGA thermograms reveal the actual concentration of B_4C in the slabs. The amount of B_4C was found slightly lower than the target composition as the filler lost is common during mixing process. This is due to the filler adhesion at the surface of hopper or feeding system. From Figure 2, as expected slab A was absent of filler.

This is followed by slab B to F with a gradual increment of filler or B_4C. However, this technique does not provide any information on the filler distribution in the sample and exhibit only information on filler concentration. Therefore this technique is coupled with the X-ray microtomography to utilise both the information on the particle distribution and the actual filler concentration.

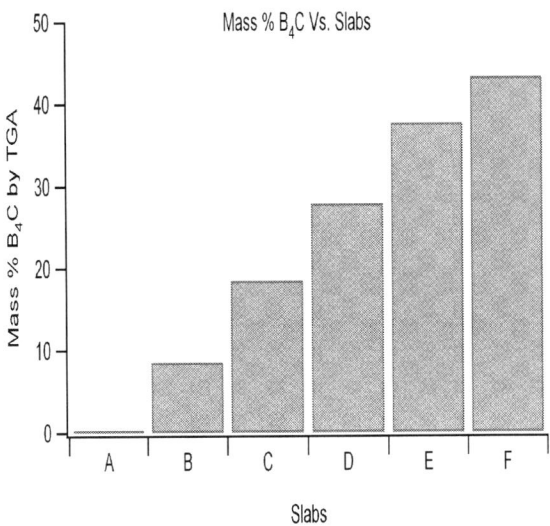

Figure 2: Average mass % of B_4C in VOI's for each of the slab

X-ray microtomography reveals the 3D volume microstructure of the sample. The density difference between the filler and the matrix enable the filler to appear darker than the matrix. Therefore particle distribution could be observed within the slices of VOI. Typical image reconstruction of 3D volume microstructure of the sample is shown in Figure 3.

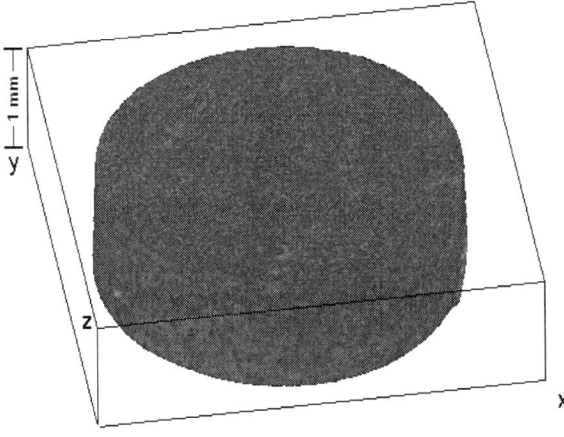

Figure 3: Typical 3D volume microstructure of VOI with 10%wt filler

Typical horizontal X-ray microtomography slices of VOI's before and after thresholded is shown in Figures 4 – 10. The image is thresholded in order to show B_4C.

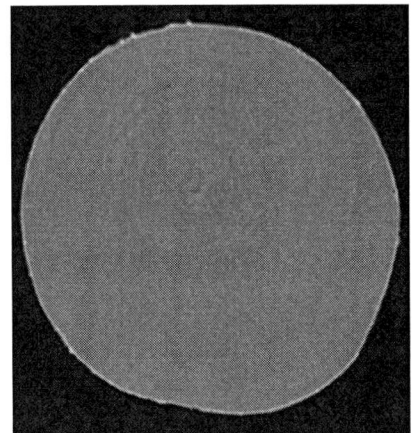

Figure 4: Typical slice of matrix without filler.

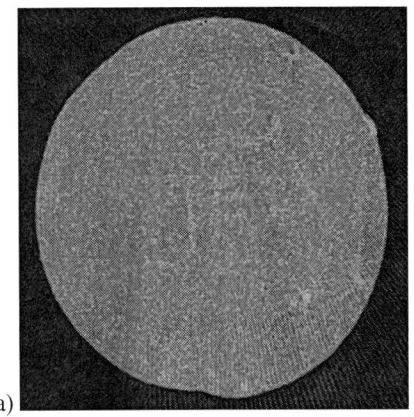

(a)

(b)

Figure 5: (a) Typical slice of matrix + 10%wt filler, (b) thresholded of image (a) and typically yield 2.515% bright areas.

The microtomography slices (Figures 5b-9b) in the thresholded images show a variation in the percentage of bright areas. The bright areas indicate the present or occupancy of filler. This is possible since the density difference between the matrix and the filler is high. The matrix density is $0.920g/cm^3$ and the filler is $2.520g/cm^3$. Thus, this has eased in the thresholding process to obtain the areas that reflect the present of B_4C. It was found that 8.514%wt by chemical analysis corresponds to 2.515% bright areas in the thresholded image as typically shown in Figure 5b. Other thresholded images give 5.331%, 8.297%, 10.640%, 13.402% bright areas and correspond to 18.4941%wt, 27.8368%wt, 37.6574%wt, 43.224%wt respectively by chemical analysis.

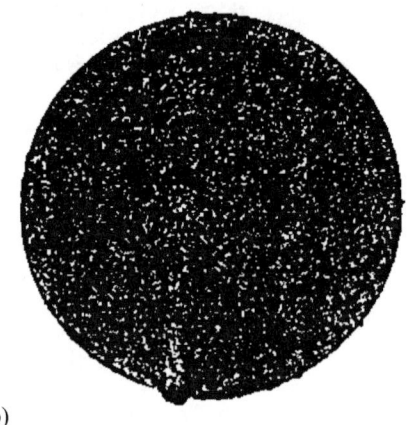

(b)

Figure 7: (a) Typical slice of matrix + 30%wt filler, (b) thresholded of image (a)   and typically yield 8.297% bright areas.

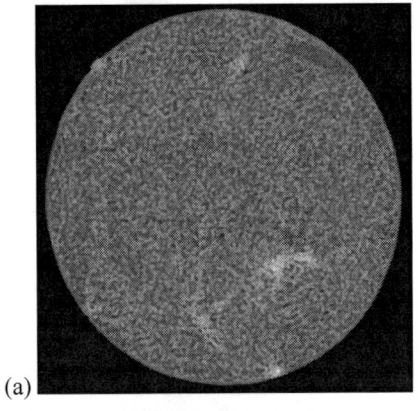

(a)

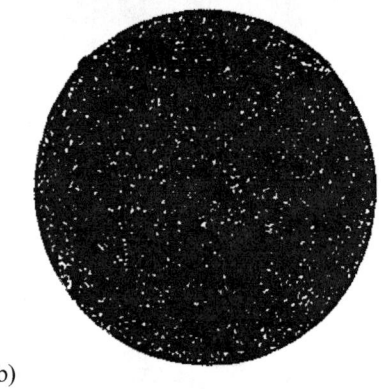

(b)

Figure 6: (a) Typical slice of matrix + 20%wt filler, (b) thresholded of image (a) and typically yield 5.331% bright areas.

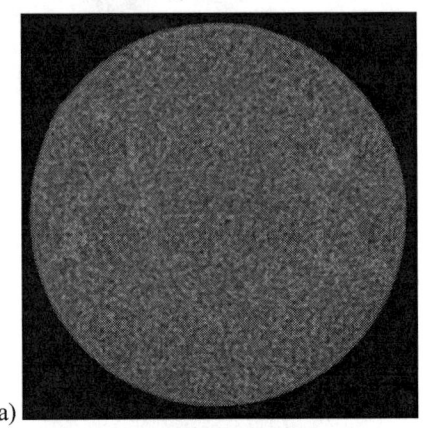

(a)

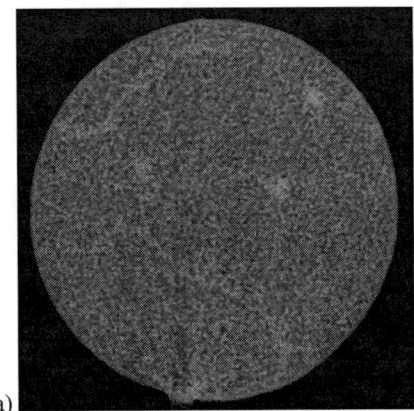

(a)

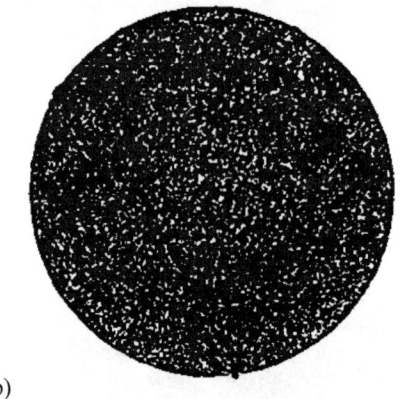

(b)

Figure 8: (a) Typical slice of matrix + 40%wt filler, (b) thresholded of image (a) and typically yield 10.640% bright areas.

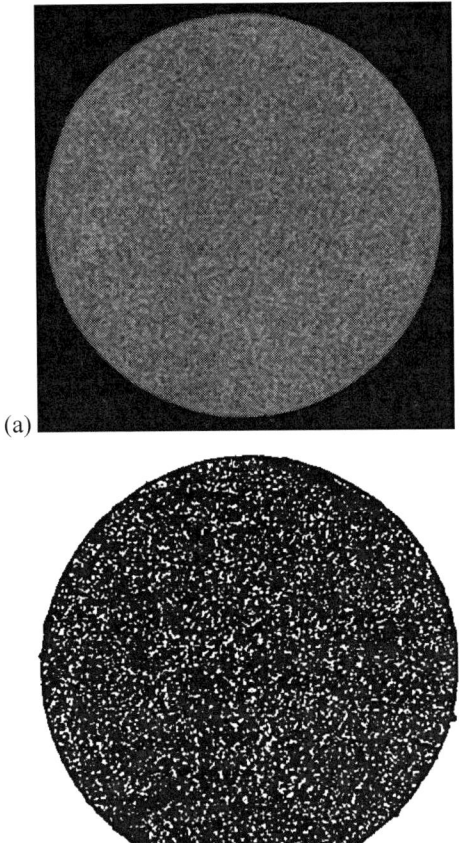

(a)

(b)

Figure 9: (a) Typical slice of matrix + 50%wt filler, (b) thresholded of image (a) and typically yield 13.402% bright areas.

A correlation factor is established to enable the filler in the sliced image being interpreted quantitatively. This is achieved by having an average value of area occupied by the filler in the sliced image. In this work, the sliced image is analysed from top to bottom of the VOI with the condition of in between 2 analysed image, there would be 9 sliced images that are abandoned from analysis to avoid misleading interpretation. The misleading is from the filler that appeared in more than 1 slice. Thus, the percentage of area occupied by the filler is compared against the result of chemical analysis. Both results are correlated by:

$$Correlation\ factor = \frac{Avarage\ vol.\ percentage\ of\ fillers\ by\ chemical\ analysis}{Avarage\ area\ percentage\ of\ fillers\ in\ the\ sliced\ image} \tag{1}$$

The existence of the correlation factor is clearly observed from the graph of average volume percentage of fillers from chemical analysis against average area percentage of fillers in the sliced image. This is shown in Figure 10.

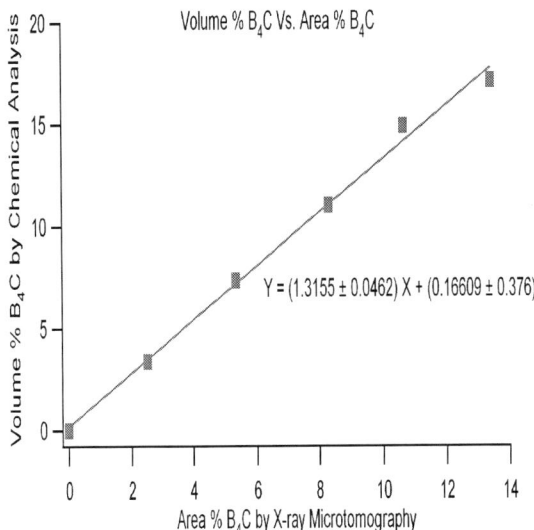

Figure 10: Volume percentage of fillers from chemical analysis against average area percentage of fillers in the sliced image

Based on Figure 10, the correlation factor is significantly the slope of the resulted graph. Based on the graph, the correlation factor was found to be approximately 1.3155. Thus, equation (1) reduced to:

$$Vol.\ \%\ of\ fillers = Avarage\ area\ percentage\ of\ fillers \times 1.3155 \tag{2}$$

Equation (2) could be used to estimate the volume percentage of B_4C for any given sample that was subjected to X-ray microtomography inspection providing any further work done using the same type of system.

Density measurements lead to the determination of amount of B_4C in the sample as the density of each of the ingredient is known. Average densities of small size cut sections that contain VOI are obtained for each slab and these results are compared between the slabs. This comparison is shown in Figure 11.

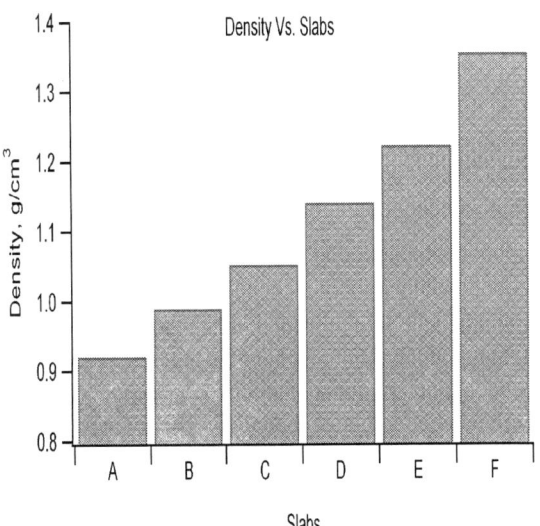

Figure 11: Avarage density measurement for each slab

From Figure 11, slab A shows the lowest value of density with 0.920g/cm³ due to the absent of filler. As the filler content is increased from slab B to F, the density also increased accordingly. The maximum achievable density is for slab F with the value of 1.354 g/cm³.

Results of chemical analysis, corrected X-ray microtomography, Archimedes method and X-ray microtomography are plotted in Figure 12.

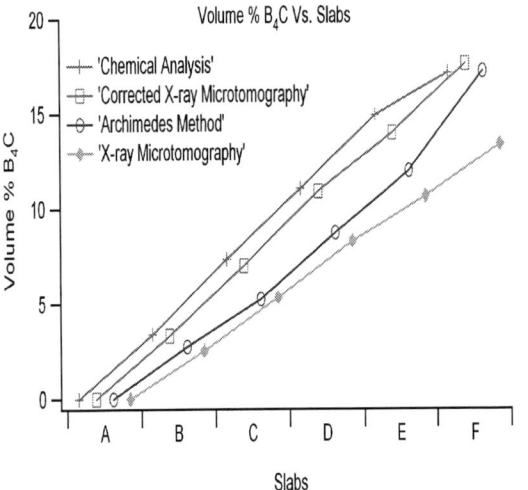

Figure 12: Comparison of volume of B_4C for each slab by 4 different methods.

Overall volume percentage of B_4C is given by 2 sets of values from chemical analysis and Archimedes method. Chemical analysis gives an accurate amount vol.% B_4C but Archimedes method suffer discrepancies if samples contain porosity or inclusions. This is shown in Figure 12 for slab C, D and E where the vol.% B_4C value obtained from density result is significantly lower than that obtained from chemical analysis. This is due to changes in density not only being affected by fillers but also by porosity or inclusions. The presence of porosity in VOI is typically shown in Figure 13 as detected by microtomography in the case of slab E.

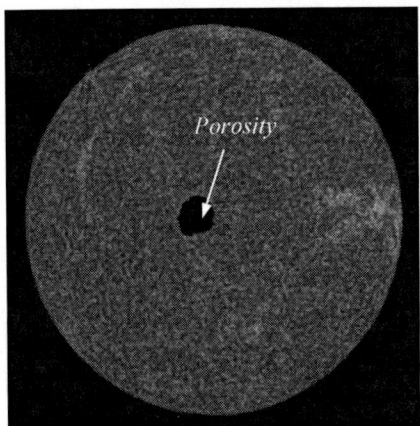

Figure 13: Present of porosity in VOI from slab E.

Although chemical analysis or specifically TGA is good in term of result accuracy and not affected by porosity, it is still considered as a time consuming analysis and destructive method. However, this technique could be used as a reference with the microtomography result in order for us to establish the correlation factor. This is true providing the relationship between the vol.% B_4C by chemical analysis and area% B_4C by microtomography is valid. As observed from Figure 12, microtomography results itself does not show the actual filler concentration. Therefore, the introduction of correlation factor is a must and in this work, the value was found 1.315. Once this value multiplied by area% of fillers obtained by microtomography, this new value or known as corrected x-ray microtomography is found close or almost identical to the actual filler concentration by chemical analysis as shown in Figure 12. Thus, quantification of B_4C in TPNR composite is possible by the use of correlation factor or corrected X-ray microtomography method. This method provide a simple mean to find the volume percent of fillers present providing the result of area% of fillers by microtomography is available and the work done using the same type of system.

CONCLUSIONS

The work could be considered as an evidence of the applicability of the X-ray microtomography to quantitatively obtain the concentration of B_4C in the TPNR composites. Results of X-ray microtomography were successfully calibrated using thermogravimetry analysis or TGA. In this work, the correlation factor was found to be 1.315. This value once multiplied with the area percentage of fillers in the sliced image would yield the concentration of fillers or B_4C in the sample. This technique enables the X-ray microtomography results to reveal the composition of B_4C in the sample in a quantitative manner with the existing capability exhibited the location or distribution of the B_4C. Further research will be focused on to establish a correlation factor for different types of filler that are typically used in the polymer composites and to obtain a set of correlation factor data for use in X-ray microtomography analysis.

ACKNOWLEDGEMENTS

The authors wish to thank Hishamuddin Husain for proofreading the paper. The excellent support from MINT and UKM throughout the research is highly appreciated.

REFERENCES

[1] T. P. Selvin, J. Kuruvilla and T. Sabu, Mechanical properties of titanium dioxide-filled polystyrene microcomposites, *Mater Lett* 58 (2004), pp. 281-289.

[2] L. Averous, J. C. Quantin and A. Crespy, Determination of the microtexture of reinforced thermoplastics by imaging analysis, *Compos Sci Technol* 58 (1998), pp. 377-387.

[3] Laura M. McGrath et al, Investigation of the thermal, mechanical, and fracture properties of alumina-epoxy composites, *Polymer* 49 (2008), pp. 999-1014.

[4] R. W. Hamilton, M. F. Forster, R. J. Dashwood, P. D. Lee, Application of X-ray tomography to quantify the distribution of TiB2 particulate in aluminium, *Scripta Materialia* 46 (2002) pp. 25-29.

[5] Dongyang Wu et al, X-ray ultramicroscopy: A new method for observation and measurement of filler dispersion in thermoplastic composites, *Composite Science and Technology* 68 (2008), pp. 178-185.

[6] Mummery PM, Derby B, Anderson P, Davis GR, Elliott JC, *J. Microsc* 177 (1995), pp. 399-406.

[7] Stock SR, *Int. Mater Rev* 44 (1999), pp. 141-164.

[8] R. H. Bossi and G. E. Georgeson, Composite structure development decisions using X-ray CT measurements, *Mater Eval* 53 (1995), pp. 1198-1203.

[9] E. Cornelis, A. Kottarand and H. P. Degischer, X-ray computed tomography characterizing carbon fiber reinforced composites, 11th Eur Conf Compos Mater (2004).

[10] Green WH, Sincebaugh P, Evaluation of complex composite structures using advanced computed tomography (CT) imaging, In: 53rd meeting of the society for machinery failure prevention technology, Virginia Beach, VA; (1999), pp. 41-50.

[11] J. Martin –Herrero and Ch. Germain, Microstructure reconstruction of fibrous C/C composites from X-ray microtomography, *Carbon* 45 (2007), pp. 1242-1253.

[12] A. Ibrahim and M. Dahlan, Thermoplastic natural rubber blends, Prog. Polym. Sci. 23 (1998), pp. 665-706

[13] A. B. Chilton, J. K. Shultis, R. E. Faw, Principles of Radiation Shielding, Prentice-Hall, Englewood Cliffs, NJ, (1984).

[14] J. K. Shultis, R. E. Faw, Radiation Shielding, Prentice-Hall, Englewood Cliffs, NJ, (1996).

[15] B. T. Price, C. C. Horten, K. T. Spinnery, Radiation Shielding, Pergamon, Elmsford, NY, (1957).

.

Diffusion of Aluminum Into Steel Substrates By Means Of Hot Dip Aluminizing

[1]Hishamuddin Hj. Husain, [1]Abdul Razak Daud, and [2]Muhamad Daud

[1] School of Applied Physics, Faculty of Science & Technology,
Universiti Kebangsaan Malaysia, 43600 Bangi, Selangor D.E.
[2]Malaysian Nuclear Agency, Bangi, 43000 Kajang, Selangor D.E.
Email <hishamuddin@nuclearmalaysia.gov.my>

ABSTRACT

Surface coating is an efficient and economical method to obtain desirable material surfaces properties. As compared to other coating techniques, hot dip coating can be considered as the most economical way to protect steel surfaces. Hot dip aluminizing technique was investigated in this study. Experiments have been conducted on the mild steel substrates with 10mm diameter. The substrates were dipped into the molten aluminum maintained at temperature 750°C for 5,10,15,20, 25 and 30 minutes. Optical microscopy, scanning electron microscope and energy dispersive X-ray spectroscopy were used in this investigation. From the microstructure observation, it showed the appearance of intermetallic layer covered by the top layer of Al on the mild steel substrate increased with the increase in dipping time. The result of EDX analysis revealed the existence of Fe and Al in form of new Al- Fe phase. This indicated the possible formation of the intermetallic layers.

Keywords: surface coating, hot dip aluminizing and intermetallic layer

INTRODUCTION

Even though new advanced materials have been developed nowadays, steel remains a major material in construction, automobiles, appliances, industrial machinery as well as in the nuclear industry. Due to steel easily corroded, a proper surface protection is required to avoid any failures and extended the life cycle of the components. Surface coating is an efficient and economical method to obtain desirable material surfaces properties. In hot dip aluminizing, when steel is in contact with a molten aluminum maintained at affixed temperature for a certain period, a reaction occurs to form a brittle interlayer of intermetallic compounds. The intermetallic layer develops between the steel substrate and the coated aluminum. The intermetallic layer grows and dissolves concurrently into the molten aluminum, which is directly associated with the loss of the steel substrate. The growth and dissolution rates of the intermetallic layer determine the thickness of the layer. The rates are closely related with the temperature of the molten aluminum and dipping time of the steel. The thickness of the layer also varies depending on chemical composition of the molten. Aluminum is very successful as a protective coating for steel because in most environments to which steel will be subjected, aluminum will act as the anode; that is it will dissolve in preference to the steel. However, the adhesion between coating and substrate by formation of the intermetallic phase is important to be investigated to produce strong and durable coating.

EXPERIMENTAL

For substrates preparation, mild steel rods with 10mm diameter were used in the experiment. Machining process was performed to remove the oxidized surface and to acquire a smooth surface of the rods. Traces of machining were removed by grinding with SiC paper prior to cleaning by ultrasonic. The composition of substrate is determined by using arc spark spectrometer. The pure aluminum (99.9% from Aldrich Chemical Company Inc) melted in a graphite crucible at the temperature maintained to 750 degree Celsius. Then, the substrates were dipped into the molten aluminum for 5, 10, 15, 20 and 30 minutes. Then, the cross section of the coated samples was analyzed by using Optical microscope, Scanning electron microscope (SEM) and Energy dispersive X-ray (EDX).

RESULT AND DISCUSSION

Table 1 shows the chemical composition of steel substrate. The carbon content in the substrate is important to be determined as it will influence the diffusion rate of atoms that will determine the thickness of the intermetallic layer. Result shows that the substrate contains about 0.17 weight %. With the percentage of carbon lesser than 0.25%,

CP1202, Neutron and X-Ray Scattering in Advancing Materials Research: International Conference – 2009
edited by A. Saat, H. A. Kassim, M. H. H. Jumali, J. M. Saleh, M. R. Othman, A. Ibrahim, F. M. Idris, and M. H. A.-R. M. Ahmad
© 2009 American Institute of Physics 978-0-7354-0739-8/09/$25.00

Table1: Chemical composition of steel substrate

Fe	C	Si	Mn	P	S	Cr	Ni	Al	Cu	W	Pb	Bi
Remaining	0.17	0.19	0.53	0.04	0.02	0.04	0.03	0.02	0.03	0.07	0.02	0.05

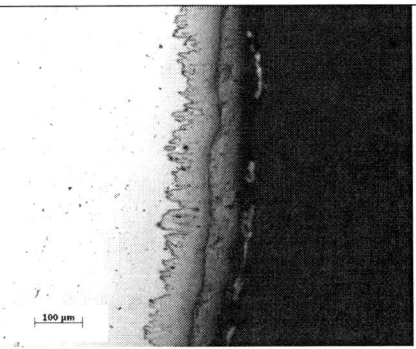

Figure 1a. Dip time: 5 minutes

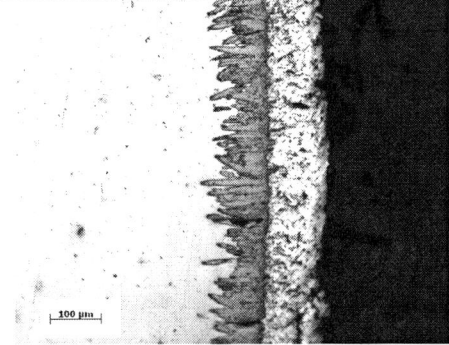

Figure 1b. Dip time: 10 minutes

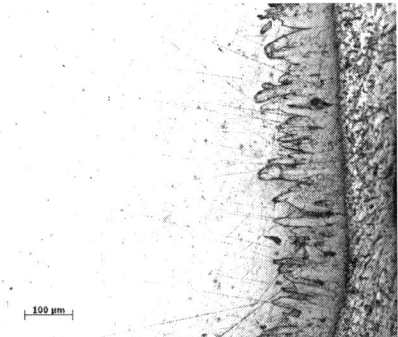

Figure 1c. Dip time: 15 minutes

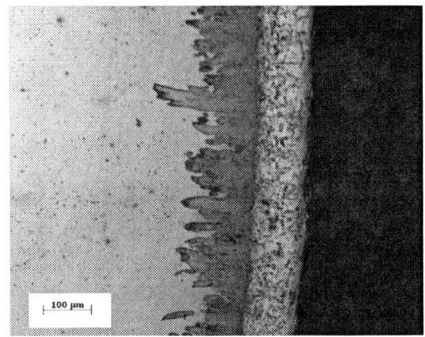

Figure 1d. Dip time: 20 minutes

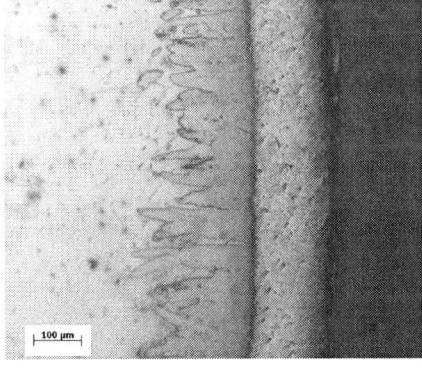

Figure 1e. Dip time: 25 minutes

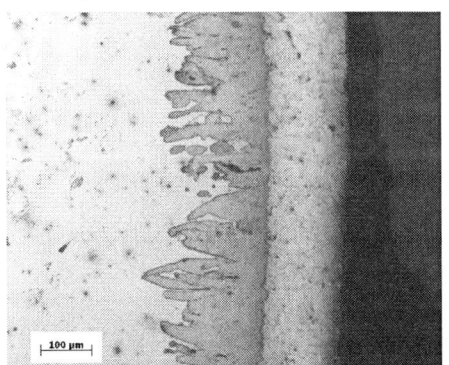

Figure 1f. Dip time: 30 minutes

Figure 1. Microstructure of aluminized samples at different times.

the substrate is categorized into low carbon steel group. It was reported that the increment of carbon atoms in the intermetallic layer lowering down the diffusion of Fe and Al. As the concentration of carbon increase, the fraction of cementites in the microstructure will increase and this will lowering down the diffusion rate of Fe and Al across the layer [Hwang et al., 2005]. The existance of Si in the composition also will affect the thickness of the intermetallic layer. High concentration of Si will retard the thickness of the intermetallic layer.

Figure 1 shows the microstructure of aluminized samples dipped at 750 °C for 5, 10, 15, 20, 25 and 30 minutes. Generally, coating layers formed at the surface of mild steel contains two different layers i.e the pure aluminum layer and the intermetallic layer. The intermetallic layers are in tounge-like shape measured between 80-200 μm.

The tounge-like shape intermetallic layer projected into the substrate with irregular projection length.

147

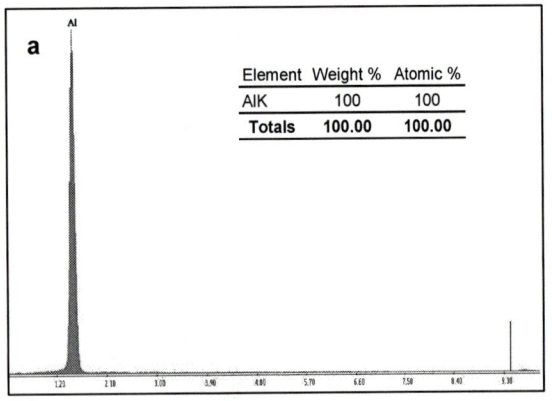

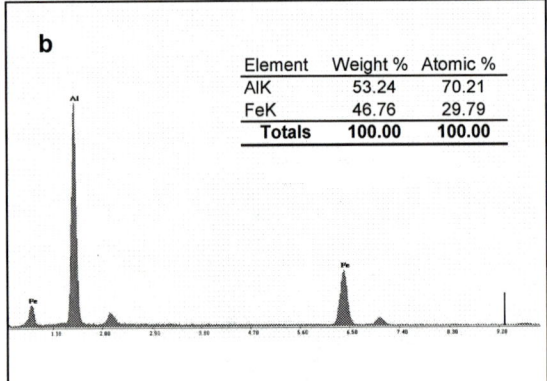

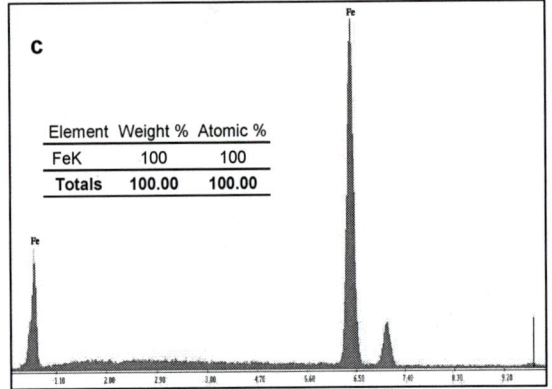

Figure 2 EDX analysis on the cross sectional area ; (a). Outer layer (b). Intermetallic layer (c). Substrate

The microstructure of the coating layers is the same even at different dipping temperatures but the morphology of the intermetallic layer is depending on temperature and dipping time [Deqing 2008]

The EDX analysis (Figure 2) shows the existence of the intermetallic phase, Fe2Al5, dominating the samples at all different dipping times. The formation of this phase in between the outer aluminum coat and the substrate is due to to its large lattice interstice favoring a rapid influx of aluminum atoms to the growth front of its crystals at high temperature. [Deqing 2008]. Although the growth of FeAl3 columnar grains was simultaneous with the growth of the Fe2Al5 layer, the coating layer developed into a single phase of Fe2Al5 due to the preferential growth of the Fe2Al5 layer [Kobayashi and Yakou 2002].

CONCLUSION

From this study it can be concluded that the tounge-like shape intermetallic layer was form in between the Aluminum coating layer and substrate. It is projected into the substrate with irregular projection length. The EDX analysis shows the existence of the intermetallic phase, Fe2Al5, dominating the samples at all different dipping times due to to its large lattice interstice favoring a rapid influx of aluminum atoms to the growth front of its crystals at high temperature.

REFERENCES

Deqing, W. 2008. Phase evolution of an aluminized steel by oxidation treatment.

Applied Surface Science 254 : 3026–3032

Hwang, S.H., Song, J.H. & Kim, Y.K. 2005. Effects of carbon content of carbon steel on its dissolution into a molten aluminum alloy. Materials Science and Engineering A 390 : 437–443

Kobayashi, S & Yakou, T. 2002. Control of intermetallic compound layers at interface between steel and aluminum by diffusion-treatment. Materials Science and Engineering A 338 : 44-53

Su, C.W. Lee, J.W. Wang, J.S. Chao, C.G. Liu, T.F. 2008. The effect of hot-dipped aluminum coatings on Fe-8Al-30Mn-0.8C alloy. Surface & Coatings Technology 202 (2008) 1847–1852

Microstructural Investigations On Ni-Ta-Al Ternary Alloys

M. NEGACHE [a,b], K.TAIBI [a], N. SOUAMI [c], Z. LOUNIS[a]

[a]; Laboratoire de Science et Génie des Matériaux Université of Sciences and Technologies Houari Boumediene, FGMGP, BP32 El Alia Bab Ezzouar 16111 Algiers
[b,] Department of metallurgy, Nuclear research center of Algiers, BP 43 Sebala/Draria , Algeria
[c] Departement of spetroscopie , Nuclear research center of Algiers; 2Bd Frantz Fanon BP399 Algiers, Algeria
mnegache1@yahoo.fr / kameltaibi@ yahoo.fr / nsouami@yahoo.fr / zlounis@yahoo.com

Corresponding author mail: kameltaibi@yahoo.fr

Abstract

The Ni-Al-Ta ternary alloys in the Ni-rich part present complex microstructures. They are composed of multiple phases that are formed according to the nominal composition of the alloy, primary Ni(γ), Ni$_3$Al(γ'), Ni$_6$AlTa(τ_3), Ni$_3$Ta(δ) or in equilibrium: two solid phases (γ'-τ_3), (τ_3-δ),(τ_3-γ), (γ-δ) or three solid phases (γ'-τ_3-δ). The nature and the volume fraction of these phases give these alloys very interesting properties at high temperature, and this makes them attractive for specific applications. We have developed a series of ternary alloys in electric arc furnace, determining their solidification sequences using Differential Thermal Analysis (DTA), characterized by SEM-EDS, X-ray diffraction and by a microhardness tests. The follow-up results made it possible to make a correlation between the nature of the formed phases and their solidifying way into the Ni$_{75}$Al$_x$Ta$_y$ (x+y=25at.%) system, which are varied and complex. In addition to the solid solution Ni (γ), the formed intermetallics compounds (γ', τ_3 and δ) has been identified and correlated with a complex balance between phases.We noticed that the hardness increases with the tantalum which has a hardening effect and though the compound Ni$_3$Ta (δ) is the hardest. The below results provide a better understanding of the complex microstructure of these alloys.

Keywords: A. multiphase intermetallics; B. phase diagram, phase identification; D. microstructure; F. diffraction, electron microscopy scanning

INTRODUCTION

The nickel-based alloys are used in various fields such as the marine industry, the nuclear industry, the aerospace and aviation for their good strength to creep and thermal fatigue at high temperatures [1-13]. This metallic alloy family is characterized by a relatively low density and good mechanical properties. Distinctively, these alloys present, by precipitation and in the Ni-rich corner, hardening and ordered phases (γ') of Ni$_3$X type (X: Al, Ti, Ta, Nb....etc) within the matrix Ni(γ) that can lead to an increase in their yield strength giving them excellent mechanical properties at high temperatures (650-1150°C) [1-3].

The existing hardening phases, coherent and ordered in the matrix, serve as barriers to the dislocations movement by enabling them to expand their field of use in order to reach temperatures up to 0.8 Tf (fusion temperature), during 100,000 operating hours. The volume fraction as well as the morphology and distribution of these phases in the whole matrix of the piece are directly linked to their thermal history and cooling mode which allow a better understanding of the relationship between microstructure and mechanical properties of these materials [1].

Most of the nickel superalloys have a very complex composition. In order to satisfy some requirements, few elements are simultaneously added to understand the specific role of each element. Some of these elements like Cr, Mo,W support austenitic structures that increase both tensile and thermal fatigue strengths at high temperatures. Other elements provide the enhancement and protection against oxydation and corrosion [14-15]. In addition, Ta, Nb and Ti are willingly added to increase the melting temperature of superalloys having compositions near to the eutectic [16]. Tantalum by its own substitution property, acts as a hardening element in the (γ') phase. It provides a greater thermal stability in this phase at high operating temperatures (760-1100°C) [3,8,10] and also improves the strength to corrosion[17]. However, an excess of this element could have adverse consequences created by certain phases very undesirable and destabilizing at the solidification such as A$_2$B type Laves phase [2].

CP1202, Neutron and X-Ray Scattering in Advancing Materials Research: International Conference – 2009
edited by A. Saat, H. A. Kassim, M. H. H. Jumali, J. M. Saleh, M. R. Othman, A. Ibrahim, F. M. Idris, and M. H. A.-R. M. Ahmad
© 2009 American Institute of Physics 978-0-7354-0739-8/09/$25.00

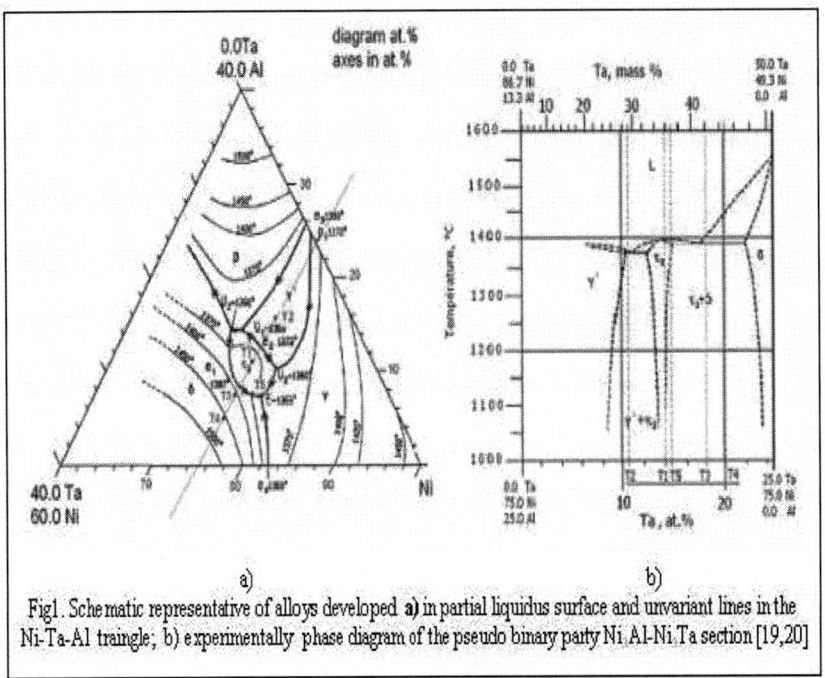

Figure 1. Schematic representative of elaborated alloys: a) in partial liquidus surface and unvariant lines in the Ni-Ta-Al traingle; b) experimentally phase diagram of the pseudobinary party Ni₃Al-Ni₃Ta section [19, 20]

According to the important role in the nickel based alloys ot these elements (Ta, Al, Ti, Nb), we started a study on some ternary alloys, in particular alloys from the pseudobinary Ni₃Al(γ')-Ni₃Ta (δ), and the ternary eutectic (Ni-Al-Ta), which generate the majority of phases of the ternary system. These alloys bring together variety of different phases with various structures that are characterized and thus the path of solidification in the ternary system model Ni-Ta-Al rich in nickel is then determined.

Obtained phases appear at high temperatures, at a range of temperatures between 1440 °C and 1323°C. In this restricted temperature gap (117°C), we distinguish the successive formation of several phases that solidify gradually as the temperature drops. The liquid becomes impoverished in element at high melting point (tantalum), thereby, providing the formation of primary phases. The Ni-Ta-Al liquidus projection in the Ni-rich corner (Figure .1) is characterized by the existence of five areas of primary (β, τ₃, δ, γ and γ'), eight equilibrium lines with three phases giving place to eutectic and peritectic binary transformations, and two invariant ternary points corresponding to reaction of the peritectic type (π) and ternary eutectic (E) in which four phases (δ/τ₃/γ/L) coexist [18-21]. The nature and the number of phases in each of these elaborated alloys mainly result on the chemical composition as well as on thermodynamic equilibrium conditions during the solidification.

The identification of the thermal history of each solidification alloy requires a projection of the alloy composition to be studied on liquidus projection ternary diagram of Ni-Al-Ta, Figure ure (1), followed by the solidification path until achieving of mono variants lines and a convergence of the solidification towards eutectic ternary point E, which is the lowest temperature in the system. It should be noted that the peritectic and the quasi peritectic reactions observed in the ternary diagram should disappear during the solidification, because of their thermodynamics transient nature. In addition, studies carried out by some writers (Durand Charre [3-4] and by P. Willemin [18]) showed that the presence of observed phase Ni₆AlTa (τ₃, denoted by Zhou.S [21]) leads to a significant rise of the partial liquidus projection of Ni₃Al (γ) and that the peritectic balance at low levels of tantalum become eutectic near the area τ₃ [3]. Thus, the existence of a wide area of the phase τ₃ prevents the formation of the ternary eutectic γ + γ '+ δ [19-21].

MATERIAL AND METHODS

The alloys were elaborated from high purity (> 99.995%) materials in an arc furnace under argon flow. The elaborated alloys were melted four times to improve their homogeneity and developed heat-treated homogenization at 1250°C for 24 hours in a vacuum furnace to avoid problems oxidation or contamination during the heat treatment [4, 11] .

Table 2: Transformation temperatures and natures of the phases obtained after DTA test.

Alloys	Nominal composition (at. %)			Transformations températures (°C)					Formed phases
	Ni	Ta	Al	Liquidus	TA	TB	TC	TD	
T1	75,00	14,00	11,00	1445	1382	1371	-	-	τ_3, δ
T2	75,00	10,50	14,50	1440	1400	1382	1373	-	τ_3, δ, γ'
T3	75,00	18,00	7,00	1445	1440	1414	1387	1373	δ, τ_3
T4	75,00	78,00	14,50	1440	1436	1405	1390	1373	τ_3, δ, γ'
T5	7,500	20,00	5,00	1405	1377	1373	-	-	δ, τ_3

Table 1: Chemical composition of studied alloys (in atom %)

Alloys	T1	T2	T3	T4	T5
Elements	Ni - Ta - Al	Ni - Ta - Al	Ni - Ta - Al	Ni - Ta - Al	Ni - Ta - Al
Composition (at. %)	75.05-13.6 -11.35	74.88-12.03-13.09	73.91-20.65-5.44	71.5 - 24.10 - 4.33	74.70 -5.17-7.97

Alloys noted T1, T2, T3, T4 and T5 are selected and displayed on the graph ternary Ni-Ta-Al and pseudo-binary Ni_3Al-Ni_3Ta of Figure (1). The composition of the T5 alloy is directly chosen from the point (E) of the ternary diagram. The chemical compositions of the four alloys noted from T1 to T4 have a constant tenure Ni (75 at. %). They put them in the pseudo binary Ni-Ta 25 at.% and Al-Ni 25 at.% sections with local balances Ni_3Al-Ni_6AlTa (γ'-τ_3), Ni_6AlTa-Ni_3Ta(τ_3-δ),(Figure .1b).

The variation of compositions Ta and Al has been studied by keeping the sum Ta (at.%)+ Al (at.%) = 25 (at.%). The results of chemical analyses obtained by EDX are given in atoms % in Table 1.

The Differential Thermal Analyses (D.T.A) tests were carried out in a SETARAM apparatus under argon atmosphere and with a rate of 5°C/min for heating and cooling. They permit to determine the transformations which proceed in alloys during cooling starting from the liquid at 1500 ± 2°C down to the ambient temperature. The transformations temperatures in the solid state were identified from cooling section. However, the solidus and the liquidus points were deducted from the heating.

The deducted points will be used to confirm the nature of the reactions and to determine the number of phases present in each area.

The observations of the microstructure were performed using a scanning electron mocroscope (SEM) type Jeol 6360 equipped with an analysing system. Quantitative chemical analyses of the largest compounds were obtained from the EDX. The contrast in the images is obtained from the backscatred electron sensitive atomic number (Z) of elements Ni, Al and Ta respectively equal to 28, 13 and 73. Apart from these, the phases in the these alloys were identified throught X-ray diffraction analysis using an X'PERT PRO Philip stype spectrometer using the Kα ray of Cu. The lattices of parameters obtained after spectrum diffraction analysis are measured and refined through a program called AFFMA. This software was developed in the laboratory, and based on the least squares method for the refinement of the lattices parameters.

The hardness and microhardness measurements of these phases were carried out using a WOLPERT and REICHERT hardness tersters a load of 98N (Hv10) to highlight the effect of hardening tantalum and measuring the hardness of different phases obtained after the solidification in differential thermal analyses tests.

RESULTS AND DISCUSSIONS

Solidification sequences and chemical composition of formed phases

The determination of the alloy solidification sequence of the alloys from the liquid state is linked to transformation temperatures of the different formed phases during solidification.

The thermogram figure (2), brings together the main points of transformations obtained during the solidification of ternary elaborated alloys. It presents a series of transformation peaks which explains the diversity of the stages and complexity of their solidification. The resulted peaks underline temperatures at the beginning of transformation and the obtained solidification interval.

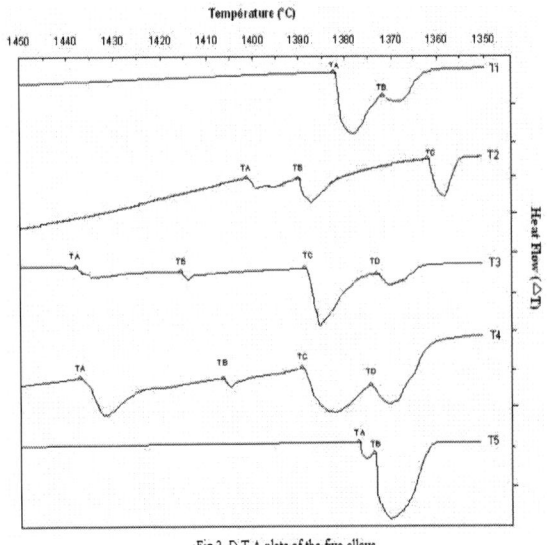

Fig 2. D.T.A plots of the five alloys

Figure 2. D.T.A plots of the five alloys

* γ : Ni, δ : Ni$_3$Ta, γ': Ni$_3$Al, τ_3 : Ni$_6$AlTa

Table 2 gives respectively, the composition of alloys, transformation temperatures and the nature of the phases obtained at the end of solidification. It shows that the alloys begin their solidification in the formation of primary phase's type: τ_3, δ, γ' or γ. Then, their solidification converges towards the eutectic ternary point E which is the lowest temperature in the system. The Beta compound of a high aluminium composition does not occur in the selected alloys.

Further on, we follow the path of solidification of each of elaborated alloys that we are correlating with the nature of formed phases.

The T1 alloy starts its solidification in the formation of the ternary compound τ_3 at 1445°C according to the reaction $L_1 \leftrightarrow \tau_3$ .The liquid carries on the way to solidification in the τ_3 fields by crystallizing it along its way and joins the mono variant line e1-E at T = 1382°C. At this point the entire liquid is transformed by a monovariant eutectic reaction giving place to two phases' τ_3 and δ before reaching the eutectic ternary point E. This eutectic reaction is well motioned by the second peak of thermo gram of D.T.A

The micrograph electronics of Figure3a illustrate the T1 alloy metallurgical structure obtained by D.T.A. It shows a biphasic structure consisting of the compound Ni$_6$AlTa (τ_3) and the eutectic Ni$_6$AlTa (τ_3)-Ni$_3$Ta (δ). This microstructure is in good conformity with the solidification path. Moreover, the T1 alloy reveals a

more important volume fraction phase τ_3 (70%) than of the phase δ (30%). That is in conformity with its double solidification, in primary, then in eutectic.

The T2 alloy emphasizes on a particularity during solidification. Indeed, this alloy begins to solidify from 1400°C by the formation of the first seeds of γ phase. The crystallization of γ' from the liquid state continues until reaching the mono variant line e2-U2 of liquidus projection of the Ni-Al-Ta system (Figure .1), where the liquid is in equilibrium with γ' and the $\gamma'+ \tau_3$ according to the reaction L2 $\leftrightarrow$ $\gamma'+ \tau_3$. Thus, the seeds of τ_3 phase start to appear and coalesce up to the U2 point where a ternary quasi-peritectic reaction $L_3 + \gamma' \leftrightarrow \tau_3 + \gamma$ with 1382 °C is then shown. Here it is noted the disappearance of the γ' phase for the benefit of the τ_3 and γ phases. Thus, the τ_3 phase becomes a major phase. The alloy ends its solidification at the eutectic ternary point E at T =1373 °C according to the following reaction: $L_4 \leftrightarrow \gamma + \delta + \tau_3$

The description of the SEM-EDX and the XRD analysis performed on this alloy have confirmed the existence of γ' phase at the end of solidification (Figure 3b, 3c and Table 3). The presence of this phase in the alloy at the end of solidification has led us to the confirmation that quasi-peritectic transformation that should undergo the compounds at the U2 point is incomplete. This is due to the fact that this point is very close to the eutectic point invariant E. The implication is that the decomposition of the primary phase γ' for the benefit of phases γ and τ_3 was only partial owing to the very short time of solidification and the lower temperature gradient between the points U2 and E. This is in conformity with the studies of Zakharov. Anatoli [20].

Alloys T3 and T4 are positioning themselves according to their nominal content (Table 1) in the primary liquidus projection corresponding to δ compound near the existing field of τ_3 ternary compound (Figure .1). These alloys begin their solidification by the formation of the first seeds of the primary δ phase at temperatures 1445 and 1440 °C, respectively. The morphology of the T3 alloy is shown in Figure 3d. Similar thing is observed in the T4 alloy as well (see Figure .3e). The solidification path of these alloys is pursuing and the formed crystals keep on growing following a curve until they cross the mono variant line e1-E to 1390 °C. From this point, the first seeds of τ_3 phase begin to appear and go on coalescing following the e1-E line based on: $L_2 \leftrightarrow \tau_3 + \delta$ reaction. The residual liquid evolves as the temperature decreases to point E. Arriving at this invariant point; the liquid crystallizes by showing a ternary eutectic constitute of τ_3, γ and δ at 1373 °C, and achieves its solidification.

The development of the T5 alloys against a chemical composition the closest to the eutectic ternary point E. The latter reveals the presence of the Ni$_3$Ta (δ) compound as being the primary phase solidification formed at 1405°C surrounded by two kinds of eutectics (Figure 3f). The first is biphasic: Ni$_3$Ta(δ)-Ni$_6$AlTa (τ_3)

formed at 1377°C and the second is majority and is formed according to the reaction of eutectic ternary point E at 1373°C.

Table3, gives the chemical composition ranges in at.% of Ni, Ta and Al elements determined by SEM-EDX for each primary phase (δ, γ' and τ₃) and give evidence that the solubility of aluminium in the phase (δ) is very small in the same way for Tantale in γ' phase. On the contrary, the γ' and τ₃ phases present considerable ranges of solubility of both aluminium and tantalum. These results are in agreement with those published by Willemin. P, Dugué. O and al [18].

Table 3. Chemical composition Ranges (in atom %.) phases formed in Ni -Al-Ta alloys rich in nickel.

| Phases | Elements | | |
	Ni	Ta	Al
Ni₃Ta (δ)	71.54 -	24.20 -	1.86 -
Ni₃Al (γ')	73.82	25.21	3.32
Ni₆AlTa (τ₃)	65.20 -	4.63	28.05 -
	66.53	- 6.12	30.06
	74.08 -	11.53 ·	8.98 ·
		14.33	13.73

Table 4: Calculated values of the lattice parameters of the main phases existing within the T1, T2 and T4 alloys

| Phases | The lattice parameters calculated (A°) | | |
	T1 alloy	T2 alloy	T4 alloy
Ni₃Ta (Tétr) δ	Not detected	a= 3.6188 ; c= 7.3938	a= 3.6122 ; c= 7.4060
Ni₃Ta (Orth) δ	a= 5.0683; b= 4.2083; c= 4.5391	a= 5.0879; b=4.1941; c=4.5692	a=5.0786; b=4.2530; c=4. 5163
Ni₃Ta (Mon) δ	Not detected	a= 5.1198; b= 4.523 c= 25.2184	a= 5.0903; b= 4.4938; c=25.3093
Ni₆AlTa (Hex) τ₃	a= 5.0872; c= 8.2094	a= 5.2406; c= 8.21031	a= 5.1220; c= 8.3289
Ni₃Al (cub) γ'	Not detected	a= 3.6012	a= 3.5684
Ni (cf c) γ	a= 3.5392	a= 3.5210	a= 3.5375

Crystallographic study of the phases

The analysis results from X-ray diffraction of the samples T1, T2 and T4 have allowed identifying the structural phases observed by SEM-EDX and affirming the suggested solidification path for each alloy cooled in D.T.A.

The diffraction spectrum of the ternary alloy T1 (Figure .4) reveals the presence of the hexagonal phase Ni₆AlTa (τ₃) which is DO₂₄ [21] and orthorhombic structure Ni₃Ta (δ) as well as the absence of characteristic lines of both Ni (γ) phase and the tetragonal Ni₈Ta phase. The absence of Ni(γ) and Ni₈Ta phases confirms the results obtained previously (Table 1and 2) by using other techniques that reveals only the existence of Ni₆AlTa and Ni₃Ta compounds at the end of solidification stage.

The diffraction spectra X-ray of samples T2 and T4 present a rich variety of phases and confirm that the solidification sequence of these alloys is achieved by a ternary eutectic reaction at point E. Indeed, the spectrum analysis reveals the presence of the hexagonal phase Ni₆AlTa, three allotropic varieties of the Ni₃Ta phase (tetragonal, Orthorhombic and monoclinic), the cubic phase Ni (γ) and finally the cubic phase Ni₃Al resulted from the residue of the quasi-peritectic T2 alloy reaction.

In table 4 are gathered lattice parameters of the main existing phases within alloys T1, T2 and T4. On one hand, it shows the existence of a variety of the crystalline structure in each sample and on the other hand, it confirms the presence of the Ni₃Al phase which was not detected by the D.T.A in the T2 alloy sample regarding its low volume fraction and its low of heat quantity released during the reaction. The calculated values of the lattice parameters coincide with those of A.S.T.M data and those published by other authors (Willimin. P and Dugué. O)[18], by Giessen. B.C [23] and Nash. P and West. D.R.F [21].

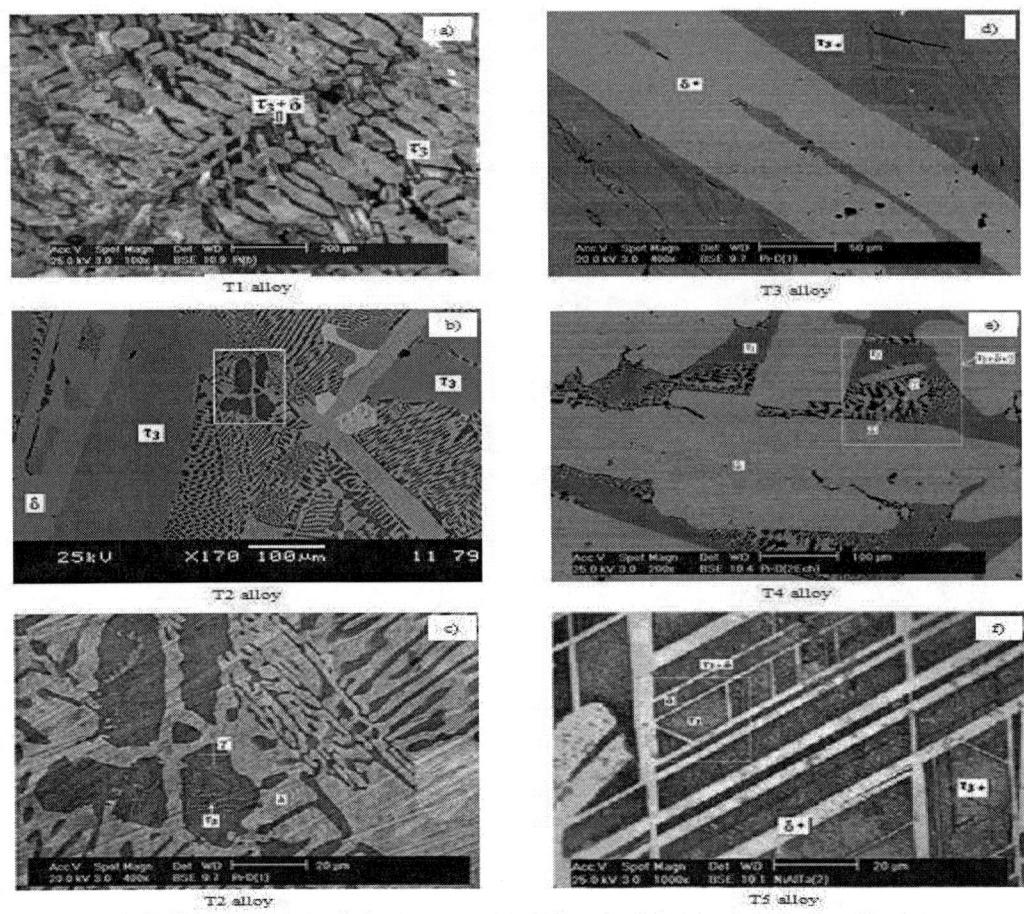

Fig. 3 SEM micrographs showing a mcrostructures of differents alloys mades up:
a) T1alloy: τ_3, δ ; b and c) T2 alloy: δ, γ', τ_3 ; d) T3 alloy: τ_3, δ ; e) T4alloy : δ, γ', τ_3 ; f) T5alloy : τ_3, δ

Figure . 3   SEM micrographs showing the microstructures of the alloys obtained after ATD test ; a) T1: τ_3, δ ; b and c) T2: δ, γ' and τ_3 ; d) T3 : τ_3, δ ; e) T4 : δ, γ' and τ_3 ; f) T5 : τ_3, δ

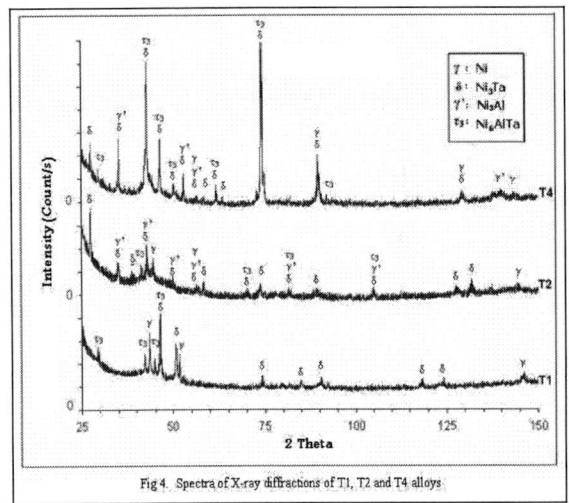

Fig 4. Spectra of X-ray diffractions of T1, T2 and T4 alloys

Figure 4. X-ray diffraction spectra of T1, T2 and T4 alloys

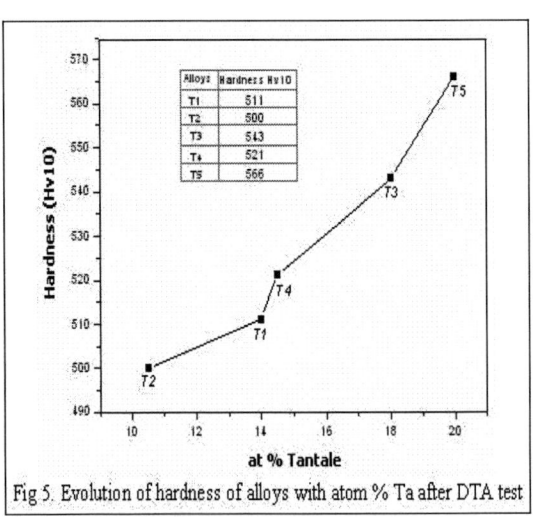

Fig 5. Evolution of hardness of alloys with atom % Ta after DTA test

Figure 5. Evolution of hardness of alloys with atom % Ta after DTA test

154

EVOLUTION OF HARDNESS

Figure 5 shows the evolution of alloy hardness provided from D.T.A tests. Hardness increases gradually with that of Tantalum concentration. The figure (8) illustrates this increase from 500 HV10 for T2 alloy which contains the lowest rate in Ta (10.5 at. %) to 566 HV10 for T4 alloy which contains the highest rate in Ta (20 at.%) i.e. an increase of 66 HV for 9.5 at.% of tantalum: corresponding to 10Hv / at. % . This increase reflects the hardening effect of this element by the formation of the (δ) phase within the alloy.

In addition, microhardness measurements made on the big stages herein show that the compound Ni_3Ta (δ) has a hardness greater than the one for compound Ni_6AlTa (τ_3) and also for the one of ($\tau_3+ \delta$) eutectic. They are about 650 Hv for Ni_3Ta (δ) of 430 Hv for Ni_6AlTa (τ_3) and finally 480Hv for ($\tau_3 + \delta$) eutectic. These values indicate that the phase Ni_3Ta (δ) is the hardest phase in these alloys for tis higt Ta (25% at) concentration.

CONCLUSION

We have approached in this work a correlation study between the ternary alloy solidification $Ni_{75}Al_xTa_y$ (x+y=25at.%) and the nature of the binary and ternary formed compounds. The experimental techniques used in the elaboration to the physicochemical and mechanical characteristic are diverse and complementary. The formed microstructures are complex with the phases: (γ), Ni_6AlTa (τ_3), Ni_3Ta (δ) and Ni_3Al (γ') which appears at very high temperatures, between 1440°C and 1323 °C as primary phase or bi-phase or three-phase eutectics. The solidification path of these alloys is explained in conformity with other studies.

We confirm the presence of three allotropic varieties of the Ni_3Ta (δ) phase (tetragonal, orthorhombic and monoclinic) phase. The obtained results on the solidification sequence and on the nature of the formed phases have been confirmed more by heating effect than by observation and physicochemical analysis. The hardness tests highlight a hardening effect of the tantalum element while the tests of micro hardness indicate that the Ni_3Ta (δ) phase constitutes the hardest structure in these alloys.

ACKNOWLEDGEMENTS

The authors would like to thank Mr S. Bouterfaia and F. Mernache of the nuclear research center of Draria and Mr A. Sari of the nuclear research center of Berine, Algeria for their help with the experiments (D.T.A and DRX tests) and for discussing the results.

REFERENCES

[1] Ati. A « Study in T.E.M of a two-phase superalloy γ/γ' deformed at high temperature: Crystal defects and parametric variation enters the phases », Thesis, USTHB / IGM, ALGERIA, 1993.

[2] Durand-Charre. M « Optimization of the microstructure superalloys for single-crystal turbine blades-Genesis and Morphology of the phases» conference: single-crystal superalloys. Toulouse Midi-Pyrénées, 22-24 March 1995. LTPCM- ENSEEG. National Polytechnic Institut of Grenoble. France

[3] M. Durand-Charre «the microstructure of Superalloys». Gordon and Breach Science Publishers. Amesterdam. 1997

[4]Ashby.M.F,Jones.D.R.H.«EngineeringMaterials1»2ndedition , Butterworth-Heineman.Oxford, 1996.

[5] Hagihara. K, Nakano. T, Umakoshi. Y. Acta Materialia 51, (2003) 2623-2637.

[6] Tsukada. Yuhki, Hasuike. Koichi, Murata. Yoshinori, Morinaga. Masahiko. Materials Science Forum Vols. 561-565 (2007) 2329-2332

[7] Seiji Miura, Yong-Myong Hong , Tomoo Suzuki and Yoshinao Mishima , Journal of phase Equilibria and diffusion Vol 22, N°3 (2001) 345-351

[8] Nunomura.Y, Kaneno.Y, Tsuda. H, Takasugi. T. intermetallics, 12, (2004) 389-399.

[9] Rusing. J, Wanderka. N, Czubayko. U, Naundorf. V, Mukherji. D, Rôsler. J. Scripta Materialia 46 (2002) 235-240

[10] Mclean.M «Directionally solidification materials for high temperature service»,The metals society-London 1983

[11] Bouterfaia. S.« Equilibrium phases liquid-solid in the multi made system up (Ni-Cr-Co-W)-Al-Ta superalloys model », Magister Thesis, C.E.N January 1986.

[12] Sins C.T, Chagel. W (eds). SuperalloysII «High Temperature Materials For Aerospace And Industrial Power», Wiley, New York, 1987.

[13] Kissinger.R.D, Deye.D.J, Anton.D.L, Cetel.A.D, Nathal.M.V, Pollok.T.M, Woodford. D.A. Superalloys, 1996, The Minerals, Metals And Materials Society, Warrendale,Pa, 1996.

[14] Djerdjar. B, Lebaili. S, Matériaux & Techniques 94, (2006) 1-12

[15] Cho.W.D, and Kim. Insoo, Metallurgical and Materials Transactions A Vol31. June (2000) 1685

[16] Baldan. A, West. D.R.F. J.Mater.Science, 16 (1981) 24-34

[17] Zhang. Wei, Arai. Katuyoshi, Qin. Chunling, Jia. Fei, Inone. Akihisa. Materials Science and Engineering A485 (2008) 690-694

[18] Willemin. P, Dugué. O, Durand-Charre. M, Davidson. J.H . Mat Scie and Tech, vol 2 (1986) 344-348

[19] Willemin. P, Dugué. O, Durand Charre. M, Davidson. J. H., MS/AIME Confr. Superalloys PA. USA (1984)367-647

[20] Zakharov. A. «TernaryAlloys: A compehensive compendium of Evaluated Constitutional Data and Phase Diagrams Ni-Al-Ta» Matport Books-Concept Series from MSIT Publisshed by VCH Verlagsgesellschaft, Weinheim, Dermany. ISBN: 3 -932120 -32 -9 . Vol 7. (1993) 483 -503

[21] Nash. P. and West. D.R.F. Metal Science- vol 17 (1983) pp 99

[22] Raghavan. V. Journal of Equilibria and Diffusion, vol.27, N°4 (2006) 405-407

[23] Giessen. B.C. and Grant. N.J. Acta Metallurgica, vol.15, (1967) 871-877.

Impurities analysis of polycrystalline silicon substrates: Neutronic Activation Analysis (NAA) and Secondary Ion Mass Spectrometry (SIMS)

A. Lounis* K. Lenouar*, Y. Gritly**, B. Abbad*M. Azzaz *, and K.Taïbi *

*Laboratory of Sciences and Material Engineering, USTHB, BP32.El Alia, Algiers, Algeria
** Unit of Silicon Technology Development, UDTS, El-Madania, Algiers, Algeria.
Corresponding author:zlounis@yahoo.com

Abstract

In this study we have determined the concentration of some impurities such as carbon, iron, copper, titanium, nickel of the flat product (polycrystalline silicon). These impurities generate a yield decrease in the photovoltaic components. The material (polycrystalline silicon) used in this work is manufactured by the Unit of Silicon Technology Development (UDTS Algiers, Algeria). The 80 kg ingot has been cutted into 16 briquettes in order to have plates (flat product) of 100 mm × 100 mm dimensions. Each briquette is divided into three parts top (T), middle (M) and bottom (B). For this purpose, the following instrumental analysis techniques have been employed: neutronic analysis (neutronic activation analysis) and secondary ion mass spectrometry (SIMS). Masses of 80 mg are sampled and form of discs 18 mm in diameter, then exposed to a flux of neutron of 2.10^{12} neutron cm^{-2} s^{-1} during 15 min. The energetic profile of incidental flux is constituted of fast neutrons ($\Phi_R = 3.10^{12}$ n.cm^{-2} s^{-1}; E = 2 Mev), thermal neutrons ($\Phi_{TH} = 10^{13}$ n.cm^{-2} s^{-1}; E = 0.025 ev) and epithermal neutrons ($\Phi_{epi} = 7.10^{11}$ n cm^{-2} s^{-1}; E > 4.9 ev), irradiation time 15 mn, after 20 mn of decrement, acquisitions of 300 s are carried out. The results are expressed by disintegration per second which does not exceed the 9000 Bq, 500 Bq and 2600 Bq, respectively for copper, titanium and nickel. It is observed that the impurities concentrations in the medium are higher. The impurities in the bottom of the ingots originate from the crucible. The impurities in the top originate from impurities dissolved in the liquid silicon, which have segregated to the top layer of the ingot and after solidification diffuse. Silicon corresponds to a mixture of three isotopes 28Si, 29Si and 30Si. These elements clearly appear on the mass spectrum (SIMS).The presence of iron and the one of nickel has been noticed.

Key Words: Multicrystalline silicon, , Photovoltaic. Impurities , NAA, SIMS, X-rays diffraction.

INTRODUCTION

In the field of photovoltaic solar energy, polycrystalline silicon is one of the most used materials. The high quality is justified by the required features for the final use of the product. It is considered that this quality is linked to the content of impurities within the metals. The impurities also reduce very largely the diffusion length of polycrystalline silicon. Carbon, oxygen, boron and transition metals are the most commonly observed in silicon because of their high mobility and solubility in polycrystalline silicon.These impurities generate a yield decrease of photovoltaic component 1-3. The material (polycrystalline silicon) used in this study is manufactured by the Unit of Silicon Technology Development (UDTS Algiers, Algeria). The aim of this work is to determine the concentration of some impurities such as carbon, iron, copper, titanium, nickel of the flat product (polycrystalline silicon). For this purpose, the following instrumental analysis techniques have been employed, namely, Neutronic Activation Analysis (NAA), Secondary Ion Mass Spectrometry (SIMS) and X-rays diffraction.

EXPERIMENTAL

The ingot has been cut into 16 briquettes in order to have plates (flat product) of 100 mm × 100 mm dimensions. The three briquettes, numbered 4, 3 and 6 representative of the ingot, will be the subject of present study.The first corresponds to the ingot corner, which has two faces in contact with the crucible. The second represents the ingot side, which has only one face in contact with the crucible. The third represents the ingot heart having no face in contact with the crucible. Each briquette is divided into three parts top (T), middle (M) and bottom (B) as indicated in Figure 1.

CP1202, Neutron and X-Ray Scattering in Advancing Materials Research: International Conference – 2009
edited by A. Saat, H. A. Kassim, M. H. H. Jumali, J. M. Saleh, M. R. Othman, A. Ibrahim, F. M. Idris, and M. H. A.-R. M. Ahmad
© 2009 American Institute of Physics 978-0-7354-0739-8/09/$25.00

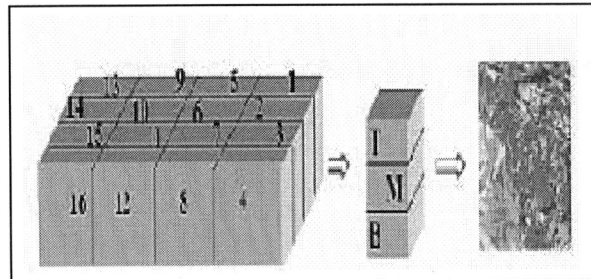

Figure 1: Selection tablets.

Regarding the neutronic analysis activation the irradiations have been realized in the thermal column of the nuclear reactor of Draria -Algiers.The samples are cleaned to eliminate the impurities. The lubricant traces and abrasive resulting from the cutting. Masses of 80 mg on average are sampled and form of discs 18 mm in diameter, then exposed to a flux of neutron of 2.10^{12} neutron $cm^{-2}.s^{-1}$.

The elements detection is accomplished by mass spectroscopy of secondary ion, the secondary ions are created by a very short pulsion of primary ions, bombarding the surface to be analyzed. For this characterization, the samples marked as 4H, 4M, 4B, 3H, 3M, 3B, 6H, 6M, 6B. To analyze the tablets by SIMS, it is indispensable to cut them into small pieces of 1 cm^2 (small port of samples). Before sawing the tablets, we immersed them into a resin bath. The samples are plunged in a hot trichloroethylene solution at 85 °C temperature during 5 min, then we rinse them successively in acetone solution in deionized water, finally dried with compressed air the surface to be analyzed. Other techniques such as the XRD (X-Ray Diffraction) and mapping has been applied to give more information for these specimen.

RESULTS AND DISCUSSIONS

The analyses achieved by ICP give us a variation coefficient (% RSD) included between 8 and 12 % which is unacceptable in atomic emission. This result reveals that considering the concentration level, it is hard to detect the impurities by this method. This is probably due to the reason that the used argon is not pure. The used standard does not respond to present matrix, consequently, the apparent concentrations and not actual values was found. However, the standard solutions stream is unknown suggesting that copper concentration into solution is less than 1 ppm. It was then decided to proceed by another strategy by two stages; first, detect the elements by mass spectroscopy of secondary ions (SIMS), then analyze the traces by neutron activation. The technique sensitivity (SIMS) seriously varies from one element to another and depends on the substrate and the effects of matrix.The mass spectrum gives the intensity (a number of the detected secondary ions) versus their indicated mass by the charge mass ratio (m/z). Silicon corresponds to a mixture of three isotopes 31 Si , 29 Si

and 30 Si, . These elements clearly appear on the spectrum (Figure 2).

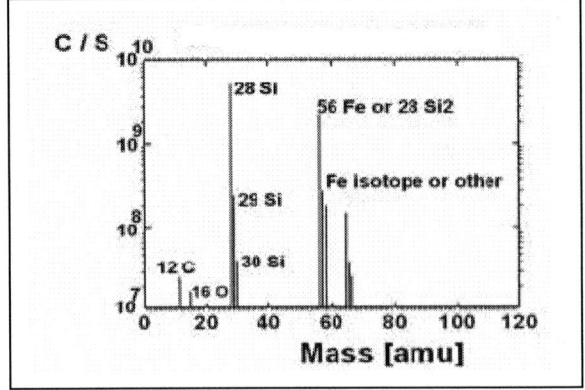

Figure 2: The mass spectrum

The presence of iron (atomic mass 55.8) and the one of nickel (atomic mass 58.76) has been noticed. Another possible mechanism for the enhanced carbon precipitation is that co-aggregation of carbon and oxygen, as observed by previous authors[26], give rise to larger precipitates, which can further result in a higher precipitation rate. Concerning the analysis by neutron activation, the choice of detection process depends on the nature of the transmitter and the complexity of the emitted spectrum.

The number of radioactive atoms N^* which accumulate in the sample during the irradiation tends towards a limit: at each instant the increase in the cores number N^* is equal to the difference between the speed of formation, considered as constant (the number of target cores N being very high) and the one of decomposition. Figures 3-5 represent the distribution of impurities in the ingot respectively for Ti, Cu and Ni

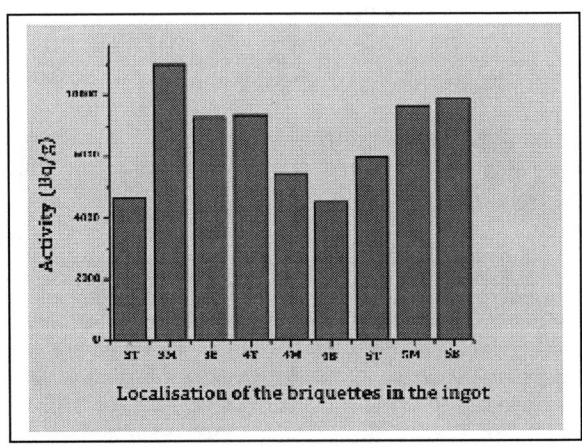

Figure 3: Distribution of titanium grades in ingot.

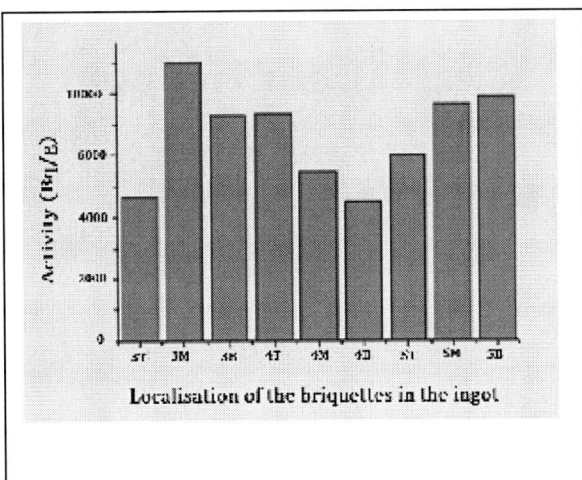

Figure 4: Distribution of copper grades in ingot.

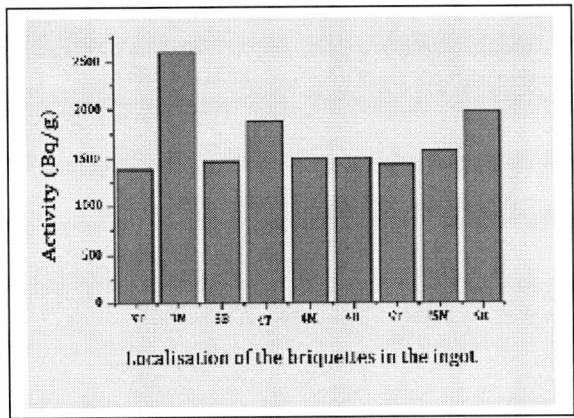

Figure 5: Distribution of nickel grades in ingot.

Table 1 : Impurities concentrations

Element	Atomic mass(g)	Concentration (at/cm^3)
Ti	47.90	$2.81 \ 10^{15}$ - $3.07 \ 10^{15}$
Cu	63.54	$1.17 \ 10^{17}$ - $2.32 \ 10^{17}$
Ni	58.76	$8.11 \ 10^{14}$ - $9.5 \ 10^{14}$

X-rays diffraction shows that we have only one phase of silicon (Figure 6).

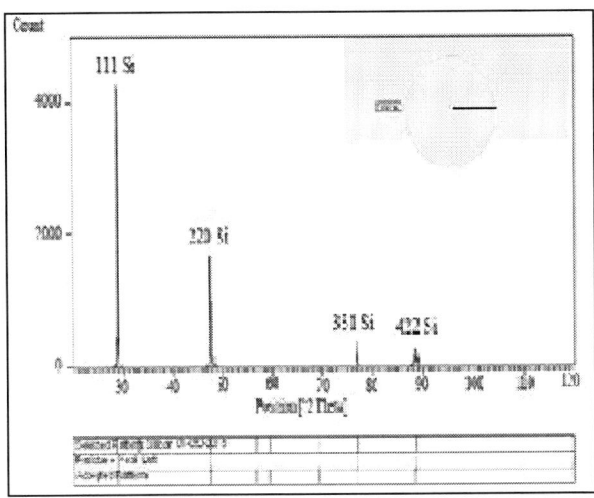

Figure 6. X-rays of silicon specimen.

We observe that the impurities concentrations in the medium are higher. The impurities in the bottom of the ingots originate from the crucible. The impurities in the top originate from impurities dissolved in the liquid silicon, which have segregated to the very top layer of the ingot and after solidification diffuse. These results practically agree with those given by Revel 25. For the analyzed elements, the concentrations found are included between two level values. This interval corresponds to the measured activity (expressed by disintegration per second) which does not exceed the 8980, 486 and 2590 Bq, respectively for copper, titanium and nickel .After calculations the concentrations are given in Table 1.

CONCLUSION

It is observed that the impurities concentrations in the medium are higher. The impurities in the bottom of the ingots originate from the crucible. The impurities in the top originate from impurities dissolved in the liquid silicon, which have segregated to the top layer of the ingot and after solidification diffuse. Silicon corresponds to a mixture of three isotopes 28Si, 29Si and 30Si. These elements clearly appear on the mass spectrum (SIMS).The presence of iron and the one of nickel has been noticed.

REFERENCES

1. C. Cabanel and J.Y. Laval, *J. Appl. Phys.*, **67**, 1425 (1990).

2. T.S. Fell, P.R. Wilshaw and M.D.D. Coteau, *Phys. Stat. Sol. (A)*, **138**, 695 (1993).

3. S.A. McHugo, *Appl. Phys. Lett.*, **71**, 1984 (1997).

4. G. Revel, N. Deschamps, C. Dardenne, J.L. Pastol, D. Hania and J.L. Nguyen Dinh, *Radional. Nucl. Chem. Lett.*, **85**, 137 (1984).

In-Situ Cold Temperature XRD of Calcium Phosphate Produced From Organic Phosphoric Acid

Meor Yusoff M. S., Wilfred Paulus and Masliana Muslimin

Materials Technology Group, Industrial Technology Division, Malaysian Nuclear Agency, 43000 Kajang, Selangor, Malaysia.

ABSTRACT

In this study, we synthesized calcium phosphate from an organic phosphoric acid, diethylhexyl phosphoric acid (DEHPA) and calcium hydroxide solution. The reaction involves a sol-gel process with a whitish gel formed. *In-situ* XRD analysis was then performed on the sample from room temperature to -140°C. At room the XRD diffractogram shows the sample as an amorphous material and as the temperature was further lowered sharp peaks begins to form indicating that the material had becomes crystalline. The peaks were identified to be that calcium hydrogen phosphate ($Ca(H_2PO_4)_2$) and this indicates that there is no hydroxyl group removal during the cooling process. The relative crystallinity values obtained for the different cooling temperatures show a slow exponential increase on the initial cooling of 0 to -100°C and at further cooling temperatures resulted fast and linear process. Also unlike the *in-situ* XRD analysis performs at high temperature no phase transformation occurred at this low temperature.

Key words: *in-situ* low temperature XRD, calcium hydrogen phosphate, DEHPA, relative crystallinity

INTRODUCTION

Calcium phosphate is one of the most common synthetic biomaterial. Hydroxyapatite (HA), a calcium phosphate compound with a chemical formula of $Ca_{10}(PO_4)_6(OH)_2$, beta tricalcium phosphate (β-TCP, $Ca_3(PO_4)_2$) and alpha tricalcium phosphate (α-TCP, $Ca_3(PO_4)_2$) are some of the common bioactive calcium phosphate materials (E. Caroline Victoria and F.D. Gnanam, 2002, A. Sinha et al., 2001). Synthesizing of these calcium phosphate compounds can be done by using the direct precipitation, solid state reaction or sol-gel methods. One of the advantages of using the sol-gel method in synthesizing calcium phosphate biomaterial is the uniform distribution of both calcium and phosphate. This can be seen from the homogeneity of the calcium phosphate product that is produced. An example of the sol-gel method is the used of triethyl phosphite with calcium nitrate or calcium ethoxide (N. Kivrak and A. Cuneyt Tas, 1998).

The transformation of the different calcium phosphate phases is much affected by the sintering temperature. Hyun-Seung Ryu et al (2005) and Kevor S. et al., 1999 have suggested the following phase transformation for calcium phosphate heated at high temperature:

Dried calcium phosphate
$$β\text{-}Ca(PO_3)_2 \dashrightarrow β\text{-TCP} + HA \dashrightarrow α\text{-TCP} + HA$$

Phase transformation from dried calcium phosphate to β-$Ca(PO_3)_2$ occurred at 450°C while when the sample was heated at 700°C it was further transformed into β-TCP and HA. Finally the stable α-TCP will be formed from the β-TCP at 900°C.

A process to produce calcium phosphate biomaterial using an organic based phosphoric acid (DEHPA) as its starting material was reported by our group earlier (Meor Yusoff et al., 2008). Following is the reported phase transformation reaction:

$$\text{Gel} \dashrightarrow β\text{-}Ca_2P_2O_7 \dashrightarrow β\text{-TCP} + HA \dashrightarrow α\text{-TCP} + HA$$

The phase transformation reaction obtained is different as compared to that obtained by Hyun-Seung Ryu et al, 2005 and Kevor S. et al., 1999 in that the initial product was β-$Ca_2P_2O_7$ instead of β-$Ca(PO_3)_2$. Another important property is the increase in the compressive strength of the calcium phosphate biomaterial when cooled below freezing temperature (Liam M. Grover, 2008).

In this paper we present the results on the use of an *in-situ* cold XRD analysis to study the effect of cooling on the synthesized calcium phosphate biomaterial. Besides determining the identity and quantity of the calcium phosphate compounds we are also interested to know the effect of relative crystallinity at the different cooling temperatures.

EXPERIMENTAL

CP1202, *Neutron and X-Ray Scattering in Advancing Materials Research: International Conference – 2009*
edited by A. Saat, H. A. Kassim, M. H. H. Jumali, J. M. Saleh, M. R. Othman, A. Ibrahim, F. M. Idris, and M. H. A.-R. M. Ahmad
© 2009 American Institute of Physics 978-0-7354-0739-8/09/$25.00

Solution of 20% (w/v) calcium hydroxide was prepared by mixing calcium hydroxide salt with distilled water. A solvent extraction process then proceeds with DEHPA used as the organic extracting agent with the above 4M HCl as the aqueous portion. The organic portion was then retained and this is mixed with the calcium solution where a brownish gel will eventually be formed. Further addition of calcium solution until the mixture reached pH10 resulted to the formation of white precipitate and this will then be collected through a vacuum filter. The whitish gel formed after drying overnight in an oven is as that shown in Fig. 1 below.

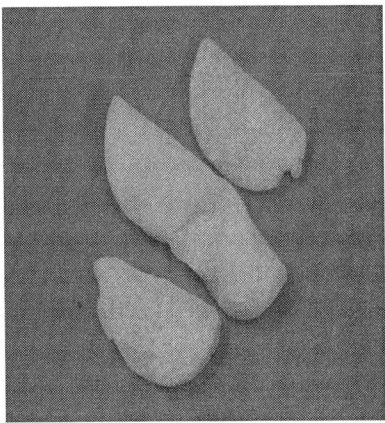

Fig.1: Calcium phosphate gel used in the study

The XRD instrument used for this in-situ study is Panalytical X'Pert PRO attached with Anton Parr high temperature attachment. Diffractograms are collected at room temperature, 0, -60, -100 and -140°C. High Score Plus software developed by Panalytical with XRD diffractogram database from Inorganic Crystal Structure Data (ICSD) was used to identify the phases present.. Relative crystallinity at the different cooling temperatures was obtained from the intensity of the most intense peak in the diffractogram.

RESULTS AND DISCUSSION

The used of an organic-based phosphoric acid as the starting material is a different calcium phosphate synthesis method compared to the commonly used precipitation or wet method. The hydrocarbon content in DEHPA tends to promote the formation of whitish gel in its early precipitation stage. This is similar to the sol-gel process where inorganic alkoxide or related hydrocarbon is used. The present of excess calcium in the gel will eventually solidified it to form white paste and upon drying a gel is formed. Fig.2 shows the XRD diffractogram of the calcium phosphate taken at room temperature. The sample seems to be in an amorphous phase. After this the same sample upon cooling to 0°C in the in-situ XRD was then analyzed again with its diffractogram showing several sharp peaks notable of which is at $2\theta = 29°$. We then proceed to cool the sample at a lower temperature of -60°C and the diffractogram obtained is almost similar to that obtained at 0°C.

The same sample was then undergoes further cooling

at -100°C and -140°C. Fig. 3 shows the diffractograms obtained at these temperatures. When the calcium phosphate sample was cooled to -100°C more sharp peaks are formed indicating that the amorphous sample had turned crystalline. As the temperature was further cooled to -140°C, the intensity of these peaks tend to increase another indication that the peak had become more crystalline than before.

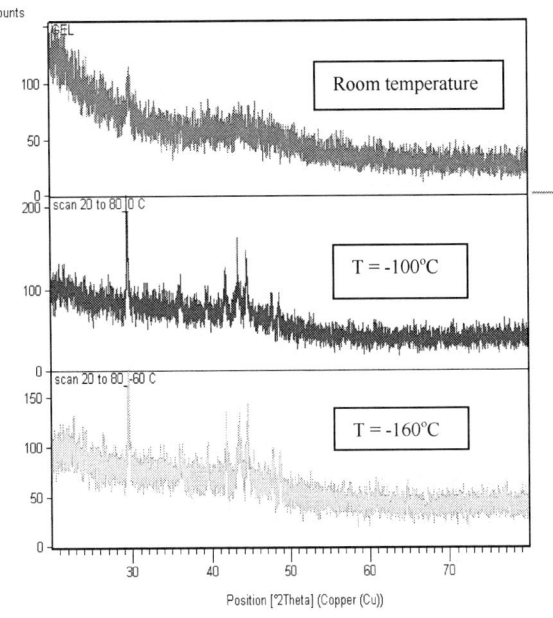

Fig. 2: XRD diffractograms of calcium phosphate sample at room temperature, 0°C and -60°C

Using the High Score Plus software with ICDD diffraction data library, we are able to identify the phase present as a single calcium hydrogen phosphate ($Ca(H_2PO_4)_2$) with reference number 00-003-0335. The present of this phase is in contrast to our earlier works (Meor Yusoff et al, 2008) on the in-situ hot XRD analysis where at the of heating of the same calcium phosphate gel will resulted to the formation of β-$Ca_2P_2O_7$ through the dehydroxylation process. In this cold stage the dehydroxylation process does not occur but what happens instead is the crystallization process.

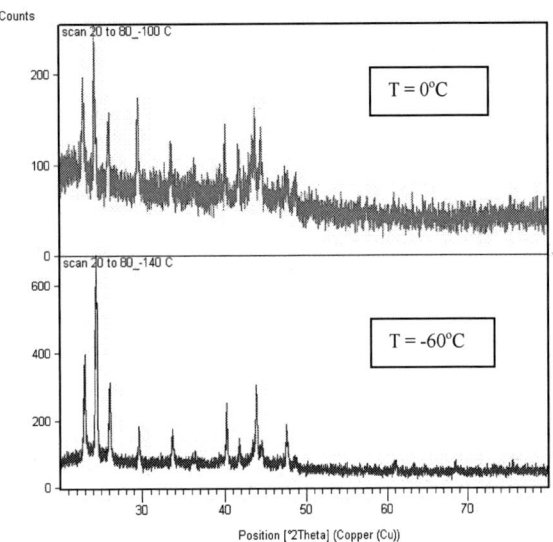

24.3% and at further cooling of -100°C the relative intensity was 35.7%. The relative intensity increments can be seen from the exponential increase of the graph as shown in Fig. 4. Further cooling of the same sample -100 to -140°C resulted to a sharp increase in the relative intensity where a linear graph was obtained.

CONCLUSION

From this study it can be concluded that cooling temperature could affect the relative crystallinity of the calcium phosphate gel. When the sample was cooled from 0 to -100°C there was a slow and exponential increase in the relative crystallinity of the sample. The sample when cooled at lower temperatures gives a fast and linear increase in its relative crystallinity. The in-situ XRD also enable to identify a single phase of calcium hydrogen phosphate through the crystallization process.

ACKNOWLEDGEMENT

The authors wish to thank the management of Nuclear Malaysia for their support on this project and also their permission in presenting this paper. We also like to thank Ms Julie Andrianny and Ms Norhaslinda Ee for carrying out the XRD analysis.

Fig. 3: XRD diffractograms of calcium phosphate sample at -100°C and -140°C

REFERENCES

E. Caroline Victoria and F.D. Gnanam, 2002, Synthesis and charaterisation of biphasic calcium phosphate, *Trends biomater. Artif. Organs.*, 16(1), pp. 12-14.

A. Sinha et al., 2001, Development of calcium phosphate based bioceramics, *Bull. Mater. Sci.*, 24(6), pp. 653-657

Hyun-Seung Ryu et al., 2005, Variations in structure and composition in magnessium incorporated hydroxyapatite/ β-tricalcium phosphate, *J. Mater. Res.*, 21(2), pp.428-436.

Liam M. Grover, Michael P. Hofmann, Uwe Gbureck, Balamurgan Kumarasami and Jake E. Barralet, 2008, Frozen delivery of brushite calcium phosphate cements, *Acta Biomaterialia*, Vol. 4, Issue 6, November 2008, pp. 1916-1923

Meor Yusoff Meor Sulaiman and Masliana Muslimin, 2008, *In-situ* High Temperature Analysis of Calcium Phosphate Biomaterial using DEHPA as the Starting Material, In Proceedings ICXRI 2008, Kota Kinabalu, Sabah.

N. Kivrak and A. Cuneyt Tas, 1998, Synthesis of calcium hydroxyappatite-tricalcium phosphate (HA-TCP) composite bioceramicpowders and their sintering behaviours, *J. Am. Ceram. Soc.*, 81(9), 2245.

X. Yang and Z. Wang, 1998, Synthesis of biphasic ceramics of hydroxyappatite and β-tricalcium phosphate with control phase content and porosity, *J. Mater. Chem.*, 8, 2233.

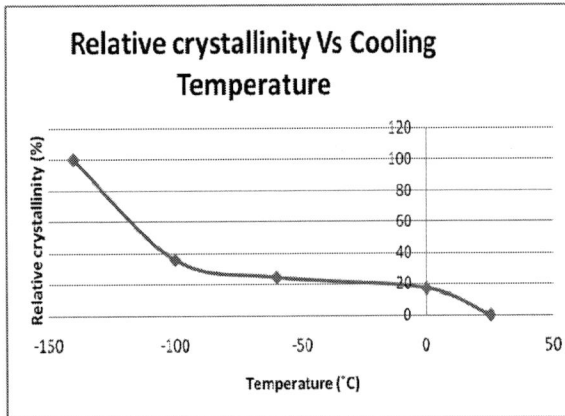

Fig. 4: Graph of relative crystallinity versus cooling temperatures

The relative crystallinity of the sample under the different cooling temperatures was measured from the intensity of the most intense peak at 2θ = 24.5°. Fig.1 shows the graph of relative crystallinity to the different cooling temperatures applied to the calcium phosphate gel. There was a slight increase in the crystallinity when the sample was cooled to temperature to 0°C registering a 17.1% relative crystallinity. When the sample was futher cooled to -60°C, the relative intensity increases to

Normalized Noise Power Spectrum of Full Field Digital Mammography System

Norriza Mohd Isa and Wan Muhamad Saridan Wan Hassan

Department of Physics, Faculty of Science, Universiti Teknologi Malaysia, 81310 UTM Skudai, Johor.
Malaysia.
email: *norriza@nuclearmalaysia.gov.my*

ABSTRACT

A method to measure noise power spectrum of a full field digital mammography system is presented. The effect of X-ray radiation dose, size and configuration of region of interest on normalized noise power spectrum (NNPS) was investigated. Flat field images were acquired using RQA-M2 beam quality technique (Mo/Mo anode-filter, 28 kV, 2 mm Al) with different clinical radiation doses. The images were cropped at about 4 cm from the edge of the breast wall and then divided into different size of non-overlapping or overlapping segments. NNPS was determined through detrending, 2-D fast Fourier transformation and normalization. Our measurement shows that high radiation dose gave lower NNPS at a specific beam quality.

Keywords: Normalized noise power spectrum, full field digital mammography, image quality, clinical radiation dose, beam quality.

INTRODUCTION

X-ray mammography is the most frequently used method for early detection of breast cancer. Sign of cancer in mammogram is difficult to detect because little difference of densities between normal tissue and cancer tissue of the breast (Goodsitt et. al., 1998). Other factors that lower the contrast resolution of mammographic images are noise from the x-ray system and x-ray beam. Noise of an imaging system is often characterized by noise power spectrum (NPS). NPS is one of the meaningful parameter and always be used to compare the noise properties of different systems (Monnin et. al., 2007).

Noise can also be characterized quantitatively based on variance of image data, σ^2 that is the sum of three components: (1) additive noise independent of exposure and mainly due to electronic noise σ_E^2, (2) photon noise which is proportional to exposure σ_P^2, and (3) bias fixed pattern proportional to square of exposure σ_B^2 (Ghetti et. al., 2008).

According to European guide line noise measurement should be performed upon acceptance testing of imaging system and after servicing of the detector or its subsystems (CEC, 2006). In practice the best method to determine NPS of digital imaging systems is still not clear. Its calculation is still complex and being addressed by number of bodies like American Association of Physicists in Medicine (AAPM), International Electrotechnical Commission (IEC) and others who have authority and expertise in this field (Dobbin et. al., 2006).

This paper describes NPS measurement work of full field digital mammography (FFDM) system at Putrajaya Hospital, Malaysia by the authors. The objectives of the study were to investigate the effect of plane fitting of the noise data (detrending) on 1D-NNPS of the FFDM, and to study the effect of segment size, sequential and random segmentations on the NNPS, with the view of choosing the best calculation method.

METHODS AND MATERIALS

EXPERIMENTAL SET-UP

The full field digital mammography system (FFDM) used in this study was Hologic Lorad Selenia M-113R (Hologic Inc., Bedford, MA 01730). The x-ray beam half-value layer (HVL) was 0.35 mm Al and was measured in the standard manner using collimated beam at 28 kVp and Mo/Mo target-filter combination (ACR, 1999).

For NPS measurement, uniform images of size 18 × 29 cm^2 were acquired at various level of incident exposure with air kerma ranging from 50 to 240 µGy using nominal focal spot of 0.3 mm. The beam quality was RQA-M2 adapted from IEC 61267-2 suited for hospital facilities (e.g. a Mo/Mo anode-filter combination, 28 kVp collimated x-ray with dedicated compression paddle, 2 mm of added aluminum filter affixed to the x-ray tube window and off anti-scatter grid used) (IEC, 2007). A 2 mm Al (99.0% purity, Keithley

CP1202, *Neutron and X-Ray Scattering in Advancing Materials Research: International Conference – 2009*
edited by A. Saat, H. A. Kassim, M. H. H. Jumali, J. M. Saleh, M. R. Othman, A. Ibrahim, F. M. Idris, and M. H. A.-R. M. Ahmad
© 2009 American Institute of Physics 978-0-7354-0739-8/09/$25.00

type 1100) added filter was placed in the beam in order to simulate the exit spectrum from breast tissue in mammography and simultaneously to minimize scattered radiations that might provide a possible noise source or artifact at low frequency in the images (Ranger et. al., 2005). HVL of the beam with the added filter was measured and its value was 0.57 mm Al. Figure 1 shows the schematic arrangement of the equipments for the measurement.

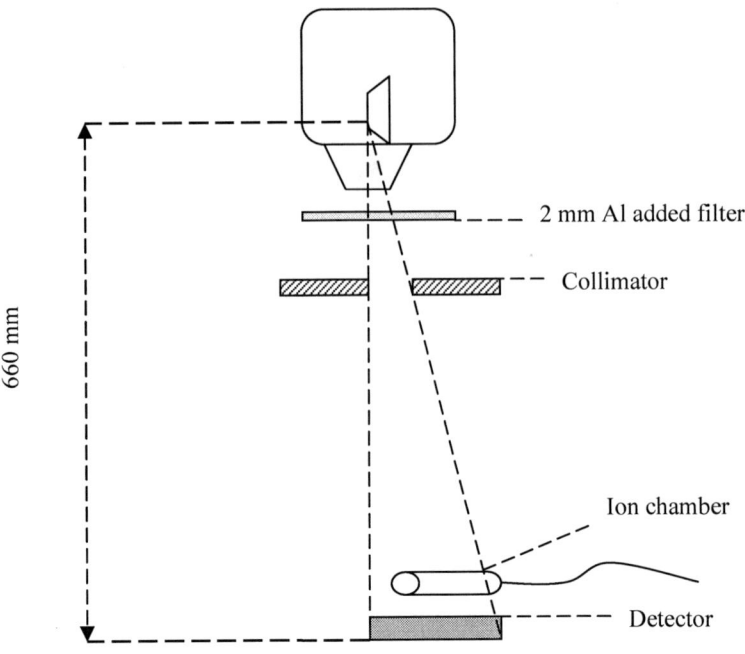

Fig. 1: Experimental set-up for image acquisition.

Three exposure levels (51.5, 143.6 and 238.4 μGy) were used in this work resulting in uniform exposure images from which the NPS were calculated. The exposure range was selected to be considerably below and above the typical digital mammographic detector exposure range which was 120 μGy in order to consider a complete characterization of the detector (Goodsitt et. al., 1998).

IMAGE ACQUISITION

A large area of the 18×24 cm^2 image was selected from each of the three exposures. For each image, different smaller areas were extracted at the centre of the image about 4 cm from the chest wall edge to avoid the low exposure region resulted from imaging conditions (Fetterly and Schueler, 2006). The region of interest (ROI) was then broken down sequentially or randomly to smaller square segments of size $L \times L$ pixels, where L is 2D fast Fourier transformation size. Three L values were considered in this work to produce three different size of segments which were 64×64, 128×128 and 256×256. For sequential selection, the segments were non-overlapping. Thus for a ROI, the bigger the segment size, the lesser the number of segments is. These segments were later fast Fourier transformed and averaged to get the NPS. For random selection, 2500 overlapping segments of size $L \times L$ were chosen randomly by the computer within the ROI, and these segments were fast Fourier transformed and averaged.

DETECTOR SYSTEM RESPONSE

Mean pixel value were determined from a 100×100 pixels region near the chest wall portion. Exposures in air were made across the clinical range of doses (20 to 300 μGy) and measured using calibrated ion chamber (Radcal model 9010, Radcal Corp., Minorva, CA) placed at 660 mm from the focal spot with the detector covered by a lead sheet to avoid any deterioration in functioning of the detector. Exposure values at the plane of the detector were then corrected by using inverse square law. For calibration purposes, the mean pixel value and exposure values were written into two columns in a computer text file. This file could then be read by our NPS computation program.

ONE DIMENTIONAL NORMALIZED POWER SPECTRUM CALCULATION

The calculation of the 1D-NNPS was based on work by Dobbin et. al., (2006, 1995); Samei and Flynn (2003) and Flynn and Samei (1999). A computer program was written to calculate 1D-NNPS from the image file. Basically the program does the following. The uniformly x-ray exposed image data are converted from digital pixel values to relative exposure values using a calibration curve (Section 2.3). Next, a 2D plane fitting is applied to the data, and the data are subtracted from the fitted plane values to get the fluctuation data. This is called detrending and is carried out to correct for low frequency background noise such as the heel effect (Goodsitt et. al., 1998).

fluctuation $(x,y) = $ image data$(x,y) - $ plane fit (x,y)

$$\ldots\ldots\ldots\ldots\ldots(1)$$

The data are then segmented either sequentially or randomly as described in Section 2.2. The segments are 2D fast Fourier transformed (FFT), the FFT amplitudes are squared and averaged. The 2D-NPS is computed using the following formula,

$$\text{2D-NPS } (u,v) = \frac{<|\text{FFT(fluctuation}(x,y)|^2>}{N_x N_y}\Delta x \Delta y$$

$$\ldots\ldots\ldots\ldots(2)$$

where $N_x = N_y = L$ (= 64, 128 or 256) is the size of FFT, and Δx and Δy = pixel size (in mm) in x and y directions.

To get the 2D normalized noise power spectrum (2D-NNPS), the 2D-NPS is divided by square of the mean value of exposure of the large area signal in order to correct for large scale non-uniformity in the image (Monnin et. al., 2007).

$$\text{2D-NNPS } (u,v) = \frac{\text{2D - NPS}(u,v)}{(\text{large area signal})^2} \qquad (3)$$

The 1D-NNPS is estimated by averaging 8 slices of values adjacent to u and v axes of the 2D-NNPS, Figure 2. It was done in order to eliminate NNPS axial contributions that are attributable to imaging device noise. Frequency of 1D-NNPS is defined as w:

$$w = (u^2 + v^2)^{1/2} \qquad (4)$$

Thus the NPS calculated by the program is 1D-NNPS(w). Sampling distance, Nyquist frequency and spatial frequency resolution for this work are as follow:

The sampling distance for this system, $\Delta x = \Delta y = 70$ μm $\qquad (5)$

Nyquist frequency, $u_{\text{Nyquist}} = \dfrac{1}{2\Delta x} = 7.14 \text{ mm}^{-1}$ $\qquad (6)$

Spatial frequency resolution, $\Delta u = \Delta y = \dfrac{1}{L\Delta x}$ $\qquad (7)$

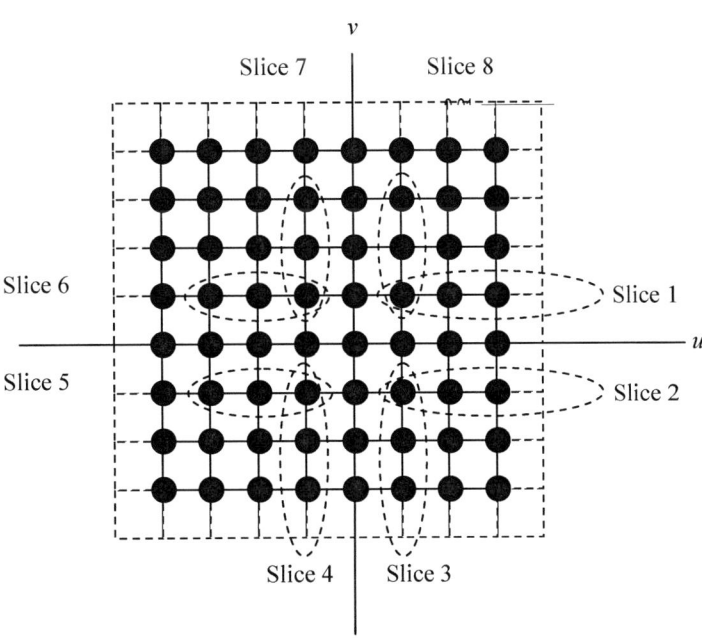

Fig. 2: *Extracting the 1D-NNPS from the 2D-NNPS. The 1D-NNPS is calculated by averaging 8 slices of values adjacent to u and v axes of the 2D-NNPS*

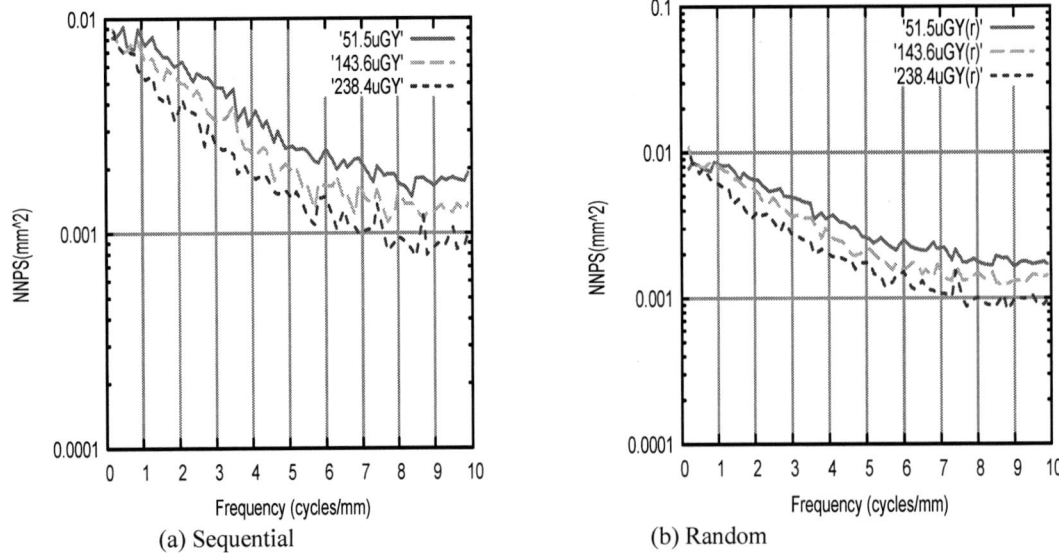

(a) Sequential   (b) Random

Fig. 2: *Extracting the 1D-NNPS from the 2D-NNPS. The 1D-NNPS is calculated by averaging 8 slices of values adjacent to u and v axes of the 2D-NNPS*

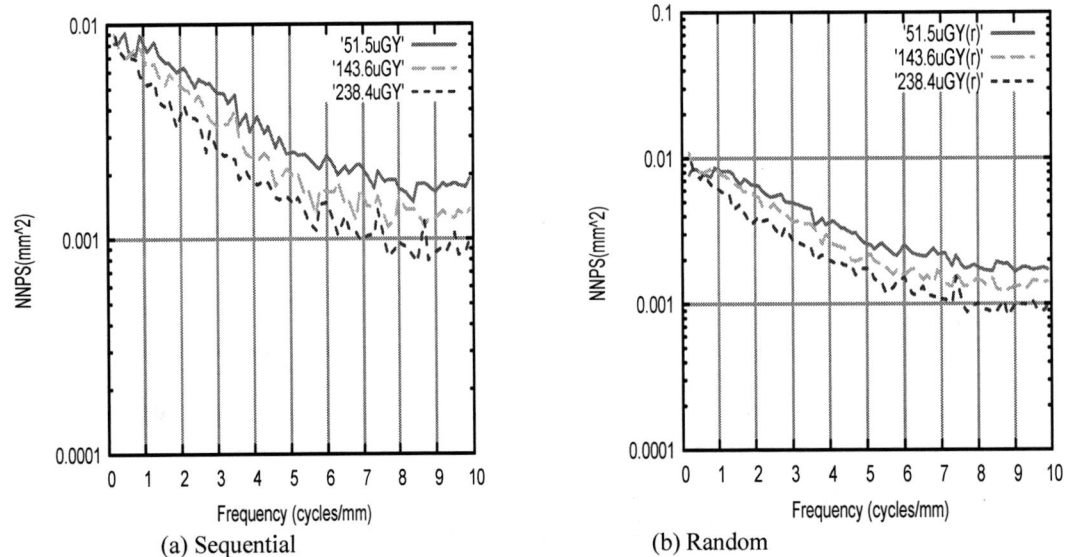

(a) Sequential   (b) Random

Fig. 3: *1D-NNPS calculated by sequential (a) and random (b) segment selection from the same region of interest, segment size is 128 × 128.*

RESULTS AND DISCUSSION

Figure 3 shows the NNPS values for two different segment selection processes namely sequential and random from the same ROI. From ROI of size 802 × 663 pixels, 30 128 × 128 segments were available for FFT by the sequential selection, while there were 2500 such segments by the random selection. It can be seen that both random and sequential selections give almost the same NNPS. However, the NNPS by the random selection has smoother curve due to larger number of segments averaged for the computation. Figure 3 also shows that 1D-NNPS values for high exposure were lower across the frequencies compared to low exposure. This reduction in noise is due to the relatively larger increase in average signal with the high exposure compared to noise. NNPS values decrease rapidly with spatial frequency at about 0.2 to 5.0 cycles/mm then decrease slowly until reaching Nyquist frequency. Beyond Nyquist frequency the NNPS are almost flat.

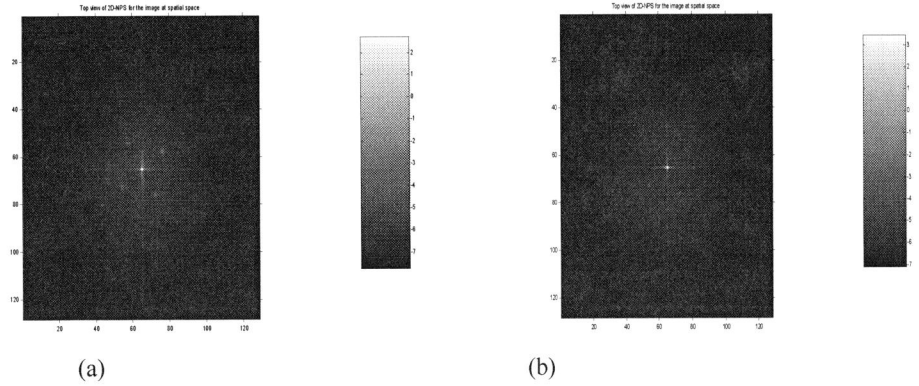

(a)   (b)

Fig. 4: 2D-NNPS for the system at (a) 143.6 μGy and (b) 51.5 μGy displayed at fixed contrast and brightness setting obtained through random segments selection.

Figure 4a and 4b show the 2D-NNPS for exposure of 143.6 μGy and 51.5 μGy respectively, for segments of 128 × 128 pixels. The two 2D-NNPS were displayed to depict the presence or absence of any off-axis noise peak in 2D-NNPS. Both 2D-NNPS show elevated noise in the horizontal and vertical direction which is higher in Figure 3b (low exposure) compared to Figure 3a (high exposure). Those were possibly due to slight structured noise pattern (system noise) in the directions. The streak artifacts at high exposure were smaller than the streak at low exposure because for high exposure, x-ray quantum noise is greater than system noise. Therefore these edge discontinuities could be attributable to imaging device noise.

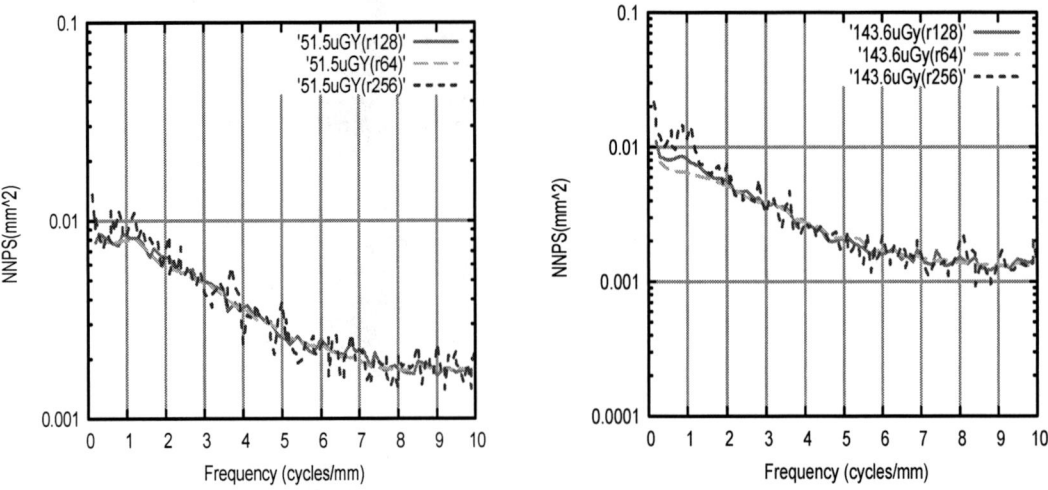

Fig. 5: 1D-NNPS for low and high exposures with various segment sizes using random segment selection. Here (r128), (r64) and (r256) mean segments of size 128 × 128, 64 × 64 and 256 × 256 pixels respectively, are randomly selected from the region of interest from NPS calculation.

Table1: Details of 4 ROIs cropped from an image, subjected to sequential and random segment selection for the NPS computation.

Cropped ROI size (pixels)	Classification of ROI (VS = very small, S = small, L = large, VL = very large)	2D FFT size or segment size	Segment selection procedure from the ROI	Number of segments obtained
657 x 657	VS			4
800 x 510	S			3
1402 x 1402	L	256	Sequential	25
1707 x1707	VL			36
657 x 657	VS			2500
800 x 510	S			2500
1402 x 1402	L	256	Random	2500
1707 x1707	VL			2500

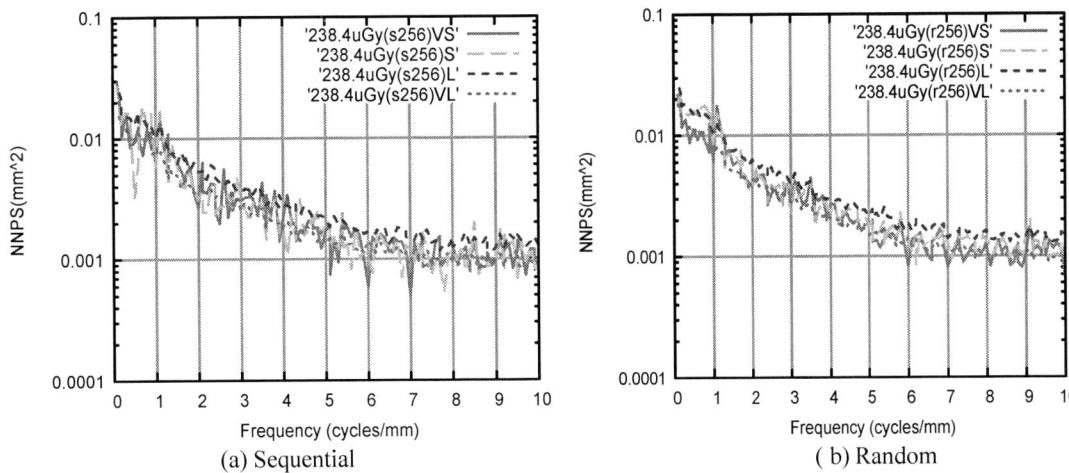

(a) Sequential   (b) Random

Fig. 6: 1D-NNPS dependence on size of ROI cropped near the chest wall edge. The segment size is kept constant at 256 × 256. Here VS, S, L, and VL stand for very small, small, large, and very large ROI size respectively as defined in Table 1.

Figure 5 shows 1D-NNPS for low and high exposure x-ray upon of the system. It can be seen that for the calculation of 1D-NNPS from the same ROI, the smaller the segment size the smoother the 1D-NNPS curve is, irrespective of the segment selection methods. For sequential segment selection, this is an expected result because the smaller the segment size, more segments are Fourier transformed and averaged for the NPS calculation thus giving smoother curve. For the random selection, the same result holds for reason yet to be explained, because the same number of segments was used in the calculation.

The high exposure (238.4 μGy) image was cropped into 4 ROIs of different sizes, segmented to 256 × 256 pixels, and subjected to both sequential and random segment selection for the NPS computation. Table 1 gives the details of the ROI. Figure 6 shows the 1D-NNPS obtained. Comparison of Figure 6a and 6b shows that random segment selection gives smoother curve than sequential segment selection. Figure 6a shows that for sequential segment selection, larger ROI gives smoother curve. The same can be said for random segment selection, Figure 6b. Comparing Figure 6a with 6b also shows that sequential selection gives less smooth curve than random selection.

1.0 CONCLUSION

A method for measuring 1D normalized noise power spectrum for FFDM is presented. Uniformly x-ray exposed image is fitted to a plane to detrend the noise data. The data is segmented to square segments randomly and with overlaps within the region of interest. The segments are fast Fourier transformed, the Fourier amplitudes squared and averaged. Normalization is made by division with square of mean signal. Eight slices of data adjacent to the u and v axes of the 2D-normalized noise power spectrum are averaged to give the 1D-normalized noise power spectrum.

The random segment selection gives smoother normalized noise power spectrum curve than the sequential segment selection. Smaller segment size and bigger region of interest also make the curve smoother.

Measurements at three x-ray dose levels showed that high radiation dose gave lower NNPS at specific beam quality. This is as expected and indicates the appropriateness of the method.

ACKNOWLEGMENT

The authors wish to thank the Head and staffs of Putrajaya Hospital Radiology Department for accessing the FFDM system and Malaysian Nuclear Agency for lending a number of test tools for the study.

REFERENCES:

ACR, (1999), Mammography Quality Control Manual, American College of Radiology, Reston, VA 20191.

CEC, (2006), European guideline for quality assurance in breast cancer screening and diagnosis, 4th Edition., Commission of the European Communities, Luxembourg.

Dobbin, J. T., Ergun, D. L., Hinshaw, D. A. and Clark, D. C., (1995), DQE(f) of four generation of computed radiography acquisition devices, *Med. Phys.* 22:1581-1593.

Dobbin, J. T., Samei, E., Ranger, N. T. and Chen, Y., (2006), Intercomparison of methods for image quality characterization. II. Noise power spectrum, *Med. Phys.* 33:1466-1475.

Fetterly, K. A. and Schueler, B. A., (2006), Performance evaluation of a computed radiography imaging device using a typical front side and novel dual side readout storage phosphors, *Med. Phys.* 33: 290-296.

Flynn, M. J. and Samei, E., (1999), Experimental comparison of noise and resolution for 2k and 4k storage phosphor systems, *Med. Phys.* 26:1612-1623.

Ghetti, C., Borrini, A., Ortenzia, O. and Rossi, R., (2008), Physical characteristics of GE Senographe Essential and DS digital mammography detectors, *Med. Phys.* 35: 456-463.

Goodsitt, M. M., Shan, H. P., Liu, B., Guru, S. V., Morton, A. R., Keshavmurthy, S. and Petrick, N., (1998), Classification of compressed breast shapes for the design of equalization filters in x-ray mammography, *Med. Phys.* 25: 937-948.

IEC, (2007), Medical electrical equipment - Characteristics of digital x-ray imaging device - Part 1-2: Determination of the detective quantum efficiency, detector used in mammography, IEC 62220-1-2. International Electrotechnical Commission, Geneva, Switzerland.

Monnin, P., Gutierrez, D. and Bulling, S., (2007), A comparison of the performance of digital mammography systems, *Med. Phys.* 34:906-914.

Ranger, N. T., Samei, E., Dobbin, J. T. and Ravin, C. E., (2005), Measurement of the detective quantum efficiency in digital detector consistent with the IEC 62220-1 standard: Practical consideration regarding the choice of filter material, *Med. Phys.* 32:2305-2311.

Samei, E. and Flynn, M. J., (2003), An experimental comparison of detector performance for direct and indirect digital radiography systems, *Med. Phys.* 30:608-622.

Aggregation In Heavy Water Micellar Dilute Solutions Of Three Nonionic Classic Surfactants: C10E7 AND C12E7 And C14E7 Study By SANS Method

Rajewska Aldona

Institute of Atomic Energy, 05-400 Swierk-Otwock, Poland
Email: aldonar@cyf.gov.pl

ABSTRACT

Three nonionic classic surfactants $C_{10}E_7$ (heptaethylene glycol monodecyl ether) and $C_{12}E_7$ (heptaethylene glycol monododecyl ether) and $C_{14}E_7$ (heptaethylene glycol monotetradecyl ether) in water solutions were investigated for concentration c=0.5% (dilute regime) at temperatures t=6°, 10°, 15°, 20°, 25°, 30° and 35 °C with two methods - tensiometric and small-angle neutron scattering (SANS) on SANS diffractometer "Yellow Submarine" at Budapest Neutron Center, Budapest (Hungary) and SANS spectrometer ("YuMO") of the IBR-2 on pulsed neutron source at FLNP, JINR in Dubna (Russia). Measurements have covered Q range from 8×10^{-3} to 0.4 Å^{-1}. The micellar solutions were prepared in D_2O since the contrast between the micelles and the solvent in neutron experiments is better with D_2O than with H_2O. It was obtained as the result that the shape of micelles changes depending on surfactant concentration and temperature. At lower concentrations and temperatures micelles are ellipsoids but at higher concentrations and temperature are rather cylinders. For calculation and approximation results from SANS experiments was used program PCG 2.0 of Glatter O. and co-workers from University of Graz (Austria).

Key words: micellar solutions, small-angle neutron scattering, nonionic surfactants, aggregation

INTRODUCTION

Three nonionic heptaethylene glycol monotetradecyl ether ($C_{14}E_7$), heptaethylene glycol monododecyl ether ($C_{12}E_7$)and heptaethylene glycol monodecyl ether ($C_{10}E_7$) were investigated in D_2O (heavy water) solutions by small angle neutron scattering method (Corti et al.,1984, Glatter and Fritz ,2000,Goyal et al.,1995, Hayter and Penfold, 1983,Porod, 1982, Toyoko Imae, 1989). All SANS measurements for surfactant $C_{14}E_7$ were performed on the time-of-flight spectrometer MURN of the IBR-2 pulsed reactor, JINR, Dubna, Russia but for surfactants $C_{12}E_7$ and $C_{10}E_7$ onSANS spectrometer "Yellow Submarine" at the Budapest Neutron Center, Budapest (Hungary). Measurements have covered Q range from 8×10^{-3} to 0.4Å^{-1}. The micellar solutions were prepared in D_2O (heavy water), since the contrast between the micelles and the solvent in neutron experiments is better with D_2O than with H_2O.The results were calculated with help of the PCG Software 1.01.02 (Austria). The information on size and form micelles was obtained from experimental data.

EXPERIMENTAL SECTION. MATERIALS

All surfactant solutions was made using D_2O as a solvent (99.9% deuterated). Heavy water was purchased in the Prikladnaya Chimia in St.Petersburg (Russia) but surfactants: $C_{14}E_7$ heptaethylene glycol monotetradecyl ether (MW = 522.77, CMC = 1.28×10^{-5}M), $C_{12}E_7$ heptaethylene glycol monododecyl ether (MW = 494.70, CMC = 0.075×10^{-3}M) and $C_{10}E_7$ heptaethylene glycol monodecyl ether (MW = 466.66, CMC = 0.96mM) were obtained from the Fluka andwere used without further purification.

Small Angle Neutron Scattering (SANS) Experiment.

All SANS measurements were performed on the time-of-flight spectrometer MURN of the IBR-2pulsed reactor, JINR, Dubna, (Russia) and on the SANS diffractometer "Yellow Submarine" at the Budapest Neutron Center, Budapest (Hungary). The characteristics of these spectrometers were described in details in Ostanievich (1988)and BNC Progress Report (2004-2005). Neutrons were used in wavelength Q range from 8×10^{-3} to 0.4 Å^{-1}. For the measurements quartz cells 2 mm of thickness, were used which were sealed to prevent evaporation during the experiment. Up to 15 of such cells were placed in a sample holder, and the temperature within the cells was kept constant in the range of ± 0.5°C by means of a thermostat. Conversion of the scattered intensities into absolute differential cross-sections was done by using an internal calibration

CP1202, *Neutron and X-Ray Scattering in Advancing Materials Research: International Conference – 2009*
edited by A. Saat, H. A. Kassim, M. H. H. Jumali, J. M. Saleh, M. R. Othman, A. Ibrahim, F. M. Idris, and M. H. A.-R. M. Ahmad
© 2009 American Institute of Physics 978-0-7354-0739-8/09/$25.00

standard (vanadium). For all samples D_2O was used as a solvent in order to achieve good contrast conditions. Background scattering was subtracted by comparison with a corresponding pure D_2O sample. The data treatment was done according to the standard procedures of Soloviev(2003).

RESULTS AND DISCUSSION

Few of scattering curves of $C_{14}E_7+D_2O$ as a function of temperature for concentration c=0.5% (dilute regime) are shown in Fig.1, for $C_{12}E_7+D_2O$ - in Fig.2, but for $C_{10}E_7+D_2O$ in Fig.3. The interparticle interactions can be neglected at this low concentration.

temperature, in particular, to a greater extent when approaching the phase boundary, assuming a threadlike or wormlike shape. The intensity of scattered radiation as a function of the magnitude of the scattering vector q is represented for monodisperse particles by (Guinier and Fournet ,1955)

$$I (q) = n\, P (q)\, S (q)$$

where $q = 4\pi / \lambda \sin (\theta/2)$, λ is the wavelength of radiation, θ is the scatterring angle, and n is the number density of particles, $P (q)$ is the particle form factor and contains the effect of particle size, shape and scattering power on scattered intensity, $S (q)$ is the structure factor and accounts for interparticle interactions, $S (q) \rightarrow 1$ for no interacting particles or in the limit of a very dilute solution.

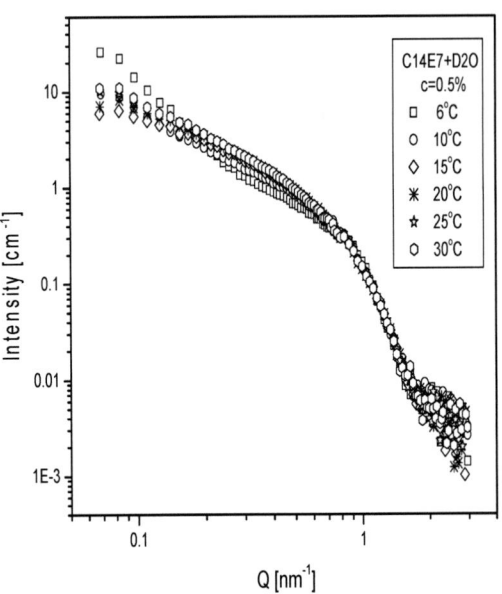

Fig.1.Intensity of neutron scattering vs scattering vector at concentration: $c_2=0.5\%$ for temperatures 6°, 10°, 15°, 20°, 25° and 30^oC. (sample $C_{14}E_7+ D_2O$)

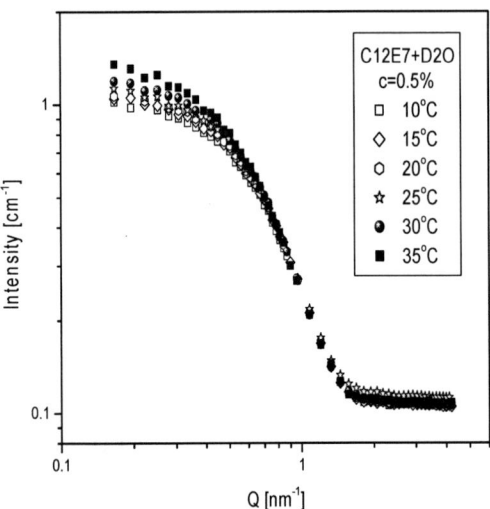

Fig.2.Intensity of neutron scattering vs scattering vector at concentration: $c_2=0.5\%$ for temperatures 10°, 15°, 20°, 25°, 30° and 35^oC (sample $C_{12}E_7+ D_2O$)

It is well - known that surfactants form micelles above the critical micellization concentration (CMC) in aqueous solution. The form and size of micelles is connected with concentration, and temperature of water solution, and with structure and size of molecule of each surfactant. Polyoxyethylene alkyl ethers have the general formula $C_nH_{2n+1}(OCH_2CH_2)_mOH$. They will be referred to as C_nE_m with n indicating the number of carbons in the alkyl chain and m being the number of ethylene oxide units in the hydrophilic moiety.

In the dilute regime the CnEm+water binary system forms an isotropic phase called the L_1 phase, which consists of micelles formed with the amphiphilic molecules and water. The system exhibits LCST (lower critical solution temperature) behavior, the micellar solutions are phase-separated into two phases at high temperatures. It is now well established that the micelles grow in size with increasing concentration and rising

The micelle shape determination for the experimental results was achieved using the Indirect Fourier Transformation method IFT (Glatter,1977a,b) and its extension Generalized Indirect Fourier Transformation method (GIFT) (Glatter, 2000). IFT is a completely model-free method, which canonly be successfully applied for the "dilute" systems where interparticle interactions can be neglected. The conventional Fourier transformation of I(q) involves the integral

$$p(r) = \frac{1}{2\pi^2} \int I(q)qr \sin(qr)dq$$

which yields the pair distance distribution function $p(r)$, where r is the distance in real space. The point, at which the p(r) falls to zero, is indicative of the particle maximum dimension. In Fig. 4 is shown the distance distribution function as determined by the Indirect Fourier Transformation (IFT) method for water solutions

$C_{14}E_7+D_2O$ (c=0.5%) for temperatures: 6°, 10°, 15°, 20°, 25° in Fig.4, the distance distribution function as determined by the Indirect Fourier Transformation (IFT) for water solutions $C_{12}E_7+D_2O$ (c=0.5%) for temperatures: 10°, 15°, 20°, 25°, 30° and 35^oC in Fig. 5 but for sample $C_{10}E_7+D_2O$ (c=0.5%) in Fig.6 for temperatures: 10°, 15°, 20°, 25°, 30° and 35^oC.

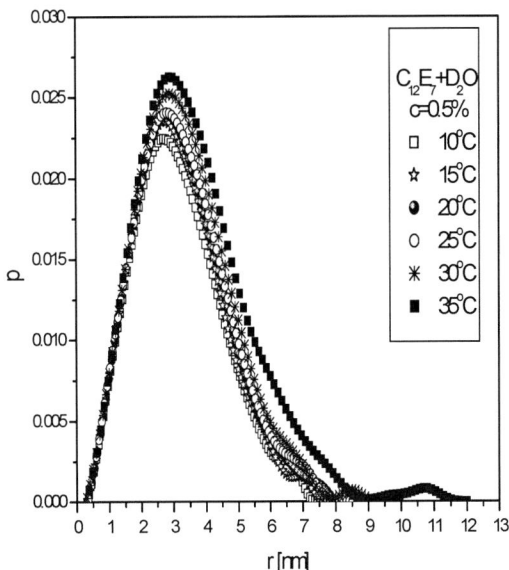

Fig.5. The distance distribution function as determined by the IFT (Indirect Fourier Transformation) method for water solutions $C_{12}E_7+D_2O$ (c=0.5%) for temperatures: 10°, 15°, 20°, 25°, 30° and 35^oC.

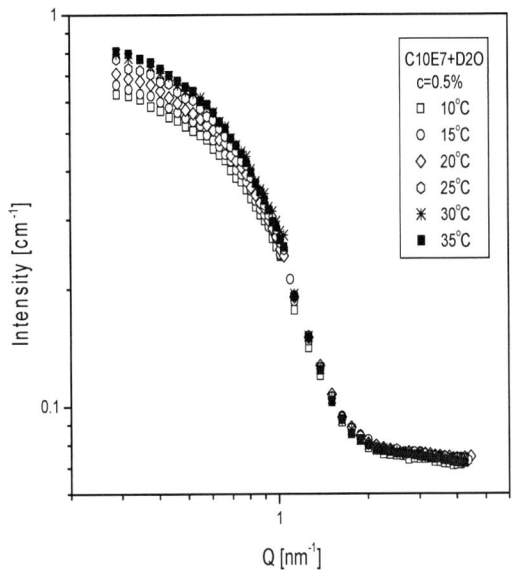

Fig.3. Intensity of neutron scattering vs scattering vector at concentration: $c_2=0.5\%$ for temperatures 10°, 15°, 20°, 25°, 30° and 35^oC (sample $C_{10}E_7+D_2O$)

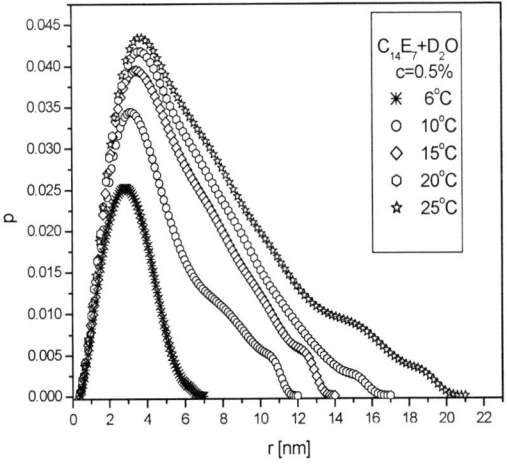

Fig.4. The distance distribution function as determined by the IFT (Indirect Fourier Transformation) method for water solutions $C_{14}E_7+D_2O$ (c=0.5%) for temperatures: 6°, 10°, 15°, 20° and 25^oC.

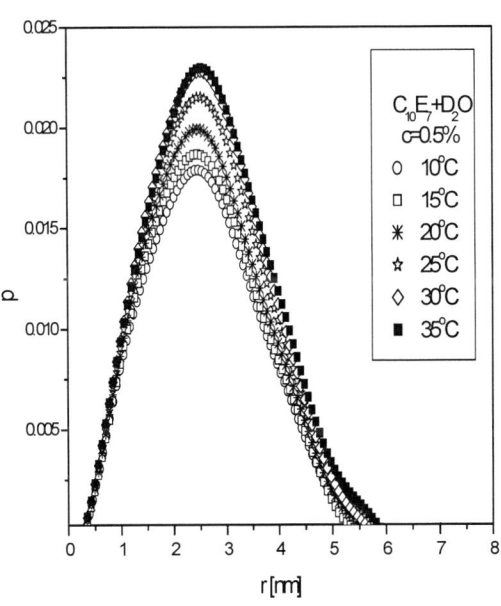

Fig.6. The distance distribution function as determined by the IFT (Indirect Fourier Transformation) method for water solutions $C_{10}E_7+D_2O$ (c=0.5%) for temperatures: 10°, 15°, 20°, 25°, 30° and 35^oC.

In case of $C_{14}E_7$ surfactant – for all temperatures at c=0.5% there are two axial ellipsoidal micelles in the investigated solutions, with longer axis 7.5 nm at 6° C and 21nm at 25^oC. For $C_{10}E_7$ surfactant at the same concentration of solution and temperature – two axial ellipsoidal micelles were observed, too. The longer axis is equal 5.2nm at temperature 10°C, but at 35° C this axis is equal to 6nm. Micelles of $C_{10}E_7$ nonionic surfactant are smaller than those of $C_{14}E_7$ surfactant in the same experimental conditions. In water solutions of $C_{12}E_7$ classic surfactant were observed micelles with the same shape as for samples $C_{14}E_7$ and $C_{10}E_7$. But micelles are smaller than for surfactant $C_{14}E_7$ and bigger than for $C_{10}E_7$. At the temperature 10°C longer axis is equal 7nm but for 35°C this axis is equal 12nm.

SUMMARY AND CONCLUSIONS

Three investigated nonionic classic surfactants are from the same homologous series with different carbon number in alkyl chain. This is the reason that size of micelles change at the same concentration of surfactants in heavy water solution. It was observed that the temperature influence on the size of micelles, too. For these three investigated surfactants the size of micelles increases with an increase of temperature and concentration.

REFERENCES:

BNC *Progress Report*, (2004-2005), ed. by M. Makai, L. Rosta (2006), Budapest, 151-152

Corti M., Minero C., Degiorgio V.,(1984), *J. Phys. Chem.* 88:309

Glatter O., (1977a) *Acta Phys. Austriaca*, 47:83; Glatter O.,(1977b) *J. App. Crystallogr.*, 10:415

Glatter O.,(2000), in: *Neutrons, X-Rays, and Light Scattering Methods Applied to Soft Condensed Matter*, P. Lindner, T. Zemb (Eds), Elsevier

Glatter O., Fritz G. et al.,(2000a), *Langmuir* 16: 8692

Goyal P.S., Menon S.V.G., Dasannacharya B.A., Thiyagarajan P., (1995), *Phys. Rev. E* 51:2308

Guinier A., Fournet G., (1955), *Small Angle Scattering of X-Rays,* John Wiley & Sons, New York

Hayter J.B., Penfold J., (1983), *Colloid and Polymer Sci.,* 261:1022

Ostanievich Yu.M., (1988) *Makromol. Chem. Macromol. Symp.* 15: 91

Porod G., In *Small Angle X-Ray Scattering,* Glatter O., Kratky O., Eds. Academic Press Inc., London, 1982, p.17

Soloviev A.G., Solovieva T. N., Stadnik A.V. et al., *Reports JINR*, Dubna, R10-2003-86

Toyoko Imae, (1989), *Journal of Coll. & Interf. Scien.*, 127: 256

Fuel Element Transfer Cask Modelling Using MCNP Technique

Rosli Darmawan, Budiman Naim Topah,

Technical Support Division, Malaysian Nuclear Agency (NUCLEAR MALAYSIA), Bangi, 43000 KAJANG, MALAYSIA

Abstract

After operating for more than 25 years, some of the Reaktor TRIGA Puspati (RTP) fuel elements would have been depleted. A few addition and fuel reconfiguration exercises have to be conducted in order to maintain RTP capacity. Presently, RTP spent fuels are stored at the storage area inside RTP tank. The need to transfer the fuel element outside of RTP tank may be prevalence in the near future. The preparation shall be started from now. A fuel element transfer cask has been designed according to the recommendation by the fuel manufacturer and experience of other countries. A modelling using MCNP code has been conducted to analyse the design. The result shows that the design of transfer cask fuel element is safe for handling outside the RTP tank according to recent regulatory requirement.

Keywords: transfer cask, MCNP, shielding model

INTRODUCTION

One of the critical situations in Reaktor TRIGA Puspati after having operated for more than 25 years is its fuel power capacity. Fuel reconfiguration exercises have to be conducted in order to maintain RTP capacity. During this exercise, some of the spent fuel has to be transferred out from the core and stored at the storage area inside the tank. In the near future, some of these spent fuels will have to be transferred out of RTP tank. Thus, a fuel transfer cask shall be provided for handling the spent fuel outside of RTP tank.

The preparation towards this requirement is being done. A fuel transfer cask is designed according to recommendation from the fuel manufacturer and experience of other countries. In order to validate the safety of the cask a shielding modeling was conducted using MCNP code.

FEUL ELEMENT TRANSFER CASK

The fuel element transfer cask is designed to transport the spent fuel element in a safe manner. It shall be adequately shielded and have rigid structure to ensure untoward accident. It shall be equipped with strong and safe lifting mechanism for handling purposes. The basics design of a fuel element transfer cask shall consists of a steel casing filled with lead shielding, lead filled plug locking at the top and bottom, and lifting hooks for handling (OSTROP, 1995).

The Design

The design for RTP fuel transfer cask is as illustrated in Figure 1. It is cylindrical in shape to accommodate the RTP fuel element shape and size. The casing is made of stainless steel 304 and the shielding material is lead with more than 99.99% purity (BSI, 1991). The thickness of the shielding was determined after referring to the manufacturer recommendation as well as rough calculation of radiation penetration. The transfer cask is equipped with top cap and release mechanism at the bottom of the cask as shown in "Detail C" in Figure 1.

MONTE CARLO TECHNIQUE

MCNP is a general purpose Monte Carlo N-Particle code that can be used for modeling neutron, photon or coupled neutron/photon/electron transport (X-5 Monte Carlo Team, 2003). MCNP differs from deterministic transport methods. While, the deterministic method solves the transport equation for average particle behavior, Monte Carlo simulates individual particles and recording particular aspects of their average behavior according to code configuration. Deterministic methods give complete information throughout the phase space of the problem. Monte Carlo only provides information about specific tallies requested by the user.

CP1202, Neutron and X-Ray Scattering in Advancing Materials Research: International Conference – 2009
edited by A. Saat, H. A. Kassim, M. H. H. Jumali, J. M. Saleh, M. R. Othman, A. Ibrahim, F. M. Idris, and M. H. A.-R. M. Ahmad
© 2009 American Institute of Physics 978-0-7354-0739-8/09/$25.00

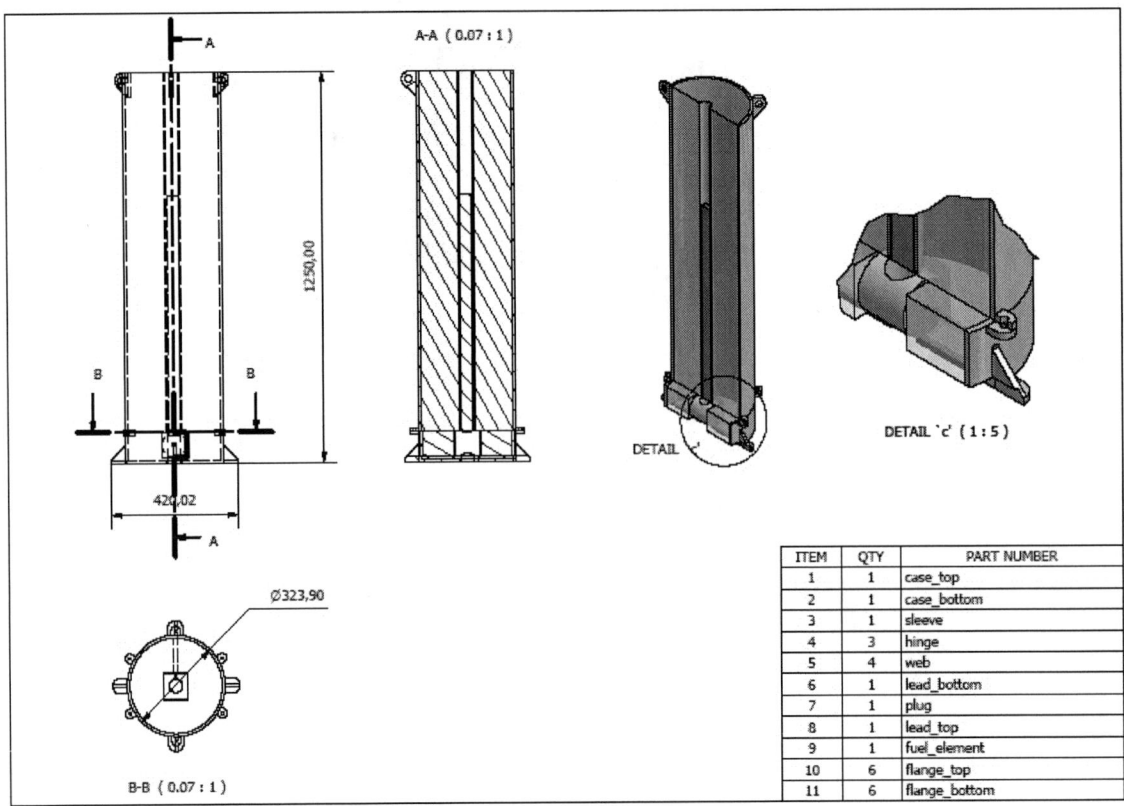

ITEM	QTY	PART NUMBER
1	1	case_top
2	1	case_bottom
3	1	sleeve
4	3	hinge
5	4	web
6	1	lead_bottom
7	1	plug
8	1	lead_top
9	1	fuel_element
10	6	flange_top
11	6	flange_bottom

Figure1: Transfer Cask Design

One of the important advantages of Monte Carlo is that it employs the point energy data so that each particle is tracked using interaction cross-section data appropriate to its actual energy. Monte Carlo is also capable to simulate sophisticated shapes and also able to import geometries from other programs. However, one critical factor in Monte Carlo is that the results depend on the number of particles scores in a cell. Therefore, to get a good variance, one has to master the right technique available in MCNP so that the number of transporting particles can be increased at particular region to simulate the actual particle transport.

MCNP is typically applied for criticality calculation, radiation dosimetry, brachytherapy calculation, stopping power ratio, wall attenuation correction and medical physics (Rogers, 2006).

The Code

The code or input file requires the user to define the geometry specification, description of materials, selection of cross-section evaluations, location and characteristics of the neutron, photon or electron source, the type of results or answer desired, as well as any variance reduction technique to be used. The geometry specification contains in both the cell cards and surface cards, and the other information such as source definition, material definition and tallies are supplied under data cards. Each block shall be separated by blank line delimiter. Users are also allow to add comments anywhere in the code with special notation such as $

characters to differentiate comments with the actual code.

Execution & Output

The calculation was conducted with MCNP version 5 which comes with Visual Editor feature. Previously MCNP uses line editor for input file which is tedious and error prone (Carter et. al, 2003). The input file was written directly using Visual Editor input menu and the execution was initiated via Visual Editor running menu. The status of the calculation was updated online in the "run window". Any warnings, errors and termination will be immediately shown during calculation. The output of the calculation is saved in an output file with inp filename. It consists of the content of input file; the interpretation and analysis of source specifications, materials specification, surface and cell specification; physical constants used; cross section tables of specified materials; the history of neutron creation, path, collision and termination; output of the tally; and finally the statistical check and variance reduction results. Graphical representation of the model and the particles created may be visualized using feature available in Visual Editor.

RESULTS AND DISCUSSION

A total of 28 runs were conducted before achieving a reliable result. The first few runs were meant to create the geometry which could be accepted by the code as

176

well as representing the actual design. The second phases of the runs were to model the source representing the fuel element. The following phase was to define the material data involved in the model and finally experimenting with tally specification.

Model Creation

The geometry created is a simplified model of the transfer cask design. It consists of one cylindrical container made of stainless steel 304 filled with pure lead. The fuel element is represented by a cylindrical shape taking into account effective length of the actual fuel element. The input file as per Figure 2 and the MCNP graphical output is shown in Figure 3. The model is to analyse the flux or dose rate just outside the container. F2 tally has been used to get the flux averaged over the outer surface of the model.

The source is defined as a fixed source emits in a cylindrical distribution with the center located at the middle of the fuel element. The Watt fission energy spectrum has been used as the source variable as it is typically modelled for fuel element (X-5 Monte Carlo Team, 2003). The schematic of the source model is shown in Figure 4. The graphical output of the source particles distribution is shown in Figure 5.

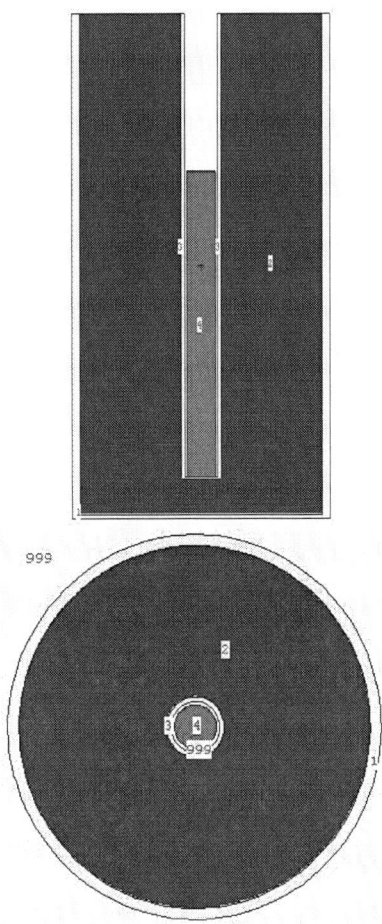

Figure 3: Graphical output of the model (not to scale)

```
1-       Simplified model of a transfer cask
2-       c
3-       c     Cylindrical cask
4-       c     fuel element
5-       c
6-        1     3  -8        (-1  2  -6   7  ):(-2  -11  7  ) $ casing
7-        2     2  -7.85     (-2  3  -6  11  ):(-3  -10  11 ) $ lead
8-        3     3  -8        (-3  4  -6  10  ):(-4  -9   10 ) $ casing
9-        4     1  -5.77      -5  -8   9 $ fuel
10-      999    0  (-6  -4   8  ):(-4  5  -8   9  ):6  :-7  :1 $ void
11-
12-       1     cz     32.39 $ Outer casing cask OD
13-       2     cz     30.48 $ Outer casing cask ID
14-       3     cz      5.1  $ Inner casing OD
15-       4     cz      4.3  $ Inner casing ID
16-       5     cz      3.754 $ Fuel element
17-       6     pz     62.5  $ Upper limit cask
18-       7     pz    -62.5  $ Lower limit cask
19-       8     pz     23.35 $ Upper limit fuel
20-       9     pz    -52.1  $ Lower limit fuel or Inner casing I
21-      10     pz    -52.5  $ Lower limit Inner casing O
22-      11     pz    -61.5  $ Lower limit Outer casing I
```

Figure 2: Input file for the geometry

177

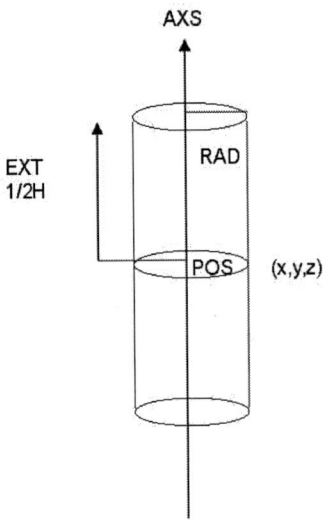

Figure 4: Schematic of the source model

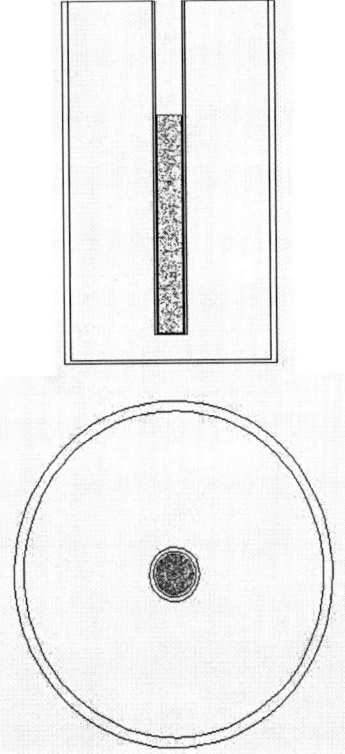

Figure 5: Graphical output of the source particles
distribution (not to scale)

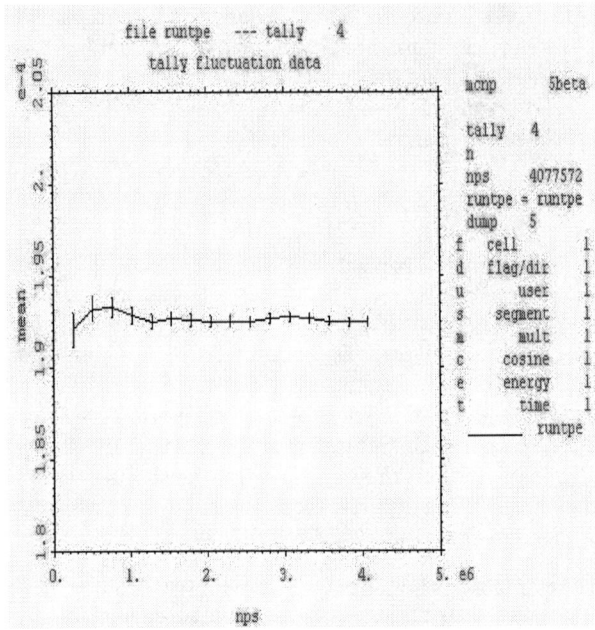

Figure 7: Output on FOM

The tally result which represents the averaged flux at the outer surface of the cask was converted into dose rate according to recommendation by NCRP and ICRP (X-5 Monte Carlo Team , 2003). The result is as per Table 1. It shows that the dose rate outside the outer casing of the cask is about 3×10^{-9} miliSievert per hour. This is far below 1 miliSievert per hour of current regulation for worker and also for member of the public at 0.025 miliSievert per hour. Manual calculation using neutron attenuation principles taking into account the effective removal cross section of shield materials has comparable result with dose rate at outer surface of 0.12×10^{-9} miliSievert per hour.

Tally result inside the other cells such as cell 3 (inside casing), cell 2 (lead shielding) and cell 1 (outer casing) have also shown similar result. The average dose rate across the volume of the three cells shows well below the regulation. Several run has been conducted and the output shows similar results as tabulated in Table 2.

The Execution

The execution has successfully passed all ten (10) standard statistical check built-in inside MCNP as shown in Figure 6. The relative error is less than 0.1 and decreasing with execution time. A check into the tally fluctuation chart shows that the Figure of Merit (FOM) is rapidly approaching a constant value indicating the calculation has converged satisfactorily (Booth, 1985).

178

Table 1: Tally results of the cask

	flux	error	energy	DFE	QF	rem/hr	m-rem/hr	mSv/hr
Cell 3	6.43E-06	0.0037	1.4609	1.32E-04	11	9.34E-09	9.34E-06	9.34E-08
Cell 2	1.91E-06	0.0042	1.0498	1.32E-04	11	2.77E-09	2.77E-06	2.77E-08
Cell 1	7.33E-07	0.0042	1.0665	1.32E-04	11	1.07E-09	1.07E-06	1.07E-08
Surface 1	2.12E-07	0.0075	1.0665	1.32E-04	11	3.08E-10	3.08E-07	3.08E-09

Table 2: Results of several runs

	flux	mSv/hr	flux	mSv/hr	flux	mSv/hr
cell 3	6.72E-06	9.76E-08	6.43E-06	9.34E-08	6.43E-06	9.34E-08
cell 2	2.02E-06	2.94E-08	1.91E-06	2.77E-08	1.91E-06	2.77E-08
cell 1	7.42E-07	1.08E-08	7.33E-07	1.06E-08	7.33E-07	1.07E-08

```
================================================================================
                    ===================

          results of 10 statistical checks for the estimated answer for the tally fluctuation chart (tfc) bin
                              of tally   2

tfc bin    --mean--    ---------relative error---------    ----variance of the variance----    --figure
                                         of merit--   -pdf-
behavior   behavior    value   decrease  decrease rate     value   decrease  decrease rate      value
                                         behavior     slope

desired    random      <0.10    yes      1/sqrt(nps)                <0.10     yes       1/nps       constant
                                         random       >3.00
observed   random      0.01     yes      yes                        0.00      yes       yes         constant
                                         random       10.00
passed?    yes         yes      yes      yes                        yes       yes       yes         yes
                                         yes          yes


================================================================================
                    ===================


          this tally meets the statistical criteria used to form confidence intervals: check the tally fluctuation chart
                              to verify.
          the results in other bins associated with this tally may not meet these statistical criteria.
```

Figure 6: Output on statistical check

CONCLUSION & FURTHER WORKS

The design for fuel element transfer cask based on RTP manufacturer recommendation and experience of other countries is found to be safe as far as dose rate level is concern. An analysis using MCNP code showed that the dose rate at the outer surface of the cask is far below the regulated level for radiation worker as well as member of the public. The dose rate obtained from conventional calculation has comparable result.

The other aspect that needs to be considered is the mechanical safety of the design. An analysis on the integrity of the design should be done to ensure its safety during handling and operation.

REFERENCES

Booth, Thomas E. (1985), *A Sample Problem for Variance Reduction in MCNP*. New Mexico, Los Alamos National Library, October 1985.pp1-2.

British Standard Institution (BSI), (1991), *Specification for Ingot Lead for Radiation Shielding, BS3909*. London, British Standard Institution. p7.

Carter, L.L. and Schwarz, R.A. (2003), *MCNP Visual Editor Computer Code Manual*. Richland,WA:Visual Editor Consultants (VEC). p5.

OSTROP. (1995), *Chapter 11: Fuel Element Handling Procedure, Ohio State University Training Reactor Operational Procedure*, Ohio State University, 1995.

Rogers, D. W. O. (2006), *Review: Fifty Years of Monte Carlo Simulations for Medical Physics*. Pys. Med. Biol. 51, p. R287-R301.

X-5 Monte Carlo Team (2003), *MCNP: A General Monte Carlo N-Particle Transport Code, Version 5, Volume 1*, New Mexico, Los Alamos National Library. April 24, 2003.pp 1-1, H1-H3.

A Study of Al-Zn-Sn Alloy Sacrificial Anode Cathodic Protection Requirements for Structure Used In Seawater

Siti Radiah Mohd Kamarudin[1], Muhamad Daud[1], Shariff Sattar[1], Abd. Razak Daud[2]

[1]Malaysian Nuclear Agency (Nuclear Malaysia),Bangi, 43000 KAJANG, MALAYSIA
[2]Faculty of Science and Technology, Universiti Kebangsaan Malaysia. 43600 Bangi, selangor, Malaysia

Abstract

The corrosion of aluminium (Al) alloys in seawater was investigated using potentiodynamic technique, complemented by Scanning Electron Microscopy (SEM), EDAX and XRD. SEM was used out to characterize the corroded surface and to observe the extent of corrosion attack on the Al alloys tested in seawater. EDAX analysis was used to identify elements present on the specimen surface. Where else XRD was to identify phase appearance. The results indicate that zinc (Zn), stanum (Sn) and copper (Cu) as alloying elements enhance corrosion behaviour of the aluminium in seawater by shifting the potential to a more negative value. In the presence of those elements, the Al alloys becomes more active, having potential of more than -1.0 V_{SCE} and showed active corrosion behaviour.

Keywords: Electrochemical behaviour, potentiodynamic, SEM, EDAX, XRD

INTRODUCTION

It is known that the addition of certain elements to aluminium, such as mercury (Hg), gallium (Ga), stanum (Sn), indium (In), magnesium (Mg), zinc (Zn), cadmium (Cd) and barium (Ba) results in alloys with more anodic potentials than unalloyed aluminium (Al) [1]. On the exposure to the atmosphere, the passive film formed on Al and retards further corrosion of the metal [2,3]. The most frequent form of corrosion of Al and Al alloys in seawater is pitting corrosion [4] due to localized breakdown of oxide film by chloride attack. This reaction serves as a first step in pitting and crevice corrosion [5]. Commercially produced Al anodes are alloyed with Zn and activators such as Hg, In or Sn [6]. This paper reports the experimental results of corrosion behaviour study of al alloys containing Zn, Sn and Cu in seawater. It will lead to a better understanding of the dissolution process occurring at the activation potential of Al alloys containing Zn, Sn and Cu and their performance in seawater.

EXPERIMENTAL AND METHOD

Pure Al (99.99%), Zn (99.9%), Sn (99.8%) and Cu (99.8%)(Aldrich Chemical) in granule form were used for samples preparation. The selected samples compositions were melt up to 800°C in a cylindrical graphite crucible in ambient environment. The molten was homogenized by stirring with a high density graphite rod and then poured into the preheated split steel mould 1.5 cm in diameter and 16 cm in height. Prior to test, samples were cut to 1.5 cm in diameter and 0.3 cm in height and mechanically ground up to 2400 grit SiC paper and washed thoroughly with distilled water followed by aceton and drying.

The electrochemical analysis was carried out by potentiodynamic method (CMS 100 software, Gamry Instrument). All potential values of 72 hours immersion in seawater are measured with respect to a Saturated Calomel Electrode (SCE) reference electrode. Following the potentiodynamic test, samples were further characterized by XRD and Scanning Electron Microscope (SEM) (FEI 400) instrument equipped with an Energy-dispersive using X-ray analyzer (EDAX)(Genesis 7000) to study morphology and elemental content.

RESULTS AND DISCUSSION

The surface chemical composition of as cast samples analysed by EDAX and measured open circuit potential (OCP) of the alloys were shown Table 1 (a) and (b).

The potential values for samples measured at room temperature are shown in Table 1(b). It shows that the potential value for each Al alloy is greater than -1.0 V_{SCE} while pure Al (sample A) has a value of -0.7 V_{SCE}. Whereas a binary alloy, Al-Zn (sample B), shows 0.22V higher potential value than pure Al. The presence of zinc and indium on Al leads to a displacement of the activation potential in approximately 0.3 V towards more negative values in seawater [7]. Al-Zn alloys with the

CP1202, Neutron and X-Ray Scattering in Advancing Materials Research: International Conference – 2009
edited by A. Saat, H. A. Kassim, M. H. H. Jumali, J. M. Saleh, M. R. Othman, A. Ibrahim, F. M. Idris, and M. H. A.-R. M. Ahmad

presence of Sn and Cu activators (sample C and D), show OCP increased significantly in the negative (active) direction for 72 hours immersion in seawater. Therefore, the addition of the alloying elements has shifted the potential towards more negative value.

Table 1. Composition (a) and OCP for samples (b)

Sample	Zn	Sn	Cu	Al
A	-	-	-	Pure
B	5.35	-	-	Bal.
C	5.40	0.97	-	Bal.
D	5.77	0.85	0.24	Bal.

(a)

Sample	Voltage (V $_{SCE}$)
A	-0.7812
B	-1.0034
C	-1.3380
D	-1.0988

(b)

The polarization curves of the alloys (Fig. 2) show corrosion potential (E_{corr}) of more than -1.0 V_{SCE} was found for the Al containing alloying elements. Anodic curves for alloys are similar to that of pure Al, but have shifting the corrosion potential towards more negative values. The more negative potential found in the case of Al alloys containing Sn and Cu may be attributed to the absence of cathodic impurities. These elements enhance the hydrogen evolution, which in turn polarized the system and shifting the surface potential to more negative values.

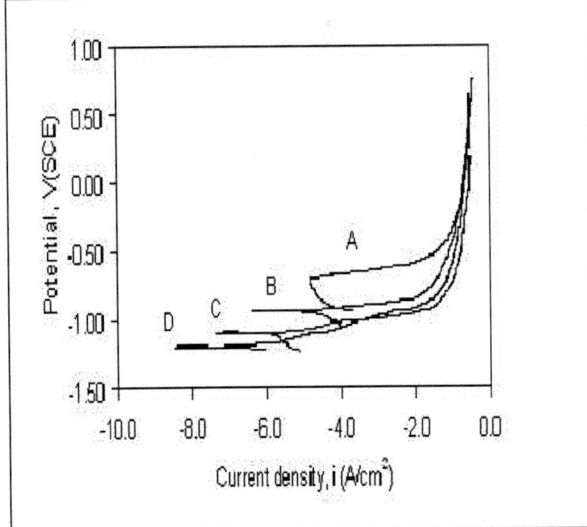

Fig. 2. Anodic potentiodynamic of specimens in aerated seawater of pH = 7.8± 0.1 at room temperature.

The relationship between morphological structure and electrochemical results was examined and particular attention paid to the influence of alloying elements on activation of Al alloys. According to Breslin et al. [8], the presence of Zn as alloying elements promotes the nucleation of $ZnAl_2O_4$ spinel, which gives rise to increase defects and cracking of protective layer. A minimum content of 5%Zn in the alloy is necessary to achieve a proper amount of intermetallic compound in order to provide the breakdown of passive layer and related anodic activation [9]. The irregular morphology (Fig.3a marked area) is believed due to non-uniform Sn precipitation. This assumption is supported by EDAX analysis that spectrum shows higher concentration of Sn (Fig.3c). Morphology of sample D (Fig.4) which containing Cu shows more severity of corrosion. According to [4] for Al alloyed with Cu, corrosion of Al is aggravated by copper corrosion product and it decreases resistance to corrosion. From the SEM surface examination, the presence of alloying elements in Al matrix were found to affect the morphology of their surface with formation of region covered with shrunken, porous and cracked surface deposits.

XRD technique (fig. 5) was carried out to evaluate the phase formation on the alloy surface after the test by scanning 2θ angle from 20° to 80°. For sample Al-5%Zn, no phase exists referring to Zn. It indicates that the only phase exit is α-Al. Mondolfo [10] saying that Zn dissolve in al matrix up to 6.2% and maintained in Al crystal. 5% Zn used will be dissolve as solid solution in Al matrix and will not show it own phase. Al-Zn alloy combination formed homogeneous solid solution and uniform grain size and grain boundaries throughout the entire crystal lattice [11]. Diffractogram result shows β-Sn peak only occur for sample with Sn addition. From the observation β-Sn phase only detected after test sample. According to [12], XPS study shows existence of Sn $^{2+}$ and Sn $^{4+}$ on the corrosion product layer. Samples before test show no indication of this phase. This suggests that dissolution process of Al alloy leaves behind Sn on the surface.

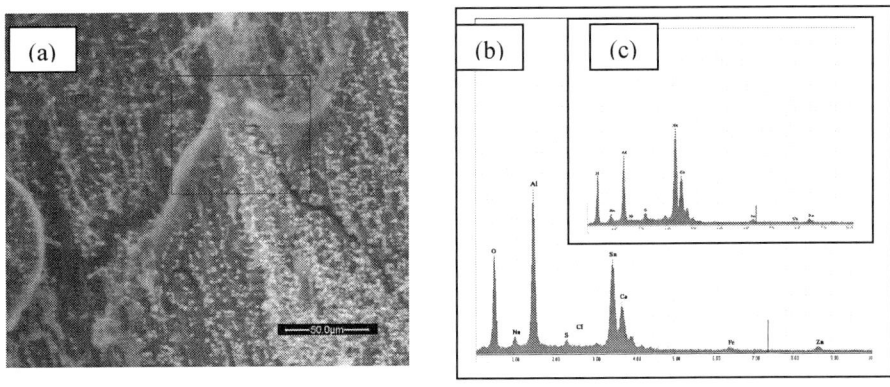

Fig.3 SEM micrograph and EDAX analysis of sample C (a & b, c)

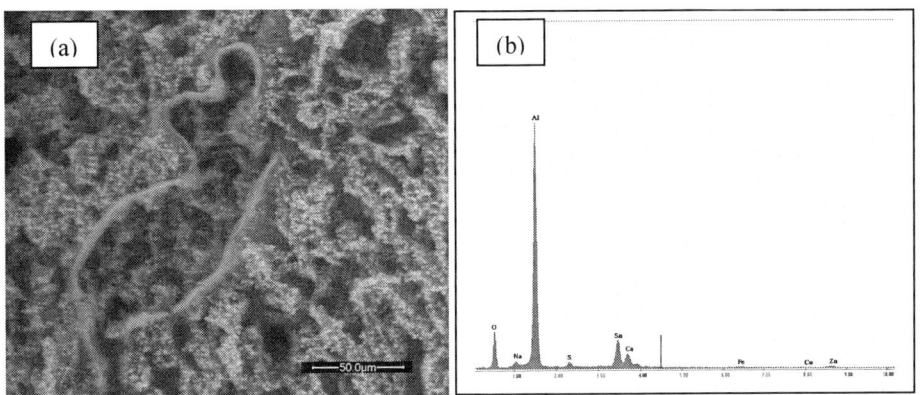

Fig. 4 SEM micrograph and EDAX analysis of sample D

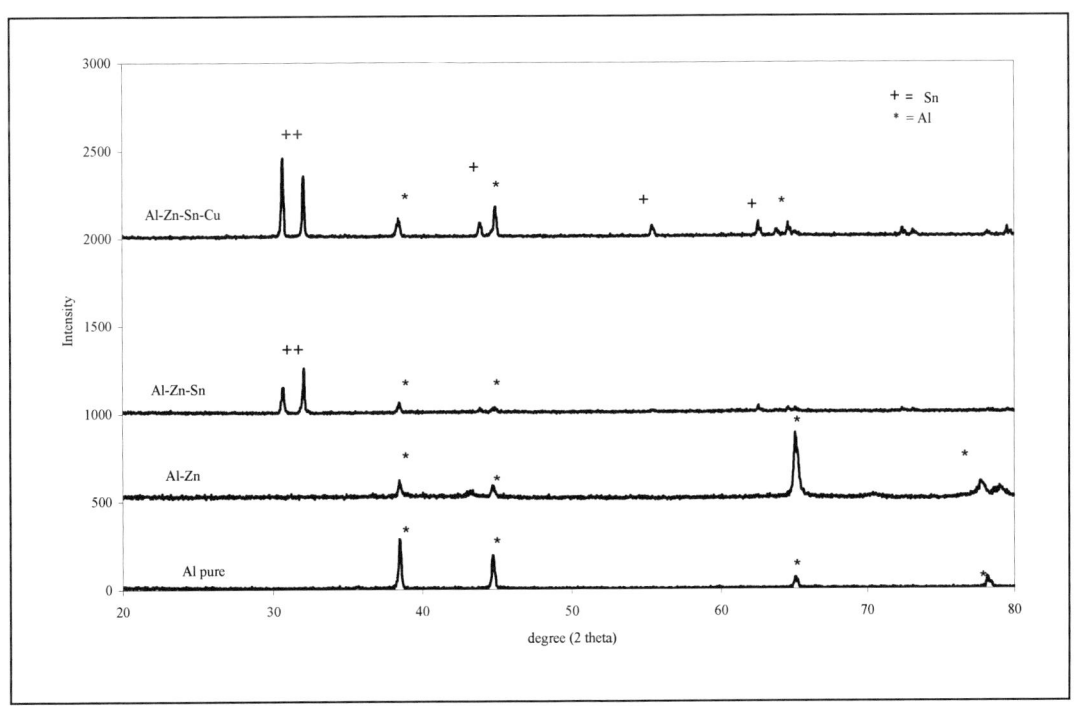

Fig.5 XRD diffractogramme of sample Al-Zn-Sn after after potentiodynamic test

CONCLUSION

Based on this study, results indicate that alloying elements in Al matrix play important role on the improving corrosion behaviour of Al alloys in seawater. The attack initiation of the Al alloys is due to higher potential achieved with the addition of Zn, Sn and Cu in Al and the corrosion resistance of Al decreased significantly with the addition of alloying elements.

REFERENCES

[1] J.T. Reding and J.J.Newport, The Influence Of Alloying Elements On Aluminum Anodes In Seawater, Dow Chemical Co.1966.

[2] H.P. Godard. An Insight Into the corrosion Behaviour of Aluminum. Plenary Lecture. National Association of Corrosion Engineers. Material Performance. (1981). 9-15.

[3] C.B.Breslin, W.M.Carroll. The Activation of Aluminum by Activator elements. Science. Vol. 35. (1993).197-203.

[4] Christian Vargel. Corrosion of Aluminum. Elsevier Ltd, Uk. (2004) 26

[5] Francis L.Laque. Marine Corrosion Causes And Prevention. John wiley & Sons, Inc. (1975) 191

[6] S.L. Wolfson. Operating Performance Of Aluminum Anodes – Results From Lab And Field Tests. Materials Performance. (1994). 22

[7] A.G. Munoz, S.B.Saidman, J.B. Bessone. Influence Of In On The Corrosion Of Zn-In Alloys. Corrosion Science 43 (2001) 1245-1265.

[8] C.B.Breslin, L.P.Friery, W.M.Carroll. Corrosion Science 36. (1994) 85.

[9] A.Barbucci, G. Cerisola, G. Bruzzone & A.Saccone. Electro. Acta.Vol.42(1997).

[10] Mondolfo, L.F. 1976. *Aluminum alloys: Structure and properties*. New York. Butterworth & Co.

[11] Jabeera, B., Anirudhan, T.S. & Shibli, S.M.A. 2005. Nano zinc oxide for efficient activation of aluminum zinc alloy sacrificial anode. *Journal Of New Materials for*

Electrochemical System 8. 291-297.

[12] Mahdi Che Isa. 2007. fabrikasi dan pencirian aloi al-Zn-Mg mengandungi unsure minor Sn dan Cu bagi perlindungan kakisan marin. Tesis Dr. Fal. Universiti Kebangsaan Malaysia.

Fractal Structures on Silica Aerogels Containing Titanium: A Small Angle Neutron Scattering Study

Widya Sari[a], Dian Fitriyani[a], Edy Giri Rachman Putra[b], Abdul Aziz bin Mohamed[c], Noorddin Ibrahim[c]

[a] Department of Physics, Andalas University, Padang, West Sumatera, Indonesia
[b] Neutron Scattering Laboratory, National Nuclear Energy Agency of Indonesia (BATAN), Kawasan Puspiptek Serpong, Tangerang, Indonesia
[c] Materials Technology Group, Industrial Technology Division, Agensi Nuklear Malaysia (ANM), Selangor, Malaysia
[d] Department of Physics, Faculty of Science, Universiti Teknologi Malaysia (UTM), Johor, Malaysia

Abstract

The fractal structure of silica aerogels containing titanium has been investigated by means of small angle scattering (SAS) technique, i.e. small angle neutron scattering (SANS). The SANS experiments were conducted using a 36 meter SANS BATAN spectrometer (SMARTer) in Serpong, Indonesia in the range of momentum transfer q, $0.006 < q$ (Å^{-1}) < 0.3. The 'power-law' for a fractal object scattering pattern $I(q) \sim q^{-D}$ observed from the all measured samples. The Fourier transform of a pair correlation model function was implemented in analyzing the structure factor from the 'power-law' scattering profiles. The results are showing that the silica aerogels containing titanium has a mass fractal where its dimension D_m is larger than pure silica aerogels. The mass fractal dimension of silica aerogels containing titanium is relatively constant between 2.2 to 2.4 with the increase of acid concentrations during a sol-gel process while the surface fractal disappeared.

Keywords: Silica aerogels, fractal structure, SANS, power-law scattering

PACS: 25.40.Dn, 82.70.Gg, 83.85.Hf

INTRODUCTION

Aerogels is a gel in which a liquid phase has been replaced by air without damaging the solid phase. It is a porous solid materials with a unique microstructures composed of nanometer size particles and pores. Silica aerogels which is the most common type of aerogels can be prepared with many different properties, for many industrial applications [1]. Silica aerogels is synthesized by a sol-gel process with a supercritical drying where its structure depends on the condition of preparation and the type of precursors. In general, during the synthesis of silica aerogels the aggregates formed and tend to exhibit fractal properties.

There are two types of fractal systems, i.e. mass fractal and surface fractal. The quantitative description of the fractal object is specified by fractal dimension D_m and the relation is given as

$$M(R) \propto R^{D_m} \qquad (1)$$

where M and R are, respectively, the object mass and its radius. This type of object is usually named mass fractal. For mass fractal objects, the fractal dimension D_m is a non-integer and varies between 1 and 3 in the three-dimensional space. Some objects possess a surface that is rough and exhibit another fractal property, i.e. surface fractal which related to a surface area S and is given as

$$S(R) \propto R^{2-D_s} \qquad (2)$$

where D_s is the surface fractal dimension. For $D_s = 2$, the particles have a smooth surface coinciding with the common view that the surface is two dimensional object. When the surface roughness increases, D_s can vary between 2 and 3.

Small angle scattering (SAS) is an important technique for studying fractal structures and other disordered systems on the scale of length from 1 to 100 nm [2,3,4,5]. Fractal concepts have been found to be extremely useful in the analysis and interpretation of small angle scattering data from many disordered systems. The intensity $I(q)$ of small angle scattering from fractal object is proportional to the negative power of the momentum transfer q

$$|\vec{q}| = q = \frac{4\pi}{\lambda} \sin(\theta/2) \qquad (3)$$

where, θ is the scattering angle and λ is the neutron wavelength. This dependence of $I(q)$ on a negative power of q is observed only when $1/\xi \ll q \ll 1/R_o$, where ξ is correlation length or average diameter of a

CP1202, *Neutron and X-Ray Scattering in Advancing Materials Research: International Conference – 2009*
edited by A. Saat, H. A. Kassim, M. H. H. Jumali, J. M. Saleh, M. R. Othman, A. Ibrahim, F. M. Idris, and M. H. A.-R. M. Ahmad
© 2009 American Institute of Physics 978-0-7354-0739-8/09/$25.00

scatter and R_0 is a radius of primary particle related to the local structure. This kind of scattering is often called 'power-law' scattering as mass fractal and surface fractal can be described by equation (4) and (5).

$$I(q) \propto q^{-D_m} \qquad (4)$$

$$I(q) \propto q^{-(6-D_s)} \qquad (5)$$

The relation between the scattering profile from the object with both mass fractal and surface fractal properties on different length scales is summarized in Figure 1.

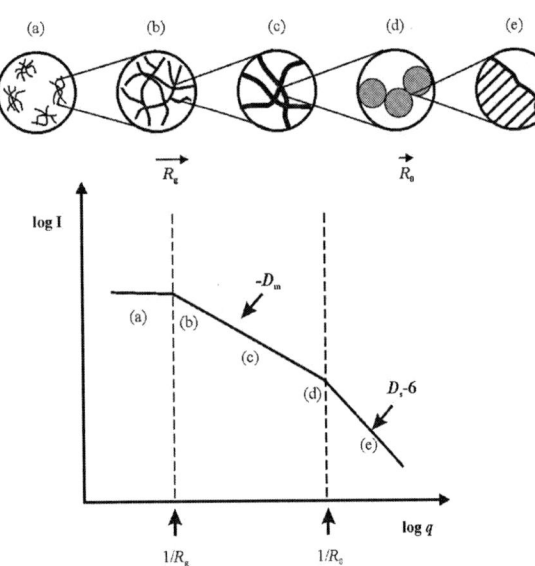

Figure 1. Schematic representation of an aggregate with mass fractal as well as surface fractal properties on different length scales and the corresponding scattering profile.

In this work we report the fractal structure of silica aerogels and silica aerogles containing titanium prepared by a sol-gel process method from small angle neutron scattering (SANS) data. The experimental data were then fitted by applying the sum of log normal sphere and aggregate fractal model calculations. A variation of acid concentrations during a sol-gel process on silica aerogels containing titanium samples was also investigated.

MATERIALS AND METHODS

Sample preparation

Silica aerogels and silica aerogels containing titanium with various acid concentrations from 1 N to 1.75 N were synthesized a sol-gel process by the Zeolite Research Group of Universiti Teknologi Malaysia, Malaysia [6]. The acid concentration of sulfuric acid was preferred in producing aerogels materials effectively. The sample was poured into 1 mm thick quartz cell and then covered with another 1 mm thick quartz cell as a sandwich with the sample thickness of 1 mm.

Small angle neutron scattering (SANS) measurement

The SANS experiments were performed using the 36 meter SANS BATAN (SMARTer) in Serpong, Indonesia [7]. The rotational speed of mechanical velocity selector was adjusted to 5000 rpm with the tilting angle 0^0 to generate a neutron wavelength λ of 3.9 Å ($\Delta\lambda/\lambda = 0.13$). Three different sample to detector distances of 1.5, 4 and 13 m have been set up covering the momentum transfer q range of $0.006 < q$ (Å⁻¹) < 0.3. During the experiment, the temperature was maintained at room temperature. The scattering data of each sample were corrected for empty cell as a background, electronic background as a noise, sample transmission and detector efficiency using a SANS data reduction software, GRAPS [8].

Data analysis

When the elementary particles are uniform, a good approximation for scattered intensity $I(q)$ can be written as the product of the form factor $P(q)$ and the structure factor $S(q)$

$$I(q) \approx \frac{d\Sigma}{d\Omega}(q) = nV^2(\rho - \rho_0)P(q)S(q) + B_{inc} \qquad (6)$$

with $P(q) = \left\langle |F(q)|^2 \right\rangle$

where n is the number concentration of scattering bodies. V is the volume of one scattering bodies, $(\rho - \rho_0)^2$ is the square of difference in neutron scattering length density between the scattering bodies and matrix or medium. B_{inc} is an incoherent background signal. The information of the geometry of primary particle can be obtained at the large scattering angles (large q, small R) and the information considering the spatial arrangements of the primary particles is present at low scattering angles (small q, large R).

For a monodisperse system of spherical and homogeneous particle of size R_0, the form factor at dilute system can be given as

$$P(q) = \left[\frac{3\sin(qR_0) - qR_0\cos(qR_0)}{(qR_0)^3} \right]^2 \qquad (7)$$

The scattering theory relates $S(q)$ via a Fourier transform with a pair correlation function $g(R)$ describing the chance to find another particle within a distance R from a certain particle. For an isotropic system in three dimensional directions, the following relation can be derived

$$S(q) = 1 + 4\pi\rho \int_0^\infty [g(R)-1]R^2 \frac{\sin(qR)}{qR} dR \qquad (8)$$

The Fourier transform of equation (7) can be described as

$$S(q) = 1 + \frac{1}{(qR_0)^{D_m}} \frac{D_m \Gamma(D_m - 1)}{\left(1 + \frac{1}{q^2 \xi^2}\right)^{(D_m - 1)/2}} \sin(D_m - 1) \tan^{-1}(q\xi)$$

$$(9)$$

with $\Gamma(x)$ is the gamma function[9]. At large scattering angles (large q, $qR_0 \gg 1$), $S(q) \approx 1$, therefore $I(q) \approx P(q)$. At small scattering angle (small q, $qR_0 \ll 1$) but large compared to the inverse finite overall size of system ($q \gg 1/\xi$), $P(q) \approx 1$ and $I(q) \approx S(q)$. This result is very often used to determine the fractal dimension of a scattering object from the scattering profile. Those model calculations are available in the Igor NIST data analysis [10] and may only be applied in the region $q\xi \gg 1$ and $qR_0 \ll 1$, regime c in Figure 1.

Results and Discussion

The log normal sphere and aggregate fractal model calculations have been applied separately to fit the corrected scattering data of silica aerogels, Figure 2. A log normal sphere model calculation is in agreement with the experimental data at high q region as $S(q) \approx 1$, Figure 2a. While at intermediate to low q region, an aggregate fractal model calculation is fit with the data as $I(q) \approx S(q)$. Here the primary particle was assumed in a spherical shape and monodisperse, Figure 2b. The sum of two model calculations was then fitted into the experimental data and shown in Figures 3 and 4.

(a)

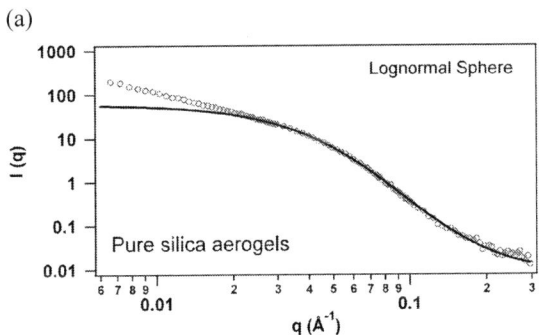

(b)

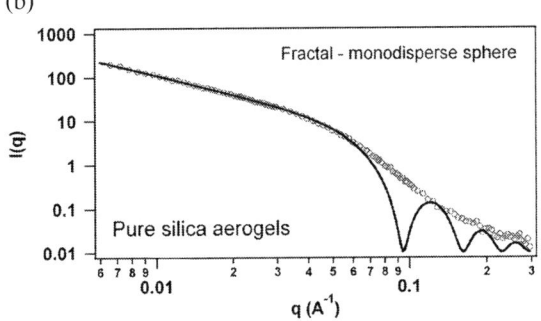

Figure 2. SANS profiles of silica aerogels fitted with two model calculations (a) log normal sphere at high q region and (b) aggregate fractals at intermediate to low q region. A symbol of circle is an experimental data and solid line is a theoretical model calculation.

Figure 3 shows the result of intensity $I(q)$ versus momentum transfer q for pure silica aerogels and silica

aerogles containing titanium at 1 N of sulfuric acid concentration. The sum of two model calculations is in agreement for both samples. The silica aerogels was assembled by their primary particles with 57 Å on diameter and the size distribution or polydispersity σ is 0.4. The mass fractal dimension of silica aerogles is 1.8. While for silica aerogles containing titanium, the particle diameter is bigger than the pure silica aerogels, i.e. 69 Å with a similar size distribution. At the same time, the mass fractal dimension also increases from 1.8 to 2.2 that indicates the addition of titanium in the silica aerogels has changed the aggregation mechanism of the primary particles. This change refers to the silica aerogels containing titanium is growing to more polymeric and branched structures with the magnitude scattering intensity $I(q) > q^{-2}$ by increasing the acid concentration. While, the surface fractal dimension D_s altered from 2, relatively smooth particles to more than 2, highly rough particles or relatively dense polymeric structure by addition of titanium[3].

(a)

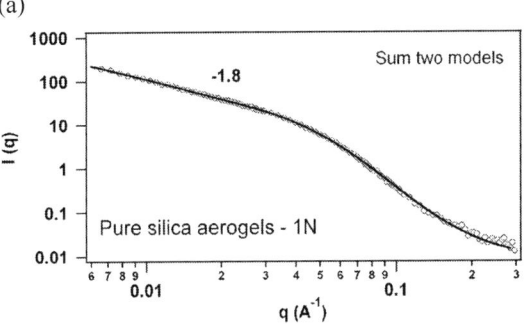

(b)

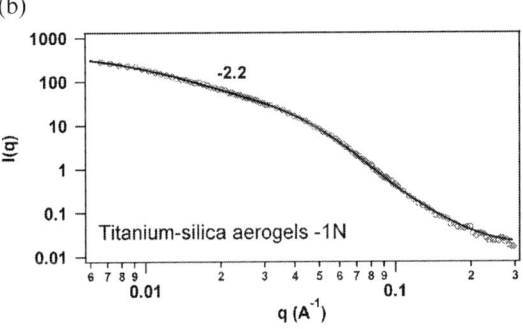

Figure 3. SANS profiles of (a) silica aerogels and (b) silica aerogels containing titanium at 1 N of acid concentration. The experimental data (symbol) were fitted by the sum of theoretical model calculations (solid line), i.e. log normal sphere and aggregate fractal.

Figure 4 shows the SANS profiles of silica aerogels containing titanium with a various acid concentrations, i.e. 1.25, 1.5 and 1.75 N. The mass fractal dimension of silica aerogels containing titanium is relatively constant between 2.3 to 2.4 with the increase of acid concentrations during a sol-gel process and formed a nanometer size of aggregates. It indicates that the mass fractal dimension of the silica aerogles containing titanium is independent on the acid concentrations. On the other hand, by increasing the acid concentration, the slope alteration of the SANS profiles at high q region disappear, regime d – e in Figure 1. This can be described that the primary particles of silica aerogels containing titanium may not be formed before they aggregate to form a large aggregate. The system has a completely 3-dimensional network structure as the formation of a primary particle is inhibited.

(a)

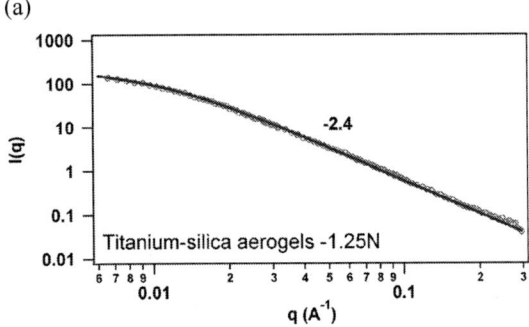

(b)

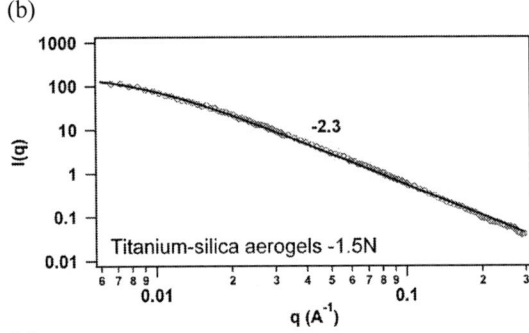

(c)

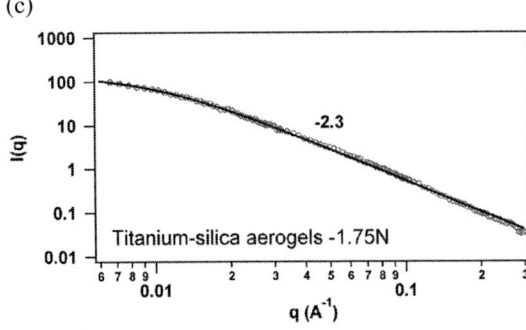

Figure 4. SANS profiles from silica aerogels containing titanium with a various acid concentrations during a sol-gel process (a) 1.25 N, (b)1.5 N and (c)1.75 N. The circular symbol is an experimental data and the solid line is the theoretical calculation.

CONCLUSION

An excellent fitting of the sum from two model calculations in revealing the fractal structures of silica aerogels and silica aerogels containing titanium has been accomplished on the SANS data taken from SANS

BATAN spectrometer (SMARTer) in Serpong, Indonesia. The mass fractal dimension of silica aerogels containing titanium sample is rather independent on the acid concentrations with the range of D_m is 2.2 – 2.4. The acid concentration affected only on the surface fractal structure as a primary particle cannot be formed by increasing the acid concentration.

ACKNOWLEDGMENT

W. Sari acknowledges the support from ICNX2009 committees for attending the ICNX2009 in Kuala Lumpur, Malaysia, June 29 – July 1, 2009.

REFERENCES

1. William, H. (1994), Transformation in Silica Gels and Zeolite Precursors, Schuit Institute of Catalysis, Laboratory of Inorganic Chemistry and Catalysis, Eindhoven University of Technology, The Netherlands.

2. Beaucage, G. (1996), *J. Appl. Cryst.* 29, 134-146.

3. Putra, E.G.R, Ikram, A., Bharoto, Santoso, E., Chiar Fang, T., Ibrahim, N., Aziz Mohamed, A. (2008), Neutron and X-ray Scattering in Materials Science and Biology, American Institute of Physics CP989, 130-133.

4. Reidy, R. F., Allen, A.J, Krueger, S. (2001), *J. Non-Crystalline Solids* 285, 181-186.

5. Schmidt, P.W. (1991), *J. Appl. Cryst.* 24, 414-435.

6. Fang, T. C. (2006), "Determination of Silica Aerogels Nanostructure Characteristics by Using Small Angle Neutron Scattering Technique", Thesis, Faculty of Science, Universiti Teknologi Malaysia, Malaysia.

7. Putra, E.G.R., Ikram, A., Santoso, E., Bharoto, *J. Appl. Cryst.* 40, 447-452 (2007); Putra, E.G.R., Ikram, A., Santoso, E., Bharoto, (2009) *Nucl. Instr. and Meth* A 600, 291-293.

8. Dewhurst, C., "GRASP: Graphical Reduction and Analysis SANS Program for Matlab", http: // www.ill. Eu/fileadmin/usersfiles/Other-Sites/Iss-grasp/graspmain. html, Institut Laue Langevin (2001-2007)

9. Teixeira, J. (1988) *J. Appl. Cryst.* 21, 781-785.

10. Kline, S. R. (2006) *J. Appl.Cryst.* 39, 895.

PHASE IDENTIFICATION AND MICROSTRUCTURE ANALYSIS OF INTERFACE Al-Si/SiC COMPOSITES

Yusof Abdullah[1], Abdul Razak Daud[2], Mohd b. Harun[1] and Roslinda Shamsudin[2]

[1]Malaysian Nuclear Agency, Bangi, Kajang 43000, Selangor, Malaysia.

[2]School of Applied Physics, Faculty of Science and Technology, Universiti Kebangsaan Malaysia, UKM-Bangi 43600, Selangor, Malaysia.

yusofabd@nuclearmalaysia.gov.my

ABSTRACT

Phase composition determined by X-ray diffraction (XRD) and microstructure analysis using scanning electron microscopy (SEM) have been performed on Al-Si/15% SiC composites prepared by stir casting method to investigate interface properties. Energy-dispersive X-ray spectroscopy (EDS) was also used to examine the product at interface where crystallisation was prominent. The interface reaction between particles and matrix was indicated that the nucleation of $MgAl_2O_4$, Mg_2Si and MgO phases growth in the composites. These Interface reaction products play an important role to ensure good physical and mechanical properties of composites.

Keywords: Phase identification, microstructure, X-ray diffraction, scanning electron microscopy, energy-dispersive X-ray spectroscopy, Al-Si/SiC composite, interface

INTRODUCTION

Recently, the research on the high performance light weight materials for automotive and aerospace industries have been received much attention. For industry application, affordable cost combine with quality and ease of fabrication is an essential factor. Metal matrix composites have received extensive studies during the past decades because of their improved properties. Among them, SiC reinforced Al matrix composites have been intensively studied since more than two decades due to its attractive properties such as high strength and high stiffness. Interface characteristics between SiC reinforcement and the Al matrix play an important role to ensure good mechanical properties. Optimization of the interfacial characteristics is meaningful by control process parameters required to obtain a desired interfacial strength. Efforts have been studied to improve the interface properties by alloying of Al matrix with Mg or Si (Mitra et al. 2004; Lee et al. 1998a) and oxidation of SiC to produce a SiO_2 layer on the surface of the SiC (Mitra et al. 2004; Urena et al. 2002). The basic principal behind alloying and oxidation methods is to reduce the Al activity by dissolving a certain amount of Si within the Al matrix, thereby suppressing the reaction between Al and SiC. The aim of this study was to investigate the particles/matrix interface reaction product by x-ray diffraction (XRD) technique and confirmation through scanning electron microscopy (SEM) equipped with energy-dispersive X-ray (EDS) analysis.

EXPERIMENTAL WORKS

Commercial α-SiC powders, magnesium alloy and aluminium-silicon alloy were used for fabrication of Al-Si/15% SiC composites. The composites were prepared by a stir casting technique in a resistance heating furnace. A graphite crucible with a hole at the bottom was used to melt aluminium with incorporation of 1.5 wt. % magnesium alloy as an alloying element. SiC powders were preheated at 1050 °C for 2 hours in a separate furnace and then poured using a funnel into the aluminium melt at temperature 750 °C. The melt was stir for about 8 minutes, prior to pouring into a cast iron mould, preheated to 150 °C. The as-cast composites were subsequently heat treated at 530 °C for 5 hours as homogenization process. The specimens were sectioned and prepared for microstructure characterization. They were mounted and polished using standard metallurgical procedures. A Fei Quanta 400 scanning electron microscopy (SEM) with a Genesis 7000 energy-dispersive X-rays (EDX) system was used to analyse microstructure while a Siemens D5000 X-ray diffraction (XRD) with Cu Kα radiation was identified the phase composition. The XRD data were analysed using Rietveld refinement program using RIETAN-2000.

CP1202, *Neutron and X-Ray Scattering in Advancing Materials Research: International Conference – 2009*
edited by A. Saat, H. A. Kassim, M. H. H. Jumali, J. M. Saleh, M. R. Othman, A. Ibrahim, F. M. Idris, and M. H. A.-R. M. Ahmad
© 2009 American Institute of Physics 978-0-7354-0739-8/09/$25.00

RESULTS AND DISCUSSION

The objective of alloying Al-Si alloy matrix with Mg alloy and oxidation of reinforcement materials during casting process is to either create a diffusion barrier to prevent the interfacial reactions or increase the diffusion distances in which to restrict the rates of reactions. The principle of alloying of the matrix is to enhance the Si activity by dissolving a certain amount of Si in the aluminium alloy, resulting in the thermodynamic suppression of the interfacial reaction between SiC and Al alloy. On the other hand, the passive oxidation of SiC is to produce a SiO_2 layer not only could be effective for prohibiting a direct contract between SiC and Al and also favors a reactive wetting between Al and SiC.

Aluminium, silicon, silicon carbide, magnesium aluminate ($MgAl_2O_4$) and magnesium oxide phases are present in the composites as shown in Fig. 1. Magnesium in the aluminium alloy can reduce the oxides present in composites and develop magnesium aluminate spinel ($MgAl_2O_4$). The formation of $MgAl_2O_4$ phase was the product of the reaction between aluminium alloy containing Mg and SiO_2 layer according to the following reaction:

$$2Al+Mg+2SiO_2 \rightarrow MgAl_2O_4+2Si \qquad (1)$$

during casting has been found effectively in restricting the formation of the Al_4C_3 phase at the interface. Al_4C_3 phase is unwanted material because of degradable material to composites. It seems that Mg alloying was segregate at particle/matrix interfaces. Mitra et al. (2004) reported that the wettability of Al melt on SiC could be improved by Mg allowing during casting. Residual of magnesium could react to form oxide layer which moved on top of liquid Al-Si matrix. Then, these oxides easily removed from the surface prior cast to form ingot.

Fig. 2 shows XRD result by Rietveld refinement using Rietan-2000 program to identify and simulate phase composition. The result indicates that simulation and experiment for Al, Si and SiC peaks are identical. But the peak intensities show differences due to orientation. The study of phase composition by Rietveld refinement can investigate overlapping peaks. The result show that magnesium siliside (Mg_2Si) and steel (Fe) are present in the composite as shown in Fig. 3. Fe was contaminated and released from degradable of propeller during stirring process. Magnesium silicide (Mg_2Si) is formed by the reaction between the silicon present in the alloy and the silicon released by the reaction of the SiO_2 particles with the magnesium present in the system (Lee et al. 1998b; Ravi Kumar et al. 2003).

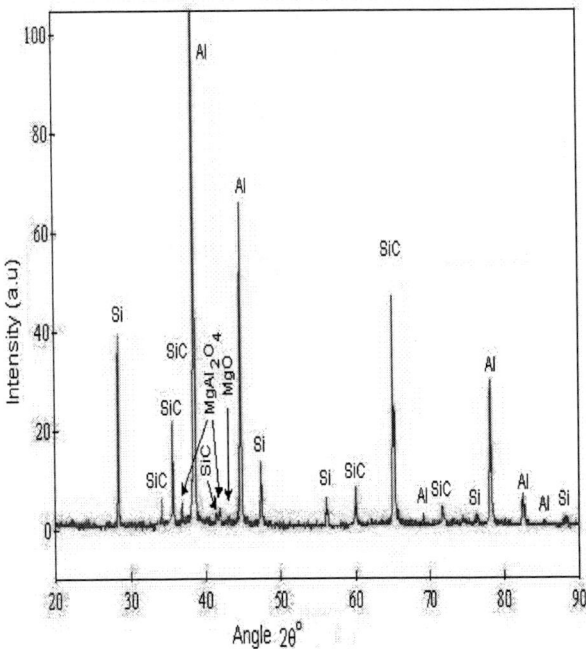

Figure 1. XRD peaks for Al-Si/15% SiC composite showing Al, Si, SiC and other phases.

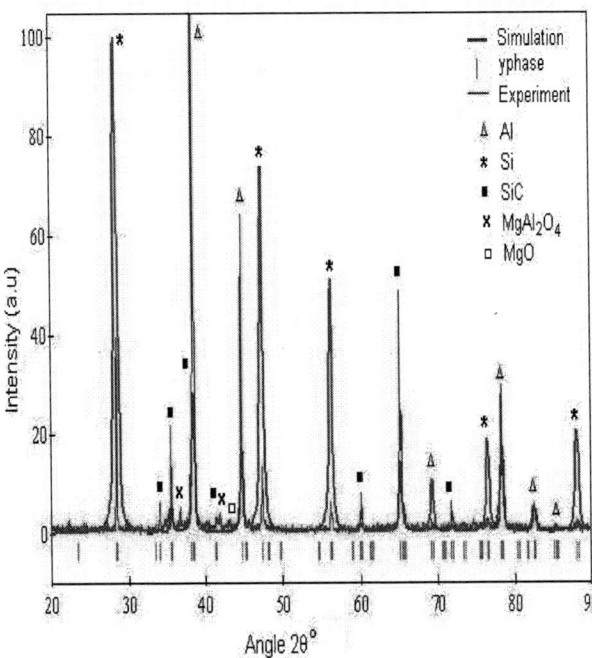

Figure 2. XRD peaks for Al-Si/15% SiC composite showing the comparison between experiment and simulation using Rietveld analysis.

The present of $MgAl_2O_4$ act as diffusion barrier between Al and SiC at the interfaces. On the other hand, alloying of Al-Si matrix with 1.5 wt. % magnesium alloy

190

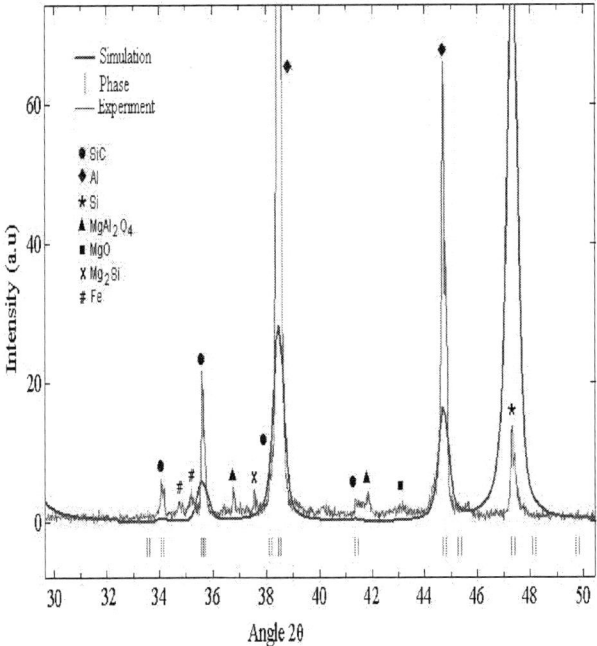

Figure 3. Rietveld refinement between 30° and 50° of 2θ angle showing the present of Fe and Mg₂Si peaks.

The XRD results indicated that aluminium chloride was not present in the composites. MgO and $MgAl_2O_4$ phases were present and believed play a role to create a diffusion barrier and inhibit Al_4C_3 phase formation. These results were same as reported by Lee et al. (1999). As described by Hashim et al. (2001) that aluminium couldn't wet the SiC particles below 900 °C, therefore, the addition of Mg alloy into Al-Si melt has resulted increasing the wettability properties during mixing process. On the other hand, the wettability between Al alloy and SiC was improved by heat treatment of SiC particles at 1050 °C. Meanwhile, Al-Si alloy which containing Si element was dissolved to enhanced the Si activity resulting in the thermodynamic suppression of the interfacial reaction between SiC and Al.

Figure 4 shows the interfacial morphology of Al-Si/SiC composites. The edge of the SiC particle was etched by the aluminium alloy as shown in Figure 4(b). The silica layer serves as a barrier and protects the direct contact between the SiC and Al matrix (Shi et al. 2001). It is suggested that the heat treatment of SiC particles before mixing into Al-Si melt could prevent the interfacial reaction between SiC particles and aluminium alloy.

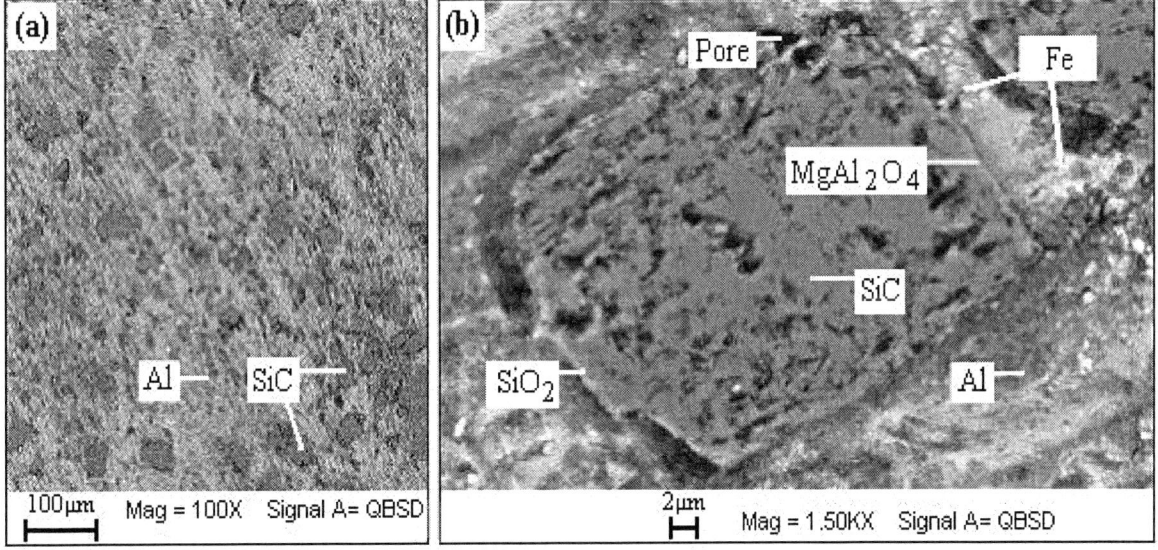

Figure 4. SEM micrographs of Al-Si/15% SiC composite showing silica, iron, magnesium aluminate and pore present in composite.

Microstructure of reinforcement/matrix interface was shown in Fig. 5(a) and (b). The microstructure and EDS spectra in conjunction of SEM analysis show the peaks for SiC, Al-Si, Mg and Fe. The results from the investigation of SEM are in excellent agreement with those derived from XRD analyses. The reinforcement/matrix interface bonding shows good properties as shown in Fig. 5(b). The good bonding between particle/matrix interface play an important role to ensure strong and stiffness of the composite during mechanical characterization.

CONCLUSION

The formation of $MgAl_2O_4$, Mg_2Si and MgO phases is due to the intermetallic compounds reaction products at interface. Heat treatment of SiC as surface reinforcement modification and the incorporation of magnesium alloying have been successful in producing good bonding at interface between particle reinforcement and matrix and enhancing the materials wettabilities.

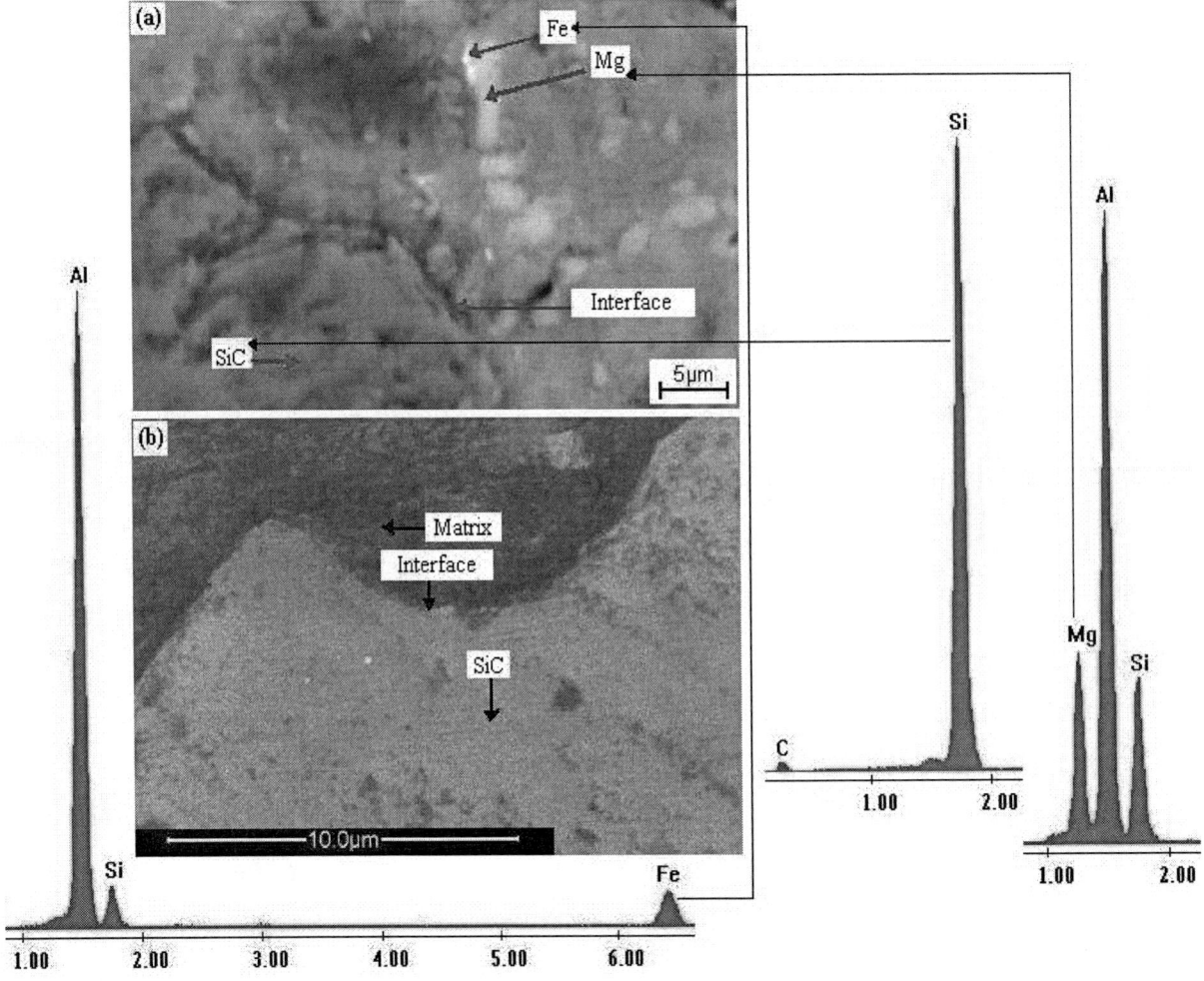

Figure 5. SEM microstructure and EDS spectrums analysis of Al-Si/15% SiC composite indicate SiC, Al-Si, Mg and Fe peaks composition.

REFERENCES

Hashim, J., Looney, L. & Hashmi, M.S.J. 2001. Journal of Materials Processing Technology 119, 324 -332.

Lee, J.C., Park, S.B., Seok, H.K., Oh, C.S. & Lee, H.I. 1998a. Prediction of Si contents to suppress the interfacial reaction in the SiCp/2014 Al composite. Acta Materialia, Volume 46, No. 8, 2635-2643.

Lee, K.B., Kim, Y.S. & Kwon, H. 1998b. Fabrication of Al-3 wt pct Mg matrix composites reinforced with Al2O3 and SiC particulates by the pressureless infiltration technique. Metal and Materials Transactions A, 29, 3087-3095.

Lee, J.C., Ahn, J.P., Shim, J.H., Shi, Z. & Lee, H.I. 1999. Control of the interface in SiC/Al composites. Scripta Meterialia 41(8): 895-900.

Mitra, R., Chalapathi Rao, V.S., Maiti, R. & Chakraborty, M. 2004. Stability and response to rolling of the interfaces in cast Al-SiCp and Al-Mg alloy-SiCp composites. Materials Science and Engineering A, 379, 391-400.

Ravi Kumar, .N.V., Pai, B.C. & Dwarakadasa, E.S. 2003. Microstructural evolution in liquid metal processed Al-alloy/SiCp composites. International Journal of Casting and Metallurgy Research, 15, 573-579.

Shi, Z., Yang, J.M., Lee, J.C., Zhang, D., Lee, H.I. & Wu, R. 2001. The interfacial characterization of oxidized SiCp/2014 Al composites. Materials Science and Engineering A303, 46-53.

Urena, A., Escalera, M.D. & Gil, L. 2002. Oxidation barriers on SiC particles for use in aluminium matrix composites manufactured by casting route: Mechanisms of interfacial protection. Journal of Materials Science 37, 4633-4643.

Instrumental Characterization Of Coir PITH By XRD, FTIR and SEM After Radium Adsorption From Aqueous Solution Under The Presence Of Humic Acids

Zalina Laili[1,2], Muhamat Omar[1], Muhamad Samudi Yasir[2], Mohd Zaidi Ibrahim[1], Mohd Yusri Yahaya[1] and Julie Andrianny Murshidi[1]

[1]*Malaysian Nuclear Agency (Nuclear Malaysia), Bangi, 43000 KAJANG, MALAYSIA*
[2]*Universiti Kebangsaan Malaysia (UKM). Bangi, 43000 KAJANG, MALAYSIA*
Liena@nuclearmalaysia.gov.my

ABSTRACT

Adsorption interactions of radium (Ra) ions onto coir pith (CP) under the presence of humic acid (HA) in the aqueous solution were investigated using X-ray diffraction (XRD), Fourier Transform Infrared Spectrophotometer (FTIR) and Scanning Electron Microscopy (SEM) techniques. XRD, IR and SEM characterization of the CP has revealed differences in the native CP and Ra ions loaded CP under the presence of HA in the aqueous solution. The X-Ray patterns showed that crystalline structure of the loaded CP exhibited a decrease in crystalline structure at around $49°-51°$ compared with unloaded CP. Characterization by IR revealed that participation of some surface functional groups during the Ra adsorption. SEM images for the morphological studies showed that there were slightly changes of the CP surfaces after the adsorption process.

Keywords: Coir pith; radium, humic acid; adsorption; XRD; FTIR; SEM

INTRODUCTION

A number of researchers have investigated the use of agricultural waste material as an adsorbent for the removal of pollutants in wastewater (Chamorthy et al., 2001; Basso et al., 2002; Villaescusa et al., 2004; Shibi & Anirudhan, 2006). The agricultural waste materials have been found to have good adsorption capacity due to inherent compound associated with cellulose such as lignin, tannin and pectin, which contains polyphenolic and aliphatic hydroxyl and carboxylic groups (Anirudhan et al., 2008). There are several mechanisms which involve in the adsorption process by the adsorbent namely Van der Waals forces, steric interaction, hydrogen bonds, hydrophobicity and polarity. Hence, the physical and chemical properties of the adsorbent are very important in controlling the adsorption process. Burg et al. (2000) reported that the selectivity and maximal quantity of pollutant adsorbed are controlled by the chemical functions at the solid sorbent surface and the solid specific area. Thus, it is very important to characterize the adsorbent before and after the adsorption process because it can help to clarify the potential mechanism involve in adsorption and provide the physical and chemical evidence about the success of adsorption.

In this study, coir pith is presented as one biosorbent alternative for the removal of radium from aqueous solution. Coir pith (CP) is an indigenously available agricultural solid waste, which is generated in the process of coir fiber from coconut husk. Raw coir pith consists of 35% cellulose, 25.2% lignin, 7.5% pentosans, 1.8% fats and resin, 8.7% ash content, 11.9 % moisture content and 10.6% other substances (Dan, 1993). Coir pith has been reported as a good adsorbent for the arsenic (V) (Anidrudhan & Unnithan (2007), chromium (Suksabye et al., 2007), nickel (Ewecharoen et al. 2008) and dyes (Sureshkumar & Namasivayam 2008). Therefore, the purpose of this study to characterize the coir pith before and after the Ra adsorption in order to evaluate the physical and chemical properties changes of coir pith. The coir pith was characterized with respect to their morphology and surface chemistry.

MATERIALS AND METHODS

Coir pith (CP) was prepared from coconut husk obtained from Bagan Datoh, Perak Darul Ridzuan. It was ground and sieved by USA Standard Sieve No. 10, 14, 18 and 35 (corresponding to 2000 μm, 1410 μm, 1000 μm, 500μm, respectively) with a sieve shaker (Fritsch model Analysette 3 Spartan, Germany). The CP was suspended in 500ml of 5% NaOH with constant stirring for about 24 h followed by a thorough washing with distilled water. It was then filtered and oven-dried at 105˚C. Humic acids (HAs) were extracted from the peat soils using the method recommended by International Humic Substances Society (IHSS) with minor

CP1202, *Neutron and X-Ray Scattering in Advancing Materials Research: International Conference – 2009*
edited by A. Saat, H. A. Kassim, M. H. H. Jumali, J. M. Saleh, M. R. Othman, A. Ibrahim, F. M. Idris, and M. H. A.-R. M. Ahmad
© 2009 American Institute of Physics 978-0-7354-0739-8/09/$25.00

Table 1 FTIR spectrum of fundamental peaks of HA, native CP (before) and after Ra adsorption (CPHARa) (cm^{-1})

Compound	O-H	C-H	C=O	C=C	O-CH$_3$	C-OH
HA	3362.8	2920.3	1705.7	1610.4	1414.9	1214.4, 1060.3*
CP (native)	3341.5	2929.1	1734.0*	1605.3	1450.4	1262.9, 1035.1
CPHARa	3329.9	2900.4	1734.6*	1604.0	1424.9	1262.6, 1032.4

* weak

modification (Ibrahim et al. 2008). Then, HAs stock solution with 0.2 mgml^{-1} was prepared by dissolving 0.1g HAs in 0.01M NaOH in a 500 ml volumetric flask. The chemicals used were of analytical grade. ^{226}Ra stock solution was obtained from Isotope Products Laboratories (Eckert & Ziegler Company). A stock solution of 245.48 Bq of ^{226}Ra (was prepared by diluting 1.7g of ^{226}Ra stock solution (361.3kBq) with 250 ml of distilled water in a 250-ml volumetric flask. ^{226}Ra working solutions with the concentrations of 16 Bqml^{-1} was freshly prepared by appropriate dilution of the stock solution prior to their usage.

Adsorption Experiments

The adsorption or Ra^{2+} were studied by adding 40 ml of 16 Bqml^{-1} ^{226}Ra working solution to 10 ml of 0.2 mgml^{-1} of HAs solutions in 50 ml centrifuge tubes containing 1.0g of CP and adjusting the pH to 9. The centrifuge tubes were sealed with screw caps. The mixtures was then shaken using a flask shaker machine at 300 oscmin^{-1} for 24 hours at room temperature. At the end of the adsorption period, the mixtures were filtered and the adsorbate loaded CP sample was dried in desiccator before they were characterized.

Characterization methods

The XRD patterns of the adsorbent were obtained with a PANalytical PW3040/60 diffractometer with Cu-Kα x-ray (λ = 1.54060 Å). The data were collected at room temperature in the range of 2θ between 5° to 60°. The surface morphology of CP before and after the Ra adsorption was examined by a scanning electron microscope (SEM) (model FEI 400). The analysis on the functional groups of native CP before and after the Ra adsorption was performed using Fourier transform infrared spectrophotometer (FTIR) (model Spectrum 2000/L183, USA) in the range 400-4000 cm^{-1}.

RESULTS AND DISCUSSION

X-ray diffraction (XRD) results

X-Ray diffraction provides a convenient and practical means for the qualitative identification of crystalline compounds (Skoog & Leary 1992). The XRD pattern of humic acid (HA), the native CP, the coir pith after treatment with NaOH (CPNaOH) and the coir pith after Ra adsorption under the presence of humic acid (CPRaHA) are shown in Fig. 1. The XRD pattern of the

native CP shows scattering angles at 2θ = 21.2, 41.9, 43.7, 49.1 and 51.1) which indicates a crystalline domain of cellulose structure. According to Anirudhan et al. (2007), in the crystalline region cellulose molecules are arranged in an ordered lattice in which –OH groups bonded by strong secondary forces. This crystalline phase indicates the porosity of the CP. The XRD pattern of the CP after Ra adsorption shows a decrease in the in the intensity around 49° - 51° which indicates a decrease in crystalline structure has occurred.

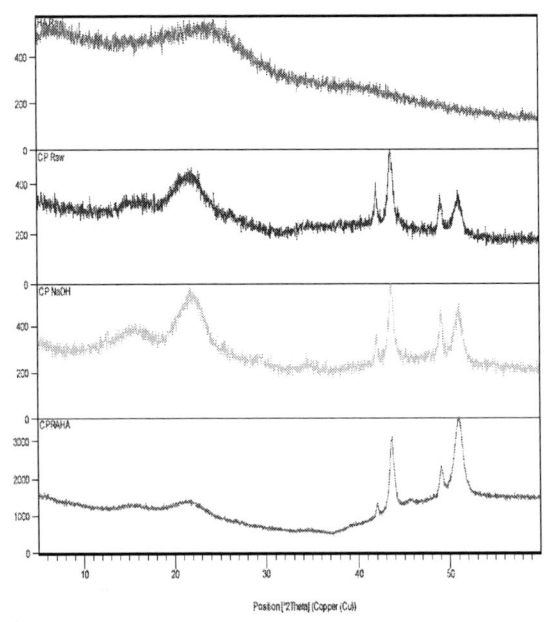

Fig. 1 XRD pattern of humic acid (HA), native coir pith (CP) before Ra adsorption, coir pith after treatment with NaOH (CPNaOH) and coir pith after Ra adsorption under the presence of humic acid (CPRaHA)

FTIR results

The FTIR spectra of the HA, native CP before and CP after the Ra adsorption are shown in Fig. 2. The FTIR spectrum for the fundamentals peaks of HA, native CP before and after the Ra adsorption are shown in Table 1. The FTIR spectrum of HA reveals the main absorption bands are at 3362.8 cm^{-1} (H-bonded OH groups), 2920.3 cm^{-1} (aliphatic C-H stretching), 1705.6 cm^{-1} (C=0 stretching of COOH and ketonic C=O), 1610 cm^{-1} (aromatic C=C and H bonded C=O) and 1250 cm^{-1} (C-O stretching and O-H deformation of COOH groups).

There are small bands at 1507.0 cm^{-1} (aromatic C=C), 1407.8 cm^{-1} (C-H deformation of CH$_2$), 1365.2 cm^{-1} (O-H deformation, CH$_3$ bending, or C-O stretching).The FTIR spectrum of CP before adsorption reveals a broad peak around 3334.5 cm^{-1}, which can be attributed to the O-H groups from cellulose structure (Anirudhan et al., 2007). The peak observed at 2929.1cm^{-1} corresponds to C-H stretching. The spectrums at 1800 - 1000 cm^{-1} were

wavenumber indicated that the interaction of -OH groups was greater after Ra adsorption in the presence of HA in aqueous solutions. It was also observed that there were a small shift of wave number for an aromatic C=C and C-OH groups. The shift of wave numbers might indicated that there were a chemical interaction between the adsorbent, HA and Ra ions. The C-H stretching groups are shifted to lower wavenumber. This is due to the

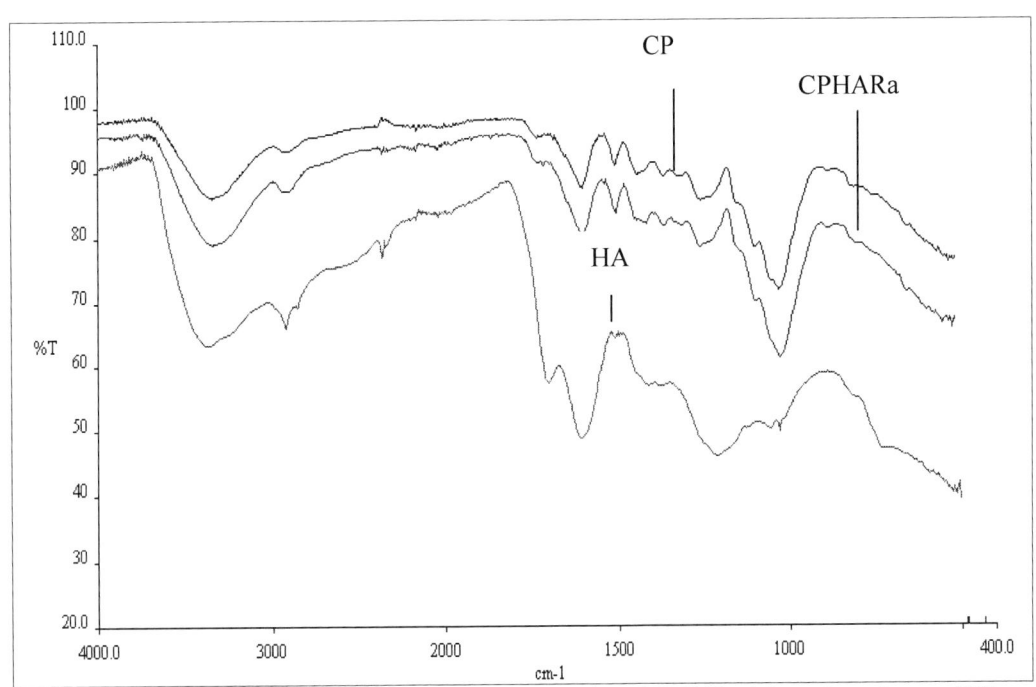

Fig. 2 FTIR spectrum of HA, CP (before adsorption) and CPHARa (after adsorption at pH 9)

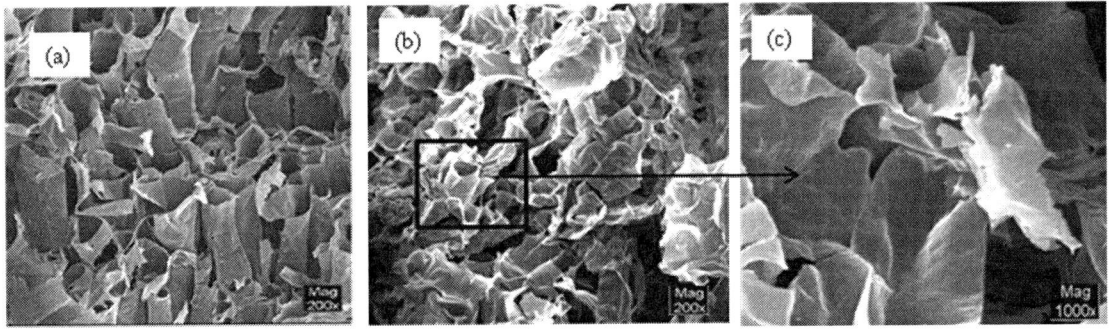

Fig. 3 SEM micrograph (a) native CP (magnification of 200x), (b) CP after the Ra adsorption under the presence of HA at pH 9 (magnification of 200x) and (c) the close- up of CP after Ra adsorption (magnification of 1000x)

fingerprint regions (Suksabye et al., 2007). These peaks are corresponding to the carbonyl stretching groups (1726.9 cm^{-1}), C=C stretching of aromatic ring vibration (1605.3 cm^{-1}, 1509.2 cm^{-1}), methoxy groups (O-CH$_3$, 1429 cm^{-1}) from lignin structure of CP (Khan et al., 2004), O-H deformation of phenolic group (1365.2 cm^{-1}) and C-OH stretching (1262.9 cm^{-1}, 145.3cm^{-1}, 1035 cm^{-1}). In the FTIR spectrum of CP after the Ra adsorption under the presence of HAs (CPHARa), it is clear that an OH stretching vibrational band shows a frequency shift to lower wave number (3341.5 to 3329.9 cm^{-1}) after the adsorption process. The shift to lower

increase in C-H bond distance, delocalization of electron density and thus the complexes are having proper hydrogen bonded complexes (Kolandaivel & Nirmala, 2004).

SEM results

SEM image in Figure 3 (a) shows the morphological of native CP before Ra adsorption at 200x magnification. At such magnification, the coir pith particles showed that the CP surface consists of close thin-walled ribbon shape cells and porous pith tissue were clearly identifiable. Figure 4(b) shows the SEM image of CP after the Ra adsorption (magnification of 200x) under the presence of HAs in aqueous solution. From the image, it could be seen that there were slightly changed of the CP surfaces after the adsorption process. The structure of porous pith tissue was disrupted and tended to coagulate after Ra adsorption.

CONCLUSION

The present study clearly shows that the characterization data have confirmed that there was a change in morphology and surface chemistry of coir pith before and after the Ra adsorption. The XRD pattern revealed that the crystalline structures are disrupted due to Ra adsorption. The FTIR and SEM characterization of the adsorbents has shown a clear difference in the native and Ra ions loaded adsorbents under the presence of HA in aqueous solution.

REFERENCES

Anirudhan, T.S., & Unnithan, M.R., (2007), Arsenic (V) removal from aqueous solutions using an anion exchanger from coconut coir pith and its recovery. *Chemosphere* 66: 60-66

Burg, P., Fydrych, P. Abraham, M.H., matt, M and Gruber, R. (2000). The characterization of an active carbon in terms of selectivity towards volatile organic compounds using an LSER approach. *Fuel* 79: 1041-1045

Dan, T.K. (1993), Development of light weight building bricks using coconut coir pith, *Ind. Coconut J.* 23: 12-19

Ewecharoen, A. Thiravetyan, P. and Nakbanpote, W., (2008), Comparison of nickel adsorption from electroplating rinse water by coir pith and modified coir pith, *Chemical Engineering Journal* 137: 181-188

IHSS. Isolation of IHSS soil fulvic and humic acids. International Substances Society, Available from:http//www.ihss.gatech.edu. [12 Mac 2007]

Khan, M.A. Ashraf, S.M., and Maholtra, V.P. 2004, Development and characterization of a lignin-phenol-formaldehyde, *J.Adhes.Adhes.* 24:485-493

Kolandaivel, P. and Nirmala, V., (2004), Study of proper and improper hydrogen bonding using Bader's atoms in molecules (AIM) theory and NBO analysis, *J. Mol. Struct.* 694(1-3): 33-38

Sureshkumar, M.V. and Namasivayam, C. (2008). Adsorption of direct red 12B and rhodamine B from water onto surfactant-modified coconut coir pith. *Colloids and Surfaces* 317:117-283

Suksabye, P. Thiraveyan, P. Nakbanpote, W. & Chayabutra, S. (2007). Chromium removal from electroplating wastewater by coir pith. *Journal of Hazardous Materials* 141:637-644

Structural And Electrical Properties oF $(La_{0.5-x}Pr_xBa_{0.5})(Mn_{0.5}Ti_{0.5})O_3$ Perovskite

Nor Hayati Alias[1,2]*, **Abdul Halim Shaari[2]**, **Wan Mohd Daud Wan Yusoff[2]**, **Che Seman Mahmood[1]**

[1]*Materials Technology Group, Malaysian Nuclear Agency, 43000, Kajang, Selangor Darul Ehsan, Malaysia*
[2]*Department of Physics, Faculty Science, University Putra of Malaysia (UPM), 43400, Serdang, Selangor Darul Ehsan, Malaysia*
**Corresponding author, email norhayati@nuclearmalaysia.gov.my*

ABSTRACT

A single phase monoclinic new perovskite based titano-manganite $(La_{0.5-x}Pr_xBa_{0.5})(Mn_{0.5}Ti_{0.5})O_3$ has been successfully prepared by ceramic solid-state technique at sintering temperature of 1300 °C. The concentration of Pr (Praseodymium), x, in molar proportion in A site has been varied as x = 0, 0.02 and 0.2. Analysis has been carried out to determine the electrical properties of the synthesized material at frequency ranging from 5 Hz to 1 MHz; and at temperature range between 25 °C to 200 °C. It is found that Pr addition promoted liquid phase sintering diffusion, porosity and agglomeration formation at 1300 °C. Dual relaxation is observed in unsubstituted Pr sample x=0 and high Pr substituted sample x=0.2. This phenomenon was a combinational contribution from a quasi dc (QDC) or low frequency dispersion (LFD), two cole-cole relaxational responses and a resistor. While low concentrated Pr substituted sampled x=0.02 shows a combinational contribution from a quasi dc (QDC) or low frequency dispersion (LFD), single cole-cole relaxational response and a resistor at room temperature. Pr substitution at x=0 (max 12000) and x=0.2 (max 16000) showed high dielectric values compared to low substituted sample x=0.02. Variation of dielectric loss tangent (tan δ) are observed for all samples at temperature ranged studied.

Keywords: perovskite, concentration, structure and electrical properties

INTRODUCTION

Recent interests in exploring new form of functional material have been renowned in the perovskite (ABO_3) structure. The structure is found to be very versatile which has many technological applications. The material which belongs in this structure group can show properties ranging from colossal magnetoresistance (CMR), superconductivity and ferroelectricity [1-5]. The ferroelectric and dielectric behaviour of this material are important in electronic industry. Hence, it is observed that there will be a continuing demand in studying new dielectric material for electronic system which is cheaper, small in size, more reliable, less loss and with the capability of increased functionality [6-7]. This diverse physical behaviour is as a result of substitution of rare earth and alkali metal cations into the polycrystalline structure; by controlling parameter x substitution concentration in A or B site. In this study a new titano manganite based material $(La_{0.5-x}Pr_xBa_{0.5})(Mn_{0.5}Ti_{0.5})O_3$ has been synthesized by solid-state reaction at temperature 1300 °C as to investigate its structural and dielectric electrical properties. The Pr (Praseodymium) substitution concentration in molar proportion in A site has been varied as x = 0.0, 0.02 and 0.2.

MATERIALS AND METHODS

Polycrystalline samples $(La_{0.5-x}Pr_xBa_{0.5})(Mn_{0.5}Ti)O_3$ have been prepared by solid-state reaction technique in molar proportion at Pr (x) concentration = 0, 0.02 and 0.2 from mixture of La_2O_3, Pr_6O_{11}, $BaCO_3$, MnO_2 and TiO_2 oxide with high purity >99.4%. Sintering process is done in air for 1300 °C for 24h. The sintered pellets are named as sample A1(x=0), A2(x=0.02) and A3(x=0.2) respectively. Structural and dielectric electrical effect of Pr substitution concentration x was investigated by using Scanning Electron Microscopy (SEM), X-Ray Diffraction (XRD) and Impedance Analyzer HP 4192A LF at frequency 5 Hz to 1M Hz in temperature ranging from 25 °C to 200 °C.

CP1202, Neutron and X-Ray Scattering in Advancing Materials Research: International Conference – 2009
edited by A. Saat, H. A. Kassim, M. H. H. Jumali, J. M. Saleh, M. R. Othman, A. Ibrahim, F. M. Idris, and M. H. A.-R. M. Ahmad
© 2009 American Institute of Physics 978-0-7354-0739-8/09/$25.00

RESULTS AND DISCUSSIONS

Figure 1 (a-c) show the SEM micrograph of Pr (x) substitution concentrations for samples A1(x=0), A2(x=0.02) and A3 (x=0.2) respectively. Unsubstituted Pr(x) sample, A1 possess clear grain-grain-boundary region. It is observed that Pr addition would enhance the liquid phase sintering at 1300 ^{0}C for sample A2 and A3. The liquid phase is dominant in A2. Both A2 and A3 samples show formation of agglomeration and porosity in the microstructure as viewed in Figure 1(b) and (c).

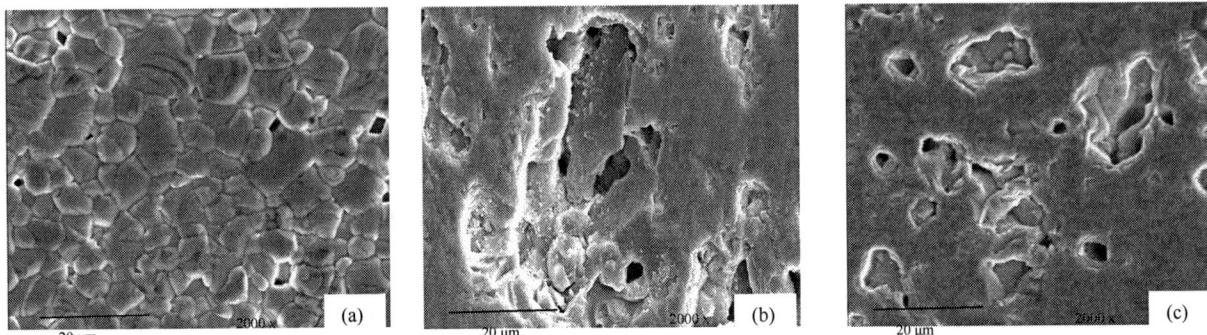

Figure 1: SEM micrographs of Pr (x) substitution concentrations: (a) A1(x=0), (b) A2(x=0.02) and (c) A3(x=0.2).

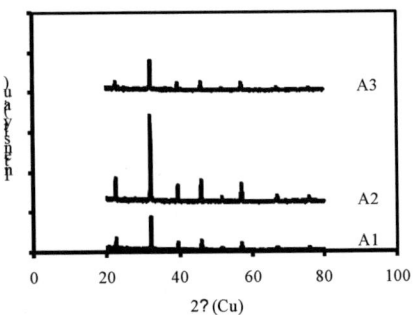

Figure 2: XRD diffractogram of samples after sintering at 1300 ^{0}C.

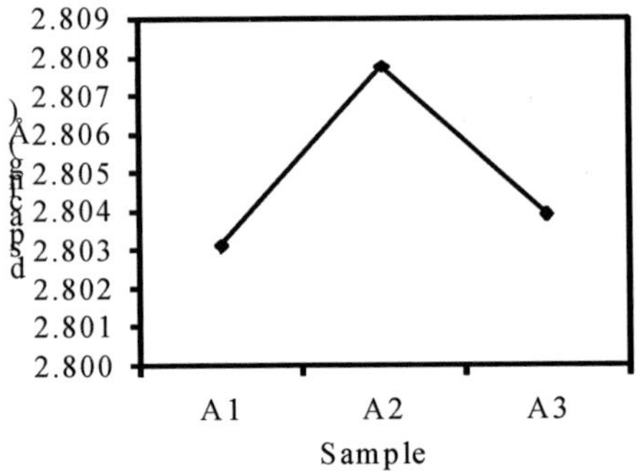

Figure 3: d spacing of sample A1, A2 and A3.

198

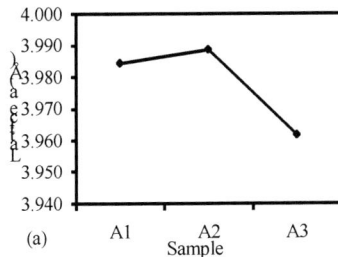 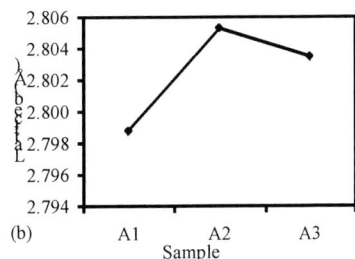 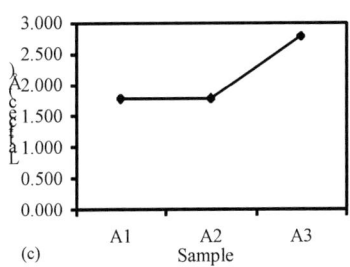

Figure 4(a)-(c): Lattice parameters of sample A1, A2 and A3.

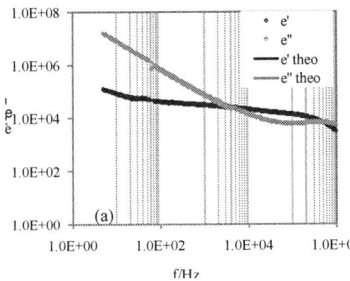 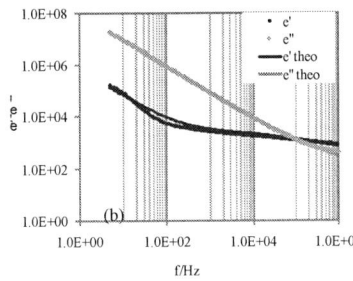 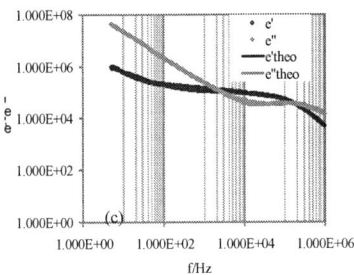

Figure 5: ε', ε'' vs frequency plot for sample (a) A1, (b) A2 and (c) A3 at 25 ℃.

The sharp and single diffraction peaks in Figure 2 of the polycrystalline $(La_{0.5-x}Pr_xBa_{0.5})(Mn_{0.5}Ti_{0.5})O_3$ compounds at Pr(x) substitution concentration x=0 in sample A1, x=0.02 in sample A2 and x=0.2 in sample A3 indicate crystal structural homogeneity and better crystallization of the samples. It is thus observed in Figure 2, there is no trace of individual peaks present;

indicating a single phase homogen crystal structure formed from the sintering temperature treatment.

This also indicates that sintering at high temperature (1300 ºC) is able to ensure the total elimination of impurity and enabling perovskite solid solution formation.

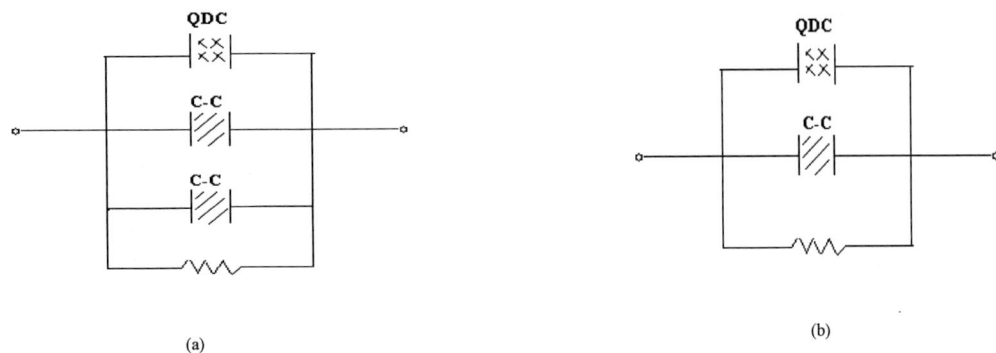

(a)
(b)

Figure 6. Suggested circuit model fitting for (a) combination of a quasi dc (QDC) of low frequency dispersion (LFD), dual cole-cole (C-C) relaxation and a resistor in A1 and A3 and (b) combination of QDC, a single cole-cole relaxation and resistor in A2.

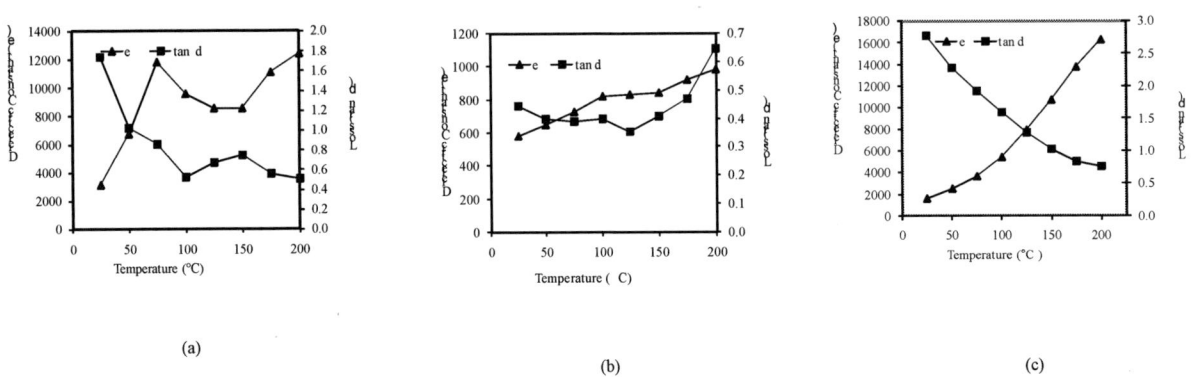

(a)
(b)
(c)

Figure 7: Plot of variation of dielectric constant (ε) with the dielectric loss tan δ vs. temperature for samples (a) A1; (b) A2 and (c) A3 at frequency 1 MHz.

Crystallographic indexing by using DIVCOL REFINEMENT (Philips Xpert HighScore Plus) indicating a single phase monoclinic P112 obtained. The error is closed to $\pm$ 0.02 delta two theta value. Figure 3 and 4(a-c) give the result of d spacing and lattice parameters calculated from the indexing respectively. In Figure 3, sample with low Pr substitution (A2) gives higher d spacing value which is 2.805Å. It can be seen that in Figure 4(a); with high concentrated concentration x=0.2 in sample A3 the lattice a parameter would decrease. The lattice b parameter seems to expand with Pr(x) substitution as depicted in Figure 4(b).

The results also indicates that the lattice parameter c of sample low substitution concentration x=0.02 in sample A2 remains unchanged within the experimental error in Figure 4(c).

Figure 5 shows the dielectric relaxational permittivity dispersion (ε') and absorption (ε') of sample A1, A2 and A3 at temperature 25°C at frequency (f) ranging from 5Hz to 10^6Hz (1 MHz). From Figure 5(a-c) dual relaxation responses are only observed for unsubstituted sample x=0 in sample A1 and high substituted sample

x=0.2. This is explained by two spectral intercept in ε' vs f and ε'' vs f spectra. Relatively parallel real and imaginary permittivity part in low spectral region observed in figure 5(a-c) is an indicative of enhance mobility of free charge carriers. This is termed as a low frequency dispersion (LFD) [8] or quasi-DC behavior (QDC) [9]. The movement of free charge carriers in these samples is enhanced by the presence of liquid phases.

The behavior of a dielectric in alternating field is examined by the approach of equivalent circuit which visualized the lossy dielectric to gain insight of the macroscopic properties of the synthesized material. The dielectric spectral had been fitted using Cole-Cole (C-C) model, Quasi Dc (QDC) and Resistor (R).

The suggested circuit model fitting for dual relaxation is viewed in Figure 6(a) is combination of a quasi dc (QDC) or low frequency dispersion (LFD), dual cole-cole (C-C) relaxation and a resistor (R) while single relaxation is viewed in Figure 6 (b) is combination of QDC, a single cole-cole (C-C) relaxation and a resistor (R).

Cole-Cole mechanism is described as [10]:-
$\chi^*(\omega) \propto \dfrac{1}{1 + (i\omega/\omega_p)^{1-\alpha}}$ and quasi-dc dispersion is

represented by $\chi(\omega) \propto \chi(0)(i\omega/\omega_c)^{-p} =$
$\chi(0)(\omega/\omega^c)^{-p} \times \{\cos(p\pi/2) - i\sin(p\pi/2)\}$ for

frequencies less than characteristic rate ω_c, and
$\chi(\omega) \propto \chi(0)(i\omega/\omega_c)^{n-1} =$

$\chi(0)(\omega/\omega_c)^{n-1} \times \{\sin(n\pi/2) - i\cos(n\pi/2)\}$ for

frequencies greater than characteristic rate ω_c [9].

Figure 7 (a-c) shows the dielectric constant and loss tangent (tan δ) values versus temperature for samples A1, A2 and A3 respectively at frequency 1 MHz. The dielectric data is described as complex permittivity ; ε*(ω) = ε'(ω) - jε' '(ω) which is real part , ε' (describing energy storage in the medium) and imaginary part, ε'' (describing energy loss in the medium). The dielectric loss tangent is described as, tan δ = ε''/ε''. Observed, only substantial amount of extrinsic defect due to porosity and liquid phase present can alter to the benefit of the dielectric constant and loss values. Higher dielectric ε values are obtained for sample A1(max 12000) and A3 (max 16000) due to dual relaxation responses. Dielectric constant and loss tan δ value is contribution from the effect of space charge polarization and/or conducting ion motion [11]. Loss tan δ value is increasing with temperature in sample low Pr substitution A2 while decreasing in sample A1 and A3.

CONCLUSIONS

Pr addition would promote liquid phase sintering (high dissolution effect) in the synthesized material. XRD results indicate a single phase monoclinic P112 structure. The dielectric electrical effect is depended on the structural and electrical responses of the synthesized material. High dielectric constant and decreased loss is achieved in sample A1 and A3.

REFERENCES

[1] V. K Wadhawan, Introduction to Ferroic Materials, (Gordon & Breach, UK, 2000).

[2] S. H. Curnoe, I. Munawar, Physica B 378–380 (2006) 554.

[3] H. Schmid, Ferroelectrics 162, (1994) 317.

[4] D.K. Pradhan, B.K. Samantary, R.N.P. Chaudhary, A.K. Thakur, Mater. Sci. Eng. B 116 (2005) 7.

[5] M. Nadeem, M.J. Akhtar, A.Y. Khan, Solid State Commun. 134 (2005) 431.

[6] P. Jha, P. Arora, A. K. Ganguli, Mater. Lett. 57 (2003)2443.

[7] M. A. Subramaniam, D. Li, N. Duan, B. A. Reisner, A. W. Sleight, J. Solid State Chem. 151 (2000)323.

[8] Jonsher, A. K. (1983). Dielectric Relaxation in Solids. London, UK: Chelsea Dielectric Press.

[9] Dissado, L. A and Hill, R. M. (1984). J. Chem. Soc. Faraday Trans. 2(80): 291-319.

[10] K. S. Cole, R. H. Cole, J. Chem. Physics 9 (1941) 341.

[11] K. Sambasiva Rao, D. Madha Prasad, P. Murali Krishna, B. Tilak and K. Ch.Vandarajulu, Material Science and Engineering B 133 (2006) 141.

Density-Functional Approximation for the Spin Dependent Quantum Transport in Magnetic Nanostructures

Khine Nyunt

Faculty of Engineering, Multimedia University Cyberjaya Campus, Persiaran Multimedia, 63100 Cyberjaya, Selangor, Malaysia.
Email: khine.nyunt@mmu.edu.my

ABSTRACT

In quasi-classical theoretical framework, the transport of electrons and holes in semiconductor devices is treated with the Boltzmann transport equation (BTE) or quantum-mechanical energy band theory - *viz.*, the effective mass approximation and the random phase approximation. On the other hand, in the mesoscopic, nanoelectronic devices, for three- and lower- dimensional structures with nanometer scaling, the wave properties, spin, charge and the interactions between spin and charge of electrons are fully utilized such as in artificial mini-Brillouin zones, quantum size effects, Coulomb blockade of single-electron tunneling and spin-polarized giant magnetoresistance (GMR) tunneling. The complexity associated with the classical quantum-mechanical formalism in the study of transport in magnetic nanostructures can be avoided by applying the so-called, Hohenberg-Kohn's density functional theory. Because of the limitations of quasi-classical theory, it is more appropriate to treat the magneto-transport problem in nanostructures by using quantum many-body theory. The starting point of the quantum trans-port theory is to take an external field as a perturbation for the many-particle system in equilibrium. This leads to a linear response and gives corresponding transport coefficients. One useful application of the Green's function techniques in quantum magneto-transport is to convert a homogeneous differential equation into an integral equation, *viz.*, as in the time-dependent Schrödinger equation. We have applied it to scattering of nanostructural defects (impurities) in the electron gas (metal) as many-body effect's model and derived an expression for its residual resistivity. Calculations of magnetic impurities in noble-metal hosts are in good agreement with the previously published results.

Key words: density functional theory, many-body problem, Green's function techniques, quantum confinement, magnetic nanostructures, magnetic impurities, metallic magnetic systems.

INTRODUCTION

Nanoengineering and nanoscience studies nanostructures, nano subsystems and nanodevices in nanoscale, which are made from atoms, molecules or molecular complexes, and the electron is considered as a fundamental particle. One of the main concerns is the behaviour of spin and charge and their interactions. Students and engineers have obtained the necessary quantum mechanics background in their mesoscopic physics classes. The behaviour of electrons and holes in microelectronics is well understood by treating them as classical particles and due to their population associated with the device operation, as in the effective mass approximation as well as in the random phase approximation for their transport [1]. The shrinking of the physical device to nanoscale and the realization of three-, two- or one-dimensional structures requires to develop a unique procedure for modeling of the

nanostructures to study their properties, processes, phenomena and the related effects. The complexity associated with many-electron wavefunctions in the classical solution of the Schrödinger equation has drastically limit the applicability of the conventional quantum mechanics.

There is a critical need to develop accurate and efficient computational procedures to perform quantum modeling and simulation of the nanoscale structures. The formulation of the quantum modeling and simulation problem can be developed by the density functional theory developed by Kohn & Sham [2] based on the Hohenberg-Kohn theory [1b] which, in principle, is an exact ground-state theory. As the name suggests, the fundamental variational parameter is the electron charge density rather than the electronic wave functions. In particular, the N-electron problem is formulated as N one-electron equations where each electron interacts with all other electrons via an effective exchange-correlation potential. These interactions are calculated

CP1202, *Neutron and X-Ray Scattering in Advancing Materials Research: International Conference – 2009*
edited by A. Saat, H. A. Kassim, M. H. H. Jumali, J. M. Saleh, M. R. Othman, A. Ibrahim, F. M. Idris, and M. H. A.-R. M. Ahmad
© 2009 American Institute of Physics 978-0-7354-0739-8/09/$25.00

using the local density approximation to exchange and correlation. Plane wave basis sets and total energy pseudo-potential methods can be used self-consistently, to solve the Kohn-Sham one-electron equations.

LOCAL DENSITY APPROXIMATION AND THE CHARGE DENSITY

The distribution of electrons in atoms was given by the statistical concept proposed, independently, by Thomas [3] and by Fermi [4]. In their consideration the following assumptions were made:(a) electrons are distributed uniformly, and (b) there is an effective potential field that is determined by the nuclei charge and the distribution of electrons. The energy analysis can be performed on electrons by assuming that they are confined in a 3-dimensional potential well. One finds the energy by making the sum of all energy levels. Hence, one can relate the total kinetic energy and the electron charge density. The approximation of distribution of electrons in an atom can be performed by using the statistical consideration. By using the so-called Local Density Approximation concept, the relation between the total kinetic energy of N electrons E and the electron density was derived.

The Thomas-Fermi kinetic energy functional is given by

$$\Gamma_F(\rho_e(r)) = 2.87 \int_R \rho_e^{5/3}(r)dr$$

(1)

and the exchange energy is found to be

$$E_F(\rho_e(r)) = 0.739 \int_R \rho_e^{4/3}(r)dr.$$

(2)

Considering the electrostatic electron-nucleus attraction and electron-electron repulsion, Thomas and Fermi derived the following energy functional by applying the electron charge density $\rho_e(r)$ for homogeneous atomic systems:

$$E_T(\rho_e(r))=2.87\int_R\rho_e^{5/3}(r)dr-q\int_R\frac{\rho_e(r)}{r}dr+\int_R\int_R\frac{1}{4\pi\varepsilon}\frac{\rho_e(r)\rho_e(r')}{|r-r'|}drdr.$$

(3)

Instead of using many-electron wave functions, Kohn [5] followed the idea and proposed to use the charge density for systems having N electrons. To perform analysis of molecular dynamics, only the knowledge of the charge density is required [7]. The charge density function (of three spatial variables x, y and z in the Cartesian coordinate system) is the one that describes the number of electrons per unit volume. One must applied the quantum mechanics and quantum modeling to analyze and understand the nanostructures and nano devices because they operate fully under the quantum mechanical effects. For example, in single electron transistor [6], signal charge is limited and elementary charge transfer is realized by a single electron using the Coulomb blockade of tunneling of electron to a quantum dot.

Under the external field, the total energy of an N-electron system is defined in terms of the 3-dimentional charge density $\rho(r)$ [1b, 2, 4]. The complexity is significantly reduced because the problem of modeling of N-electron, Z-nucleus molecular systems becomes equivalent to the solution of equation for single electron. Assume that the electron charge density $\rho(r)$ is smoothly distributed in R. Thus, the total energy is given as:

$$E(t,\rho(r)) = \Gamma_1(t,\rho(r)) + \Gamma_2(t,\rho(r)) + \int_R \frac{e\rho(r')}{4\pi\varepsilon|r-r'|}dr',$$

(4)

where the third term is for the potential energy due to interaction (repulsion) of the electron in R; the first and second terms are the interacting (exchange) and non-interacting kinetic energies of the single electron in an N-electron, Z-nucleus system:

$$\Gamma_1(t,\rho(r)) = \int_R \gamma(t,\rho(r))\rho(r)dr,$$

(5)

$$\Gamma_2(t,\rho(r)) = -\frac{\hbar^2}{2m}\sum_{j=1}^{N}\int_R \psi_j^*(t,r)\nabla_j^2\psi_j(t,r)dr;$$

(6)

$\gamma(t,\rho(r))$ is the parameterization function.

It should be emphasized that the Kohn-Sham electronic orbitals [2] are subject to the following orthonormal constraints:

$$\int_R \psi_i^*(t,r)\psi_j(t,r)dr = \delta_{ij}$$

(7)

where δ_{ij}, the Kronecker delta is 1 for i = j and 0 for i ≠ j.

The state of substance (media) depends largely on the balance between the kinetic energies of the particles and the interparticle energies of attraction.

The expression for the total potential energy is justified easily. The third term in Eqn. (4) represents the Coulomb interaction in R, and the total potential energy is a function of the charge density $\rho(r)$.

The total kinetic energy due to interactions of electrons and nuclei, and electrons is integrated into the equation for the total energy. The total energy, as given by Eqn. (4), is stationary with respect to variations in the charge density. The charge density is found by taking note of the time-dependent Schrödinger equation. The first-order Fock-Dirac [7] electron charge density matrix is

$$\rho_e(r) = \sum_{j=1}^{N} \psi_j^*(t,r)\psi_j(t,r).$$

(8)

The 3-dimensional electron charge density is a function in three spatial variables (x, y and z in the Cartesian co-ordinate system). The charge of the total number of electrons N can be obtained by integrating the electron charge density $\rho_e(r)$.

Thus,

$$\int_R \rho_e(r)dr = Ne.$$

(9)

Hence, $\rho_e(r)$ satisfies the following properties:

$$\rho_e(r) > 0, \quad \int_R \rho_e(r)dr = Ne,$$

$$\int_R |\sqrt{\nabla\rho_e(r)}|^2 \, dr < \infty,$$

and

$$\int_R \nabla^2 \rho_e(r)dr = \infty.$$

(10)

For the nuclei charge density, we have

$$\rho_n(r) > 0 \quad \text{and} \quad \int_R \rho_n(r)dr = \sum_{k=1}^{z} q_k.$$

(11)

There exist an infinite number of anti-symmetric wave functions that give the same $\rho(r)$. The minimum-energy concept (energy-functional minimum principle) is applied. The total energy is a function of $\rho(r)$, and the so-called ground state wave function ψ must minimize the expectation value $\langle E(\rho)\rangle$.

THE DENSITY FUNCTIONAL

The searching density functional $F(\rho)$, which searches all ψ in the N-electron Hilbert space H_ψ to find $\rho(r)$ and guarantee the minimum to the energy expectation value, is expressed as

$$F(\rho) \le \min_{\Psi \to \rho} \langle \Psi | E(\rho) | \Psi \rangle,$$

(12)

$$\Psi \in H_\Psi$$

where H_ψ is any subset of the N-electron Hilbert space.

Using the variational principle, we have

$$\frac{\Delta E(\rho)}{\Delta f(\rho)} = \int_R \frac{\Delta E(\rho)}{\Delta\rho(r')}\frac{\Delta\rho(r')}{\Delta f(r)}dr' = 0,$$

(13)

where $f(\rho)$ is the non-negative function.

Thus, $\dfrac{\Delta E(\rho)}{\Delta f(\rho)}\Big|_N = const.$

(14)

The solution to the high-fidelity modeling problem is found using the charge density as given by Eqn. (8).

ONE-PARTICLE DENSITY OF STATES AND THE GREEN'S FUNCTION

It is well known that many-body problems cannot be solved exactly. Therefore, one is obliged to resort either to statistical methods or to perturbation theory, and in many cases to both. Because of the complexities of condensed matter, one has to resort to modern techniques such as Feynman diagrams[9a-9c] and Green's functions[9d-9h] to solve a particular solid state problem. In condensed matter or solid state theory, the Green's function (GF) technique is widely used in both classical or statistical mechanics and quantum mechanics, e.g., in the quantum theory of many-body systems, viz., such as Kubo's quantum transport theory[10]. In principle, it is merely a particularly unified formulation of the usual quantum mechanical problem. One useful application of the Green's function technique is to convert a homogeneous differential equation into an integral equation, viz., as in the time-dependent Schrödinger equation. Mathematically, it provides a simplified method of solving the Schrödinger equation. The physical meaning of the GF used here is just so-called *propagator*, the wavefunction at r excited by the potential and associated wavefunction at r'. From a fundamental point of view, the GFs are related to the probability amplitude of a quantum state by bilinear expressions and are therefore connected with probabilities, rather than with probability amplitudes. Thus, in many cases, they have more physical significance than the 'not-directly-observable' probability amplitude.

The importance of Green's functions in condensed matter theory arises from the fact that exact expressions for many physical properties of interest (e.g., specific heat [11], resistivity [9g], mobility[12],optical conductivity[12], magnetic or spin susceptibility [9h][11]) and GMR [8] may be written in terms of them. For example, it can readily be shown that the one-particle density of states $\rho_d(\omega)$, could be written classically as

$$\rho_d(\omega) = \frac{2}{\hbar}\int\frac{d^d k}{(2\pi)^d}\delta\left(\omega - \frac{E_k}{\hbar}\right),$$

(15)

204

where d is the dimensionality of the system, and the factor s=2 is for electron-spin where the opposite spin do not raise the spin degeneracy or represents a summation over the diagonal spin indices of the Green's function, $G_{rs}^r(k,\omega)$. The local, static one-particle conductivity tensor σ_{ij} is given by

$$\sigma_{ij} = \frac{8\pi e^2}{\hbar} \sum_{nm} \int\int < \frac{\partial G(k,\omega)}{\partial k_i} \frac{\partial G(k,\omega)}{\partial k_j} > \delta(E_n - E_m)\frac{\partial f}{\partial E_n}\delta k_i \delta k_j$$

(16)

where f is the Fermi-Dirac distribution function and the brackets < > denotes an ensemble average.

If we re-write Eq.(1) in terms of the non-interacting Green's function, we obtain

$$\rho_d(\omega) = -\frac{s}{\pi\hbar} \int \frac{d^d k}{(2\pi)^d} \text{Im}\{G_{rs}^r(k,\omega)\} = ...$$

$$........................\frac{s}{\pi\hbar} \int \frac{d^d k}{(2\pi)^d} \text{Im}\{G_{rs}^a(k,\omega)\}$$

(17)

Hence, the density of states is related to the spectral density function, $A(k,\omega)$, through the retarded $G_{rs}^r(k,\omega)$, or advanced $G_{rs}^a(k,\omega)$, Green's function:

$$\rho_d(\omega) = \frac{s}{2\pi\hbar} \int \frac{d^d k}{(2\pi)^d} A(k,\omega). \qquad (18)$$

with

$$\int_{-\infty}^{\infty} A(k,\omega)d\omega = 1. \qquad (19)$$

In general, the decay of excitations is much strongly disordered (e.g., large impurity scattering and diffusive transport) system than it is in the pure system. This is of concern, since most mesoscopic (low-dimensional) systems are treated as disordered systems because of the randomness introduced by the impurity scattering. Now, by one-particle density of states, one doesn't mean the quantity for a single electron, but rather the density of states that is appropriate for a single electron in the sea of other electrons and impurities. We say that the Green's function is a temperature or thermal Green's function if the system is in a grand canonical ensemble at non-zero temperature. The so-called, temperature or Matsubara Green's function [13] can be recognized through the reverse of the analytic continuation procedures, which was

$$\rho_d(\omega) = -\frac{s}{\pi\hbar} \int \frac{d^d k}{(2\pi)^d} \text{Im}\{G_{rs}(k,i\omega \to \omega + i\eta)\},$$

(20)

and this may be related to the non-interacting Green's function and the self-energy by Dyson's equation as

$$\rho_d(\omega) = -\frac{s}{\pi\hbar} \int \frac{d^d k}{(2\pi)^d} \text{Im}\left\{\frac{1}{[G_{rs}^0(k,\omega_n)]^{-1} - \Sigma(k,i\omega_n)}\right\},$$

(21)

where the self-energy includes those parts from impurities and the electron-electron interaction, which itself may be mediated by the impurity scattering in the diffusive limit. The result for the interacting GF is essentially the same for zero-temperature Green's function, arises from the properties of the linear perturbation expansion and resummation into Dyson's equation itself, and not from any particular properties of the Green's function. Thus, the spectral density in the interacting system represents the entire density of states including both the single-particle properties and their modification (e.g., in the presence of strong impurity scattering) that arises from the self-energy corrections arising from the presence of the interactions.

The transport properties of quasi-low-dimensional system is an emerging area of research on nanostructures, such as in a quantum wire (QW), where the motion of electrons along the axis of the wire is free-electron like. If we restrict ourselves to carrier transport <u>parallel</u> to confining potentials, then we can discuss the homogeneous transport properties of such low-dimensional systems in the same context as homogeneous transport in bulk systems, i.e., in terms of macroscopic phenomenological transport parameters such as carrier mobility, diffusion coefficient, conductivity, thermo-power, and etc. In the case of transport <u>perpendicular</u> to the confining barriers, the transport in this context is quite different, in which quantum mechanical reflection and transmission from the confining barriers themselves play a central role.

In the discussion of transport properties of 1-D systems, in particular, many interesting predictions of many-body behaviour have been suggested, *e.g.*, the Luttinger liquid formation [14] or charge-density-wave ground state's presence due to a lattice distortion [15].Due to energy broadening effects of disorder in present systems, it leads to the normal Fermi liquid-like behaviour.

SCATTERING MECHANISMS IN LOW-DIMENSIONAL SYSTEMS

For many scattering mechanisms, the scattering rate in a low-dimensional system differs from that in 3-D system in the different initial and final states, which become increasingly restricted as the dimensionality is reduced. In a quasi- 1-D system, in particular, there are only two possible mechanisms of scattering, forward or backward with respect to wire axis. The density of states tends to be reflected in the scattering rates, being increased especially at the sub-band edges, as the dimensionality is reduced. In 1-D system, the scattering rate can be divergent at low energies because the singularity in the 1-D density of states (DOS) of the ideal system at the sub-band minimum [18]. New mechanisms that are not present *per se* in bulk systems, such as, surface- or boundary-roughness scattering, interface states, interface phonons, remote impurities, etc., also must be included. In addition, the nature of the allowed (acoustic and polar optical) phonon modes themselves are modified by the inhomogeneous structure associated with low-dimensional systems. The connection of these scattering mechanisms in quasi-, low-dimensional systems to the phenomenological transport parameters of the homogeneous systems may be found in [18].

IMPURITY LATTICE VIBRATION MODES

Lattice vibrations are affected by two kinds of impurities. Out of the two kinds; mass-difference defect and force-constant defect, they are inter-connected in reality. For illustrative purposes, we need to consider the simplest case, for a perfect crystal with a single atom per primitive cell. The algebraic equations of motion describing the stationary states of lattice vibrations is

$$M_s \omega^2 u_{ls\alpha} = \sum_{l's'\beta} V_{ls\alpha,l's'\beta} u_{l's'\beta}$$
(22)

$$\omega^2 u_l - \frac{1}{M} \sum_{l'} \Phi_{ll'}.u_{l'} = 0$$
(23)

in terms of the displacements u_l and the force constant Φ_{ll}. The displacement is transformed into where U_k is normalized. Because l runs from 1 to N, Eq.(17) represents a set of N vector equations. We can define a $3N \times 3N$ force constant matrix Φ,

$$D(\omega^2) = \omega^2 I - (1/M)\Phi$$
(24)

The Eqn.(22) is rewritten in terms of u, a $1 \times 3N$ column matrix standing for the set of displacement u_l. By supposing, for simplicity, that the mass is at the site $l=0$ is changed to M_o, the dynamic equations becomes

$$\omega^2 u_l - \frac{1}{M}\sum_{l'}\Phi_{ll'}.u_{l'} + \frac{\delta M}{M}\omega^2 u_o \delta_{lo} = 0$$
(25)

where $\delta M = M_o - M$. This can be written as

$$D(\omega^2)u + \delta(\omega^2)u = 0.$$
(26)

The mass defect leads to a perturbation of the dynamic matrix of the perfect crystal.

We now use the classical technique using the Green's function, G, and re-write Eqn.(25) in an algebraically equivalent form

$$(1 + G\,\delta D)u = 0,$$
(27)

where G is the matrix inverse of the dynamic matrix D satisfying

$$G(\omega^2)D(\omega^2) = 1.$$
(28)

After Eqn.(22) is solved, the eigenfrequencies, ω_k and eigenvectors, u_k are known, because they satisfy

$$\omega_k^2 u_k - \frac{1}{M}\Phi u_k = 0.$$
(29)

We can now write an expression

$$G^{-1}u_k = Du_k = (\omega^2 - \omega_k^2)u_k,$$
or
$$Gu_k = \frac{1}{(\omega^2 - \omega_k^2)}u_k.$$
(30)

Multiplying by $u_k{}^*$ on both sides, and then summing over the wavevectors k to include all the vibrational modes, we finally obtained the reduced form of the matrix $G(\omega^2)$:

$$G_{ll'}(\omega^2) = \frac{1}{N}\sum_k \frac{U_k U_k^*}{(\omega^2 - \omega_k^2)}e^{ik.(l-l')}.$$
(31)

Because we assume that the mass defect at the site $l=0$ provides only a highly localized perturbation, substitution of Eqn.(31) into Eqn.(27) gives an equation involving only u_o, the vector displacement on this site. We only need to consider $G_{oo}(\omega^2)$, i.e.,

$$\frac{\delta M}{NM}\sum_k U_k U_k^* \frac{\omega^2}{(\omega^2 - \omega_k^2)} = 1.$$
(32)

or

$$\frac{\delta M}{NM}\sum_k U_{k\alpha}U_{k\beta}^* \frac{\omega^2}{(\omega_k^2 - \omega^2)} = \delta_{\alpha\beta}.$$
(33)

This is in the form of a 3x3 matrix expression, written in the αth direction of the component of U_k. Assuming that the frequency and polarization of each normal modes of the perfect lattice are known, all frequencies of the normal, localized modes of the crystal with the defect can be found from Eqn.(32).

The graphical solution of the normal mode frequencies, which are the roots of Eqn.(33), for a cubic crystal having a defect with cubic symmetry. We look in such the plot where the explicit function $f(\omega^2)$ [part of Eq.(33)] intersects the horizontal line at $(-M/\delta M)$. δM is either positive (negative) for heavy (light) impurity, then each root of the solution must lie below (above) the pole of ω_k^2 of $f(\omega^2)$, respectively. We notice that the highest root is not constrained, but can move away from the top of the band by a finite amount [17].

FRIEDEL-ANDERSON ANALYSIS

One of the important areas of strongly-correlated electron states involves magnetic impurities in metals, where the representative many-body effect's model applies as an electron gas. The effect of magnetic moment formation and its influence on the resistivity of heavy electron (heavy fermions) metals leads to the Kondo problem[20]. Our attention here is to focus on the Friedel-Anderson Hamiltonian of a magnetic impurity in metals for its simplicity, as well as its importance in many-body theory[21][22]. Then, we will be ready for a full treatment related to the residual resistivity calculation using the GF techniques.

The Friedel-Anderson model is famous for its validity in achieving the results and, also for its flexibility in handling similar problems and its easiness to get the quantitative results. Thus, it becomes one of those important theories to handle many-body effects. The fundamental point of the Friedel-Anderson model is the use of a localized description for the impurity and a delocalized description for the electrons in the host metal. In the simplest case, the single-orbital, spin-up d-electrons on the impurity site, the Hamiltonian may be written as

$$H_d = \sum_\sigma \varepsilon_d n_{d\sigma} + U n_{d\uparrow} n_{d\downarrow}.$$

(34)

The atomic d-level is split, polarized by on-site Coulomb repulsion and is broadened by hybridization with s-electrons in the conduction band of the host, whose Hamiltonian is written as

$$H_s = \sum_{k\sigma} \varepsilon_k n_{k\sigma}.$$

(35)

Further introduction of an s-d hybridization interaction term with strength V_{kd}

$$H_{sd} = \sum_{k\sigma} V_{k\sigma}[c_{k\sigma}^* c_{d\sigma} + c_{d\sigma}^* c_{k\sigma}].$$

(36)

The Friedel-Anderson Hamiltonian is just the sum of the above three terms,

$$H = H_d + H_s + H_{sd}.$$

(37)

The energy for single occupation of the d-level is ε_d, while that for double occupation (will cost energy U, the Hubbard energy) is $\varepsilon_d + U > \varepsilon_F$. Due to the interaction between the localized d-state and the Bloch states of the conduction electrons of the host metal, s-d hybridization takes place as a perturbation of the d-level. This hybridization of the atomic d-level and the s-band of the host due to quantum mechanical tunneling can be interpreted as formation of a virtual bound state by resonant scattering of conduction electrons. The transition rate between the d-level and s-band can be described by the Fermi golden rule, in which the density

of states in the s-band at the d-level $\rho(\varepsilon_{d\sigma})$ make the d-level extend to some resonance width with $\Gamma_\sigma = \pi V^2 \rho(\varepsilon_{d\sigma})$; thus, the sharp level at $\varepsilon_{d\sigma}$ will be replaced by a spectral density function of a Lorentzian shape

$$A_{d\sigma}(\varepsilon) = \frac{1}{\pi} \frac{\Gamma_\sigma}{(\varepsilon - \varepsilon_{d\sigma})^2 + \Gamma_\sigma^2}.$$

(38)

A slight shift of the center can be added with no particular physical significance. The consequence of the broadening of the d-levels by resonance is that the occupancy of the $d\uparrow$ level

$<n_{d\uparrow}>$ becomes <1, while $<n_{d\downarrow}>$ becomes >0, so whether the local moment is still retained on the impurity site demands further theoretical research. We may evaluate now

$$<n_{d\uparrow}> = \frac{1}{\pi} \int_{-\infty}^{\varepsilon_F} \frac{\Gamma_\sigma d\varepsilon}{(\varepsilon - \varepsilon_{d\sigma})^2 + \Gamma_\sigma^2} = \frac{1}{\pi} arc \cot \frac{\varepsilon_d - \varepsilon_F + <n_{d\downarrow}>U}{\Gamma_\sigma}.$$

(39)

Of course, we have a comparable expression for $<n_{d\downarrow}>$, so

$$<n_{d\downarrow}> = \frac{1}{\pi} arc \cot \frac{\varepsilon_d - \varepsilon_F + <n_{d\uparrow}>U}{\Gamma_\sigma}.$$

(40)

Self-consistent solutions of these equations are plots of graph of the two equations with chosen range of parameters. The formation of the local magnetic moment is a cooperative phenomena and is due to the second term in the Hamiltonian [Eqn.(34)]. Through this analysis, Anderson explained [17] why the iron group elements (transitional impurities: V, Cr, Mn, Fe and Co) show local magnetic moments when they are dissolved in Cu and Ag but showed no local magnetic moments when dissolved in Al.

THE ELECTRON SPIN DENSITY OSCILLATION AND RKKY INTERACTION

In dilute solid-solution of a magnetic ion in a non-magnetic, metal crystal, the exchange coupling between the ion and the conduction electrons has important consequence. Before considering consequences of this interaction for dilute alloys, let us discuss its role in producing indirect coupling between local moments, i.e., the scattering of conduction electrons by the impurity moment can be spin-dependent.

By taking the electron-spin into account, the eigenstate of a conduction electron is

$$\psi_{k\sigma}(r) = \Omega^{-\frac{1}{2}} e^{ik.r} |\sigma\rangle,$$

(41)

where spin index $\sigma = \uparrow$, or $\downarrow$. Consider a local moment with spin $\mathbf{S}$ located at the site R, which affects the spins of conduction electrons s_j described by a contact interaction

$$H' = -J \sum_j s_j . S \delta(r-R)$$

(42)

Then the spin of each conduction electron experiences an effective field coming from the impurity moment

$$H_{eff}(r) = -\frac{J}{g_L \mu_B} S \delta(r-R),$$

(43)

where g_L and μ_B are the Landé factor and Bohr magneton, respectively. For such a field the response of an electron gas is determined by its susceptibility $\chi(\mathbf{q})$. Since the Fourier transform of this field is

$$H_{eff}(r) = -\frac{J}{g_L \mu_B} S,$$

(44)

we get the spin density at r as,

$$s(r) = \frac{J}{g_L^2 \mu_B^2 \Omega} \sum_q \chi(q) e^{iq.r} S,$$

(45)

by assuming $\mathbf{R} = 0$. For a free-electron gas, the spin susceptibility $\chi(\mathbf{q})$ is given by

$$\chi(q) = \frac{3 g_L^2 \mu_B^2}{16 E_F} \frac{N}{\Omega} \left[1 + \frac{k_F}{q} (1 - \frac{q^2}{4 k_F^2}) \ln \left[\frac{2 k_F + q}{2 k_F - q} \right] \right],$$

(46)

where N is the total number of electrons in the system. To evaluate the sum over q in Eqn.(45) we convert it into an integral, using the integral representation

$$\ln \left[\frac{2 k_F + q}{2 k_F - q} \right] = 2 \int_0^\infty \frac{dx \, \sin(2 k_F r) \sin(qr)}{x}.$$

(47)

The result is

$$\chi(q) = \frac{3 g_L^2 \mu_B^2}{16 E_F} \frac{N}{\Omega} \left[1 + \frac{k_F}{q} (1 - \frac{q^2}{4 k_F^2}) \int_0^\infty \frac{dx \sin(2 k_F r) \sin(qr)}{x} \right]$$

(48)

The Fourier transform of it is

$$\chi(r) = \frac{k_F^3}{8 k_F r} \int_0^\infty dx \, x \, \sin(2 k_F r) \chi(q).$$

(49)

From Eqn.(45), it becomes

$$\frac{1}{\Omega} \sum_q \chi(q) e^{iq.r} = \frac{3 g_L^2 \mu_B^2 k_F^3}{128 \pi E_F} \frac{N}{\Omega} \frac{\sin 2 k_F r - 2 k_F \cos 2 k_F r}{(k_F r)^4}.$$

(50)

When this expression is substituted into Eqn.(45), the spin density for k_F r $\gg$ 1 is

$$s(r) = \frac{3 \pi n^2}{64 E_F} JS \frac{\cos 2 k_F r}{(k_F)^3}.$$

(51)

It is clear that when a localized moment is introduced into a metal, the spins of the conduction electrons develop an oscillating polarization in the vicinity of the moment. This is the Rudermann-Kittel-Kasuya-Yoshida (RKKY) oscillation for electron spin density [23]. These spin density oscillation is in analogy with the Friedel electron-charge density oscillations which appear when an electron gas screens out a charge impurity.

If there is another localized spin S' at r, this interaction leads to an effective interaction between the localized spins of the form

$$H_{RKKY} = -\frac{J^2}{g_L^2 \mu_B^2 \Omega} \sum_q \chi(q) e^{iq.r} S.S'.$$

(52)

Rudermann and Kittel [23a] showed its role for a broadening of the nuclear magnetic resonance absorption. It is also the origin of the exchange coupling between the localized moments in the rare-earth metals and its oscillatory nature leads to 'helimagnetism'. Because of the nature of this interaction which depends on the conduction electron density through the Fermi momentum, there should be the possibility to change the magnetic order continuously by alloying.

KONDO SCREENING AND RESIDUAL RESISTIVITY

We now return to the dilute alloys and consider what effect a local moment, assuming that it exists, has upon the resistivity. The related problem is so called the Kondo lattice problem or, the problem of the coexistence of conduction electrons and local magnetic moments. Kondo[20] was the first to recognize that the resistance minimum in certain alloys is a direct consequence of the presence of a local magnetic moment. We know that, in this case, the local moment is screened at low temperatures by conduction electrons and the ground state is singlet. The formation of this singlet state is a non-perturbative process which affects electrons very far from the impurity.

Kondo lattices, empirically, resemble metals with very small Fermi energies. It is well known that conduction and localized electrons in Kondo lattices hybridize at low temperatures to create a single narrow band. However, the details of this process is not fully understood. The most interesting problem is how the localized electrons contribute to the volume of the Fermi sea, especially, in systems, the so-called Kondo insulators, with one conduction electron and one spin per init cell. The conservative approach to Kondo insulators would be to calculate their band structure treating the on-site Coulomb repulsion U as a perturbation.

In this case the insulating state, in 1-D model of the Kondo lattice, forms not due to a hybrid-ization of conduction electrons with local magnetic moments, but as a result of strong anti-magnetic fluctuations. The presence of these strong anti-magnetic fluctuations has been confirmed by modeling approach [20] which demonstrate a strong, sharp enhancement of the staggered susceptibility in 1-D Kondo spin-liquids.

The relevant energy scale $\Delta_S(J)$ formally resembles the expression for the Kondo temperature

$$T_K = \sqrt{J}\exp(-2\pi/J)$$

$$(53)$$

$m(J)=JS$ is always larger due to the presence of the large logarithm. Therefore, at small J the anti-ferromagnetic exchange by the conduction electrons (the so-called Rudermann-Kittel-Kasuya-Yoshida (RKKY-) interaction) plays a stronger role than the Kondo screening. Thus the contributions to the low energy dynamics come from anti-ferromagnetic fluctuations [17].

$$\Delta_S(J) = \pi\upsilon/\xi \approx Jg^{-1}\exp(-2\pi/g)$$

$$(54)$$

where $m=JS$, the dimensionless coupling constant $g = i(\text{пб})$ is

$$g = \frac{\pi\upsilon}{M} = \frac{\pi}{\sqrt{M^2 + \pi^2/2J^2\ln(1/JS)}}$$

$$(55)$$

and a correlation length ξ of the disordered ground state model [18],

$$\xi = a^{-1}g\exp(2\pi/g) \qquad (56)$$

Depending on even or odd value of $|M-2S|$, the topological term is non-essential (even), equal to $2\pi i$, or essential (odd). In this essential case, the model becomes critical and its low energy behaviour is the same for the spin-1/2 anti-ferromagnetic Heisenberg chain and the energy scale [Eqn.(54)] marks the crossover to the critical regime where n is the same correlation functions as staggered magnetization of the Heisenberg chain. The

specific heat is linear at low temperatures $T\ll\Delta_S(J)$ and comes from gapless spin excitations [11].

Kondo lattices, empirically, resemble metals with very small Fermi energies. It is well known that conduction and localized electrons in Kondo lattices hybridize at low temperatures to create a single narrow band. However, the details of this process is not fully understood. The most interesting problem is how the localized electrons contribute to the volume of the Fermi sea, especially, in systems, the so-called Kondo insulators, with one conduction electron and one spin per init cell. The conservative approach to Kondo insulators would be to calculate their band structure treating the on-site Coulomb repulsion U as a perturbation.

He demonstrated this correlation by calculating the spin-flip scattering of a conduction electron from a localized magnetic moment to second order in the Born approximation [17]. He proposed that the resulting resistivity contains a term proportional to $J^3\log T$ and the exchange coupling constant J, when it is negative, leads to in increasing resistivity at low temperatures, which, when combined with the decreasing phonon contribution, produces a resistance minimum [9g].

RESIDUAL RESISTIVITY

By use of the GF operator, the iteration which represents the scattering of electrons firstly from k to k' and then successively to k'' and k''' and back to k again is

$$\langle k|G|k\rangle = G_0(k) + G_0(k)\langle k|V|k\rangle G_0(k) + G_0(k)\langle k|V|k'\rangle G_0(k')\langle k'|V|k\rangle G_0(k) + ...$$

$$(57)$$

If no intermediate state is identical to initial state, we shall have a set of terms whose sum is defined as

$$G_O(k)M(k)G_O(k) \text{, where}$$

$$M(k) = \langle k|V|k\rangle + \sum_k\langle k|V|k'\rangle G_0(k')\langle k'|V|k\rangle + \langle k|V|k'\rangle G_0(k')\langle k'|V|k''\rangle G_0(k'')\langle k''|V|k\rangle + ..$$

$$(58)$$

which is often called the proper self-energy. The Dysonian is

$$M(k,\varepsilon) = \langle k|V|k\rangle + \sum_{k'}\frac{|\langle k|V|k'\rangle|^2}{\varepsilon - \varepsilon_0(k') + i\eta}.$$

$$(59)$$

By truncating it at second tern, we have a finite, real and imaginary part of $M(k,\varepsilon)$. In particular,

$$\text{Im}M(k,\varepsilon) = -\pi\sum_{k'}|\langle k|V|k'\rangle|^2\delta[E(k)-E(k')]$$

$$(60)$$

which is an expression for the probability of an electron being scattered out of the state k. Then Eqn.(57) becomes

$$\sum_{k'} Q(k,k') = \frac{2n_i}{\hbar} \operatorname{Im} M(k,\varepsilon).$$

(61)

As a result of substitution in a simple momentum relaxation time in closed form, we get

$$\frac{1}{\tau(\varepsilon)} = \frac{2n_i}{\hbar} [-\operatorname{Im} M(k,\varepsilon)].$$

(62)

Rewriting Eqn.(57) gives

$$\langle k|G|k \rangle = G_0(k)[1 - M(k)G_0(k)]^{-1}.$$

(63)

After manipulating, we get

$$\operatorname{Im} M(k,\varepsilon) = \operatorname{Im} M(k,\varepsilon)^{-1}.$$

(64)

Consequently, the spin averaged collision frequency becomes

$$\frac{1}{\tau(\varepsilon)} = \frac{2n_i}{\hbar} \left[\frac{1}{2} \sum_\sigma \operatorname{Im} G_{kk}^\sigma(\varepsilon)^{-1} \right].$$

(65)

Therefore, the residual resistivity, according to the Drude theory, becomes

$$\Delta\rho = \frac{n_i}{n} \frac{m^*}{e^2 \hbar} \sum_\sigma \operatorname{Im} G_{kk}^\sigma(\varepsilon_F)^{-1}$$

(66)

As the impurity atom has an unfilled or partly filled d-shell and the conduction band states of the matrix are not d-like (Ni,Fe), say s-like (Cu,Be) or s-p mixture (Al-Si), the Anderson model is applicable. We applied the Friedel-Anderson model here. By manipulation from GF matrix elements, we have the conduction-electron Green's function as

$$(\omega - \varepsilon_k) G_{kk}^\sigma = 1 - \sum_m \frac{|V_{mk}|^2}{\omega - \varepsilon_k} G_{mm}^\sigma ,,$$

(67)

$$\therefore G_{kk}^\sigma = (\omega - \varepsilon_k)^{-1} - \sum_m \frac{|V_{mk}|^2}{\omega - \varepsilon_k} G_{mm}^\sigma$$

(68)

where V_{mk} is the matrix element of the mixing interaction and G_{mm}^σ is the Green's function for the localized electron (d-electron, for example) whose orbital is specified by '0'; a five-fold orbital degeneracy is assumed. To the lowest order,

$$G_{kk}^\sigma(\omega)^{-1} = [\omega - \varepsilon_k - \sum_m |V_{mk}|^2 G_{mm}^\sigma.$$

(69)

At the Fermi surface

$$G_{kk}^\sigma(\varepsilon_F)^{-1} = -\sum_m |V_{mk}|^2 G_{mm}^\sigma.$$

(70)

$$\therefore \operatorname{Im} G_{kk}^\sigma(\varepsilon_F)^{-1} = -\sum_m |V_{mk}|^2 \operatorname{Im} G_{mm}^\sigma$$

(71)

But from the definition of the occupation number of the localized state, it is defined as

$$n_m^\sigma = \int_{-\infty}^{\varepsilon_F} d\varepsilon \, \eta_m^\sigma(\varepsilon)$$

$$..... = \int_{-\infty}^{\varepsilon_F} d\varepsilon \frac{\Delta}{(\varepsilon - \varepsilon_{m\sigma})^2 + \Delta^2}$$

$$..... = \sum_{m\sigma} \cot^{-1} \frac{\varepsilon_{m\sigma} - \varepsilon_F}{\Delta}$$

(72)

$$\cot^2 \pi n_m^\sigma = 1 - \csc^2 \pi n_m^\sigma = \sum \frac{(\varepsilon_{m\sigma} - \varepsilon_F)^2}{\Delta^2}$$

$$\frac{1}{\sin^2 \pi n_m^\sigma} = \sum \frac{(\varepsilon_{m\sigma} - \varepsilon_F)^2}{\Delta^2} + 1$$

$$\therefore \sin^2 \pi n_m^\sigma = \frac{\Delta^2}{(\varepsilon_{m\sigma} - \varepsilon_F)^2 + \Delta^2}$$

(73)

By using the expressions for the localized density of states,

$$\sin^2 \pi n_m^\sigma = \eta_m^\sigma \Delta \pi$$

$$\therefore \frac{\sin^2 \pi n_m^\sigma}{\pi \eta} = \eta_m^\sigma \langle v^2 \rangle \pi$$

$$\therefore \frac{\sin^2 \pi n_m^\sigma}{\pi \eta} = -\sum_m |V_{mk}|^2 \operatorname{Im} G_{mm}^\sigma$$

(74)

Substituting in Eqn.(62) gives

$$\operatorname{Im} G_{kk}^\sigma = \frac{\sin^2 \pi n_m^\sigma}{\pi \eta}$$

(75)

Therefore, Eq.(62) becomes

$$[\tau(\varepsilon_F)]^{-1} = \frac{n_i}{\hbar} \sum_{m,\sigma} \frac{\sin^2 \pi n_m^\sigma}{\pi \eta} \tag{76}$$

Thus, we have

$$\Delta \rho = \frac{n_i}{n} \frac{m^*}{\pi e^2 \hbar \eta} \sum_{m,\sigma} \sin^2 \pi n_m^\sigma. \tag{77}$$

The calculated results associated with the present formulation, applied to Be:Ni,Cu:Ni, and Al:Ni magnetic alloy systems has been published as where[1c].

CONCLUSION

All knowledge about a system, in principle, can be usually obtained from the quantum mechanical wave function. At atomic level, microelectronics can be studied using the wave function solving the Schrödinger equation for N-electron systems. However, this problem cannot be solved even for simplest nanostructures which consist of a couple of molecules. The density functional theory was then developed, and the electron charge density is used as the fundamental variational parameter rather than the electron wave functions. The N-electron problem is formulated, in this formalism, as N one-electron equations where each electron interacts with all other electrons via an effective exchange-correlation potential. These interactions are then calculated using the local density approximation to exchange and correlation. Plane wave basis sets and total energy pseudo-potential methods are then used to solve the Kohn-Sham one electron equations.

Because of the limitations of semi-classical theory, it is more appropriate to treat the transport problem in nanostructures by using quantum many-body theory. The starting point of the quantum transport theory is to take an external field as a perturbation for the many-particle system in equilibrium. This leads to the induced charge density, or change in charge density, caused by the external potential. The induced charge density contributes an additional potential that combines with the external potential to give the net additional potential in the system. The importance of the Green's Function Technique in condensed matter physics arises from the fact that exact expressions for many physical properties of interest may be written in terms of them. The Green's function formulation gives a full treatment for electron gas and a proper calculation of scattering self-energy for the electron-electron single-particle scattering. From a fundamental point of view, the Green's Functions are related to the probability amplitude of a quantum state by bilinear expressions and therefore connected with probabilities rather than with probability amplitudes. In many cases, they have more physical significance than the probability amplitude, which is not a directly observable parameter. Many have made contributions to the many-body problems in condensed matter theory, this way.

ACKNOWLEDGEMENTS

The author would like to thank Assoc. Prof. Dr. Yow Ho Kwang, Dean of the Faculty of Engineering, MMU, for the encouragement and support during the preparation of the manuscript.

REFERENCES

1. For example: Density Functional Theory (a) W. A. Goddard III (Ed),

   Handbook of Nanoscience, Engineering and Technology, CRC Press, 2003,

   §23-18; (b) R. M. Driezler and J. da Providencia, Density Functional Methods in Physics,

   Plenum, NY, 1985 (c) P. Hohenberg and W. Kohn, Phy. Rev. 136, 3B (1964) 864;

   Random Phase Approximation (c) C. Kittel, Quantum Theory of Solids, Wiley, NY, 1963, 102

   &113; (d) D. Pines in Solid State Physics 1 (1955);

   Effective Mass Approximation (e) P. T. Landsberg, ed., Solid State Theory-Methods and

   Applications, Wiley, NY, 1969, 114.

2. W. Kohn and L. J. Sham, Phys. Rev. 140A (1965) 1133.

3. L. H. Thomas, Proc. Cambridge Phil. Soc. 23 (1927) 542.

4. E. Fermi, Z. Physik 48 (1928) 73.

5. W. Kohn and R. M. Driezler, Phys. Rev. Lett. 56 (1986) 1993.

6. K. K. Likharev, IEEE Trans. Magn. 23 (1987) 1142.

7. P.A.M.Dirac, Lecturers on Quantum Field Theory, Academic

   Press, N.Y., 1966.

8. J. Binder, P. Zahn and I. Mertig, Jour. Appl. Phys. 87 (2000) 5182.

9. For example: Feynman's diagrammatic interpretation (a) Feynman Diagrams in the Many-body Problem, McGraw-Hill, NY, 1976;(b) R.P. Feynman, Phys. Rev. 76(1949)749; (c) K. Nyunt, "Many-Body Approach to Transport in Quantum Confined Nanostructures: A Review on the Green's Function Techniques" ICMN'08, Kuala Lumpur, Malaysia, 13-15 May, 2008, E2, 224; Green's Function Techniques (d) P.T. Landsberg, ed., Solid State Theory-Methods and Applications, Wiley, NY, 1969, 407; (e) S. Raimes, Many-Electron Theory, North-Holland, Amsterdam, 1972; (c) D.J. Thouless, Quantum Mechanics of Many-Body Systems, Academic Press, NY, 1972; (f) D. J. Thouless, Phys. Rep. 13(1974)93;(g) K. Nyunt & Sann M. Htwe, Rangoon University(unpublished), 1989; (h) K.Nyunt & Kyi K. Hlaing, Rangoon University (unpublished), 1987.

10 For example: Kubo's quantum transport theory: R. Kubo, Phys. Rev. 87(1952) 568.

11. J. Feder *et al*., Phys. Rev. 168, 2(1968) 649.

12. D. Controzzi et al., Phys. Rev. Letters. 86, 4(2001)680; F.H.L. Essler *et al*., Phys. Rev. B 65(2002) 115117-1.

13. T. Matsubara *et al*., Progr. Theoret. Phys. 26(1961) 739.

14. J.M. Luttinger, Phys. Rev. 121(1961) 924.

15. C. Kittel, Quantum Theory of Solids, Wiley, 1963; R.E. Peierls, Quantum Theory of Solids, Clarendon, Oxford, 1955.

16. F. W. Byron and R. W. Fuller, Mathematics of Classical and Quantum Physics, Addison-Wesley, NY, 1969.

17. D. Feng & G. Jin, Introduction to Condensed Matter Physics, Vol.1, World Scientific, NJ, 2005.

18. D. K. Ferry & S. M. Goodnick, Transport in Nanostructures, Cambridge, 1997.

19. G.D. Mahan, Many-Particle Physics, Plenum Press, NY, 1995.

20. For example:KONDO lattice problem : (a) J. Kondo, Progr. Theoret. Phys. 32(1964) 37; (b) J. Kondo, J. Appl. Phys. 36(1966) 1177.

21. For example: Friedel-Anderson analysis:(a) J. Friedel, Metallic Alloys, Nuovo Cimento 7 Suppl, 287 (1958);(b) J. Friedel, Phil. Mag. Suppl 3 (1954) 446-507; *ibid* 43(1952) 153; (c) J. Friedel, in *The Physics of Metals*, Vol.1 Electrons (J.M. Ziman, ed.) Cambridge, 1971, 30; (d) P. W. Anderson, Phys. Rev. 86(1952) 694; (e) P. W. Anderson, J. Appl. Phys. 37(1966) 1194.

22. C. Zener, Phys. Rev. 81(1951) 440.

23. For example: RKKY interaction: (a)M. A. Rudermann & C. Kittel, Phys. Rev. 149(1966) 491; (b) T. Kasuya, Progr. Theoret. Phys. (Kyoto) 16 (1956) 45; (c) K. Yoshida, Phys. Rev. 106 (1957) 893.

24. D. Pines, Elementary Excitations in Solids, Benjamin, NY, 1963.

25. For example Landau liquid theory: (a) L.D. Landau, Zh. Eksper. Teor. Fiz., 30 (1956) 1058 (Soviet Phys,-J.E.T.P,3(1957) 920); (b) *ibid*, 32 (1957) 59 (Soviet Phys,-J.E.T.P,5(1957) 101;

(c) *ibid*, 35, (1958) 97 (Soviet Phys,-J.E.T.P,8(1959)70.

26. For example: Edwards, Sherrington & Bhagavan GF Methods (a) M.R. Bhagavan, NORDisk Institute for Theoretisk Atomfysik (NORDITA Lectures), Stockholm, 1967; (b) S.F. Edwards & D. Sherrington, Proc. Phys. Soc., 90(1967) 3; (c) M.R. Bhagavan, & S.F. Edwards, Proc. Phys. Soc., 90(1967) 953.

Calculation of Ion Charge State Distributions After Inner-Shell Ionization in Xe Atom

Adel M. Mohammedein, Adel A. Ghoneim, Kandil M. Kandil, Ibrahim M. Kadad

Applied Sciences Department, College of Technological Studies P.O. Box 42325, Shuwaikh 70654, Kuwait
e-mail : admohamed@yahoo.com, am.mohammedein@paaet.edu.kw

ABSTRACT

The vacancy cascades following initial inner-shell vacancies in single and multi-ionized atoms often lead to highly charged residual ions. The inner-shell vacancy produced by ionization processes may decay by either a radiative or non-radiative transition. In addition to the vacancy filling processes, there is an electron shake off process due to the change of core potential of the atom. In the calculation of vacancy cascades, the radiative (x-ray) and non-radiative (Auger and Coster-Kronig) branching ratios give valuable information on the de-excitation dynamics of an atom with inner-shell vacancy. The production of multi-charged ions yield by the Auger cascades following inner shell ionization of an atom has been studied both experimentally and theoretically. Multi-charged Xe ions following de-excitation of K-, L_1-, $L_{2,3}$-, M_1-, $M_{2,3}$- and $M_{4,5}$ subshell vacancies are calculated using Monte-Carlo algorithm to simulate the vacancy cascade development. Fluorescence yield (radiative) and Auger, Coster- Kronig yield (non- radiative) are evaluated. The decay of K hole state through radiative transitions is found to be more probable than non-radiative transitions in the first step of de-excitation. On the other hand, the decay of L, M vacancies through non-radiative transitions are more probable. The K shell ionization in Xe atom mainly yields Xe^{7+}, Xe^{8+}, Xe^{9+} and Xe^{10+} ions, and the charged X^{8+} ions are the highest. The main product from the L_1- shell ionization is found to be Xe^{8+}, Xe^{9+} ions, while the charged Xe^{8+} ions predominate at $L_{2,3}$ hole states. The charged Xe^{6+}, Xe^{7+} and Xe^{8+} ions mainly yield from $3s_{1/2}$ and $3p_{1/2,3/2}$ ionization, while Xe in $3d_{3/2,5/2}$ hole states mainly turns into Xe^{4+} and Xe^{5+} ions. The present results are found to agree well with the experimental data.

Keywords: Auger cascade, multiple charged ions, charge state distributions

INTRODUCTION

The vacancy cascades following initial inner-shell vacancies in a single and multi-ionized atoms often lead to highly charged residual ions. The inner-shell vacancy produced by ionization processes may decay by either a radiative or non-radiative transition. In addition to the vacancy filling processes, there is an electron shake off process due to the change of core potential of the atom. In the calculation of vacancy cascades, the radiative (x-ray) and non-radiative (Auger and Coster-Kronig) branching ratios give valuable information on the de-excitation dynamics of an atom with inner-shell vacancy. The production of multi-charged ions yield by the Auger cascades following inner shell ionization of an atom has been studied both experimentally and theoretically.

Krause et al. [1] and Carlson et al. [2-5] measured the multicharged ions formed as a result of relaxation of inner-shell vacancies in rare gases using characteristic x-ray. Holland et al. [6] studied the partial cross-section for multiple photoionization in rare gases using a time-of-flight (TOF) mass spectrometer and synchrotron radiation. The multiple photoionization of Xe in the L- and M- subshells regions using synchrotron radiation are measured [7-9].

There are two major theoretical methods used to calculate the vacancy cascades originating from initial inner-shell vacancies. The first model is based on straightforward construction of de-excitation decays [10-17]. The second method is based on Monte Carlo technique to simulate possible de-excitation pathways including the radiative and non-radiative transition processes [18-22].

CP1202, *Neutron and X-Ray Scattering in Advancing Materials Research: International Conference – 2009*
edited by A. Saat, H. A. Kassim, M. H. H. Jumali, J. M. Saleh, M. R. Othman, A. Ibrahim, F. M. Idris, and M. H. A.-R. M. Ahmad
© 2009 American Institute of Physics 978-0-7354-0739-8/09/$25.00

In the present work, Mont Carlo technique is adapted to calculate the multiply charged ions following K-, L- and M- subshells vacancies creation in Xe atom. The radiative (x-ray) and non-radiative (Auger and Coster-Kronig) branching ratios for ionization are evaluated using Multiconfiguration Dirac Fock wave functions from Grant et al. [23] and Dirac Fock Slater wave functions from Lorenz et al. [24], respectively. The results of multicharged Xe^{i+} ions are compared with the available theoretical and experimental data.

CALCULATION METHOD

Monte- Carlo technique is used to simulate vacancy cascades following K- L- and M- subshells in Xe atom. Radiative and non- radiative transitions and electron shake off processes are considered. The details of the calculations are described in El-Shemi et al [21] and Abdullah et al. [22]. A brief description of the method is given in the following.

The simulation of each cascade begins with the implementation of atomic data for all possible radiative and non-radiative transitions for a single ionized Xe atom. The radiative and non-radiative branching ratios which give valuable information on the de-excitation dynamic of an atom with a core vacancy, are calculated as following:
fluorescence yield:

$$\omega(C_i \to C_f) = \frac{\Gamma_R(C_i \to C_f)}{\Gamma(C_i)} \quad (1)$$

And Auger yield:

$$a(C_i \to C_f) = \frac{\Gamma_A(C_i \to C_f)}{\Gamma(C_i)} \quad (2)$$

Where the initial configuration C_i will decay into final configurations C_f. The sum of partial widths $\Gamma(C_i)$ of all radiative $i \to j$ and non-radiative $i \to jk$ transitions is given by:

$$\Gamma(C) = \sum_{i,j} \Gamma_{ij}(C_i \to C_f) + \sum_{i,jk} \Gamma_{ijk}(C_i \to C_f) \quad (3)$$

The partial widths of radiative and non-radiative transitions for single ionized atom are calculated as follows:

$$\Gamma_R(n_i l_i \to n_f l_f) = \frac{4k^3}{3g_i} \sum_{\substack{L,S,J,M \\ L',S',J'M'}} \left| \langle n_i l_i LSJM_J | D | n_f l_f L'S'J'M'_J \rangle \right|^2 \quad (4)$$

$$\Gamma_A(n_i l_i \to n_j l_j, n_k l_k) = \frac{2\pi}{g_i} \sum \left| \langle n_i l_i, \varepsilon_A l_A, L'S'J'M' | H^{ee} | n_j l_j, n_k l_k, LSJM \rangle \right|^2 \quad (5)$$

where $n_i l_i$, $n_j l_j$ and $n_k l_k$ denote the atomic subshell involved in the transition. g_i is the statistical weight of the initial state $n_i l_i$, the value k is equal to the x-ray transition energy divided by light velocity c, D is the dipole operator, and H^{ee} is the operator of electron- electron interaction operator.

An analysis of each cascade started with consideration of all possible electron transitions, which may fill an initial core vacancy, then one transition is selected. The selection is random in the interval (0,1) based on the branching ratios of all possible transitions. In the second step, a new configuration of vacancies appears after the occurrence of the selected transition. The consideration of this configuration is similar to that at the first step. The random selection of the next transitions in the given cascade has been realized with allowance for the relative probabilities of possible transitions in all available vacancies. Successive vacancy configurations appear until all the vacancies reach the outer shell leading to the production of highly charged ions have been produced. After finishing with one cascade, the same initial hole is created again in the inner - subshell and the cascade is simulated again.
Finally after multiple simulations 10^5 times, the final charge state distribution (CSD) in outer shells and average charge state, $< i >$, of ions are calculated. The probabilities of charged ions state distributions, $p(i)$, and the average charged ions, $< i >$, are calculated for ith charged ions as:

$$p(i) = \frac{\sum_{1}^{n} i}{10^5} \quad (6)$$

The mean charge state of ions is given by:

$$< i > = \sum p(i) i \quad (7)$$

Where i denote the number of ejected electrons.

Radiative (x-ray) and non-radiative (Auger and Coster-Kronig) transitions rates for single ionized atoms are calculated using Multiconfiguration Dirac Fock wave function and Dirac Fock Slater wave functions respectively. The radiative and non-radiative transition rates for multi-ionized atoms are evaluated using a statistical weighting procedure from Jacobs et al. [25]. The transition probabilities of multi-ionized atoms change proportional to the number of electrons in the subshells. The radiative transition rates for multi-ionized atoms are given by:

$$a_r(n_1 l_1^{N_1}, n_2 l_2^{N_2} \to n_1 l_1^{N_1-1}, n_2 l_2^{N_2+1}) = N_1 \frac{(4l_2 + 2 - N_2)}{(4l_2 + 2)} A_r(n_1 l_1 \to n_2 l_2) \quad (8)$$

and the non-radiative transition rates are given by:

214

$$a_a(n_1l_1{}^{N_1}, n_3l_3{}^{N_3}, n_4l_4{}^{N_4} \rightarrow n_1l_1{}^{N_1-1}, n_3l_3{}^{N_3+1}, n_4l_4{}^{N_4+1}) =$$

$$N_1 \frac{(4l_3+2-N_3)}{(4l_3+2)} \frac{(4l_4+2-N_4)}{(4l_4+2)} A_a(n_1l_1 \rightarrow n_3l_3, n_4l_4) \quad (9)$$

If both of the final vacancies occur in the same principle shell, the non-radiative transition rates are given:

$$a_a(n_1l_1{}^{N_1}, n_3l_3{}^{N_3} \rightarrow n_1l_1{}^{N_1-1}, n_3l_3{}^{N_3+2}) =$$

$$\frac{N_1}{2} \frac{(4l_3+2-N_3)}{(4l_3+2)} \frac{(4l_3+1-N_3)}{(4l_3+1)} A_a(n_1l_1 \rightarrow n_3l_3{}^2) \quad (10)$$

where A_r and A_a are the radiative and non-radiative transition rates for single ionized atom, respectively, a_r and a_a are the transition rates for atom with various spectator inner-shell vacancies, n_1l_1 is initial state and n_3l_3, n_4l_4 are a the final state, and N_i is the number of vacancies in the n_il_i sub-shell.

RESULTS AND DISCUSSIONS

The fluorescence yield (radiative branching ratios) of an atomic shell or subshell configurations is defined as the probability that a vacancy in that shell or subshell is filled through a radiative transition. The Auger yield (non-radiative branching ratios) is the probability that a vacancy in the atomic shell and subshell is filled through a non-radiadive transition by an electron from a higher shell or subshell. Table 1 presents the radiative (fluorescence) yield and non-radiative (Auger and Coster - Kronig) yield for single ionized Xe atom. The present results are compared with those published by other authors [10,26]. The decay for K shell vacancy in Xe atom through radiative transitions is more probable than the non-radiative transitions. The total branching ratio of

radiative yield is 88.8% while the branching ratio of non-radiative yield is 11.2% as shown in Table 1. On the other hand, the decay of L- and M- subshells is more probable through non-radiative transitions (Auger and Coster- Kronig transitions).

Fig.1 shows the probability of final charge state distributions for Xe^{i+} ions after de-excitation of K vacancy.

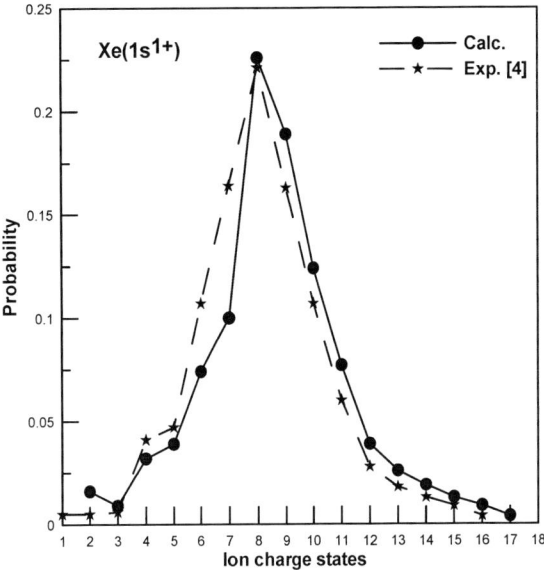

Figure1. Final charge state distribution of Xe^{i+} following de-excitation decay of initial K vacancy in Xe atom.

The calculation and experimental Xe^{i+} ion charge states have a maximum at i=8 and a shoulder at i= 4. This may be attributed to the initial $K-L_{2,3}$ radiative transitions which have a probability (88.8%) as shown in Table 1. The decay through radiative transition $K-L_{2,3}$ replace the vacancy to

	K	L$_1$	L$_2$	L$_3$	M$_1$	M$_2$	M$_3$
Table 1: Total radiative and non-radiative branching ratios of xenon							
RADIATIVE YIELD Radiative yield[10] Radiative yield[26]	88.8	4.97 5.43	8.79 7.97	8.80 7.31	0.05 0.05 0.05	0.01 0.01 0.01	0.05 0.07
C K YIELD CK yield[10] CK yield[26]		49.80 43.25	16.10 12.37 15.40		95.20 94.83 94.30	89.60 89.84 89.20	88.10 87.61
Auger yield Auger yield[10] Auger yield[26]	11.2	45.20 51.32	75.10 79.65 76.90	91.20 92.69	4.80 5.11 5.70	10.75 10.70 10.70	11.80 12.32

the $L_{2,3}$ subshells as a first decay step in the cascade. Then the resulted $L_{2,3}$ vacancies may decay through L-NN, L-NO and L-OO Auger transitions leading to the production of Xe^{4+} ions. Ions mainly produced from Xe in K shell vacancy state are found to be Xe^{7+}, Xe^{8+}, Xe^{9+} and Xe^{10+}. The number of ejected electrons result from de-excitation decay of inner-shell vacancies in xenon forms an asymmetric peak. In comparison with the available experimental data [4] the present calculations agree well.

The charge state distribution of Xe^{i+} ions yield after de-excitation decay of an initial L_1-shell vacancy is shown in Fig. 2. The charged Xe^{9+} ions predominate in the charge state distributions. In the L_1 vacancy state, the intensity of Xe^{8+} ions decreases gradually, while that of Xe^{9+} ions increases. The spectrum of ejected electrons after an initial L_1 shell vacancy forms asymmetric peak. The spectrum of charged Xe^{1+}, Xe^{2+} and Xe^{3+} ions is weak and has no significant effect on the charge state distributions. The results of ion charge state distributions after de-excitation of an initial L_1-shell vacancy are very close to Carlson and Krause [4] experimental results.

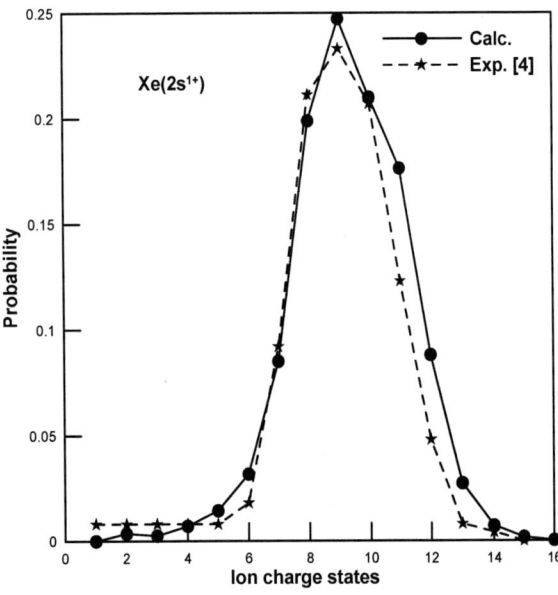

Figure 2. Final charge state distributions of Xe^{i+} ions following after de- excitation decay of initial L_1 vacancy in Xe atom

Fig. 3 shows the experimental and the theoretical charge state distributions of ions after de-excitation decay of an initial L_2- and L_3 subshell vacancies. The charged Xe^{8+} ions predominate in the charge state distributions. The shoulder appears at i=4 in both experimental data and calculations. This behaviour is previously observed at the K hole state. The reason of appearance of this shoulder is related to the first decay step of L_2 and L_3 subshell vacancies that may occur through L α radiative transitions. These transitions lead to vacancy movement to a higher $M_{4,5}$ subshells without changing the number of vacancies producing a maximum abundance of Xe^{4+} ions. The probability of charge state distributions are maximum in the middle of the distribution, while it decreases at the lower and higher values of ion charge states. The present results are compared with theoretical calculations [10] and experimental data [4,9,19, 27] and the agreement is reasonable as indicated in Fig. 3.

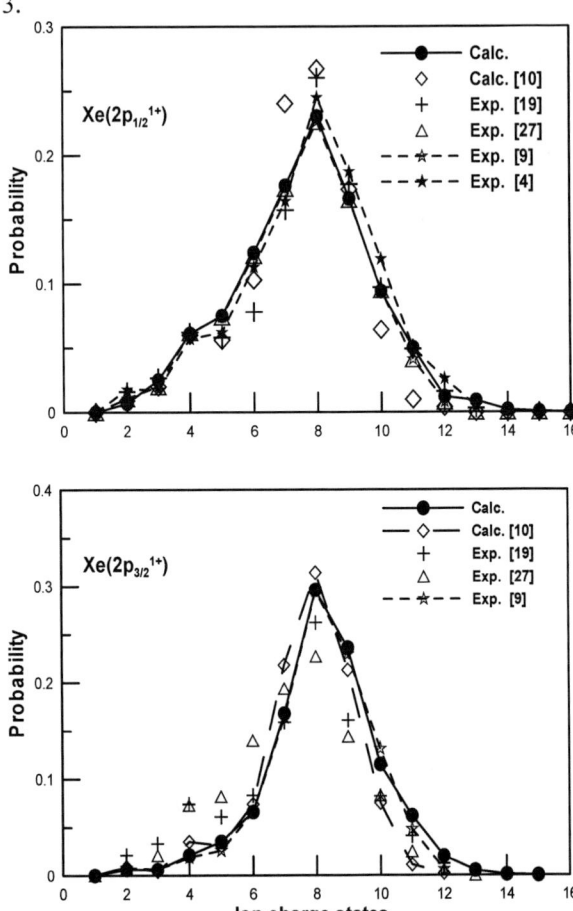

Figure 3. Final charge state distributions of Xe^{i+} ion result after de-excitation decay of initial L_2 and L_3 vacancies in Xe atom

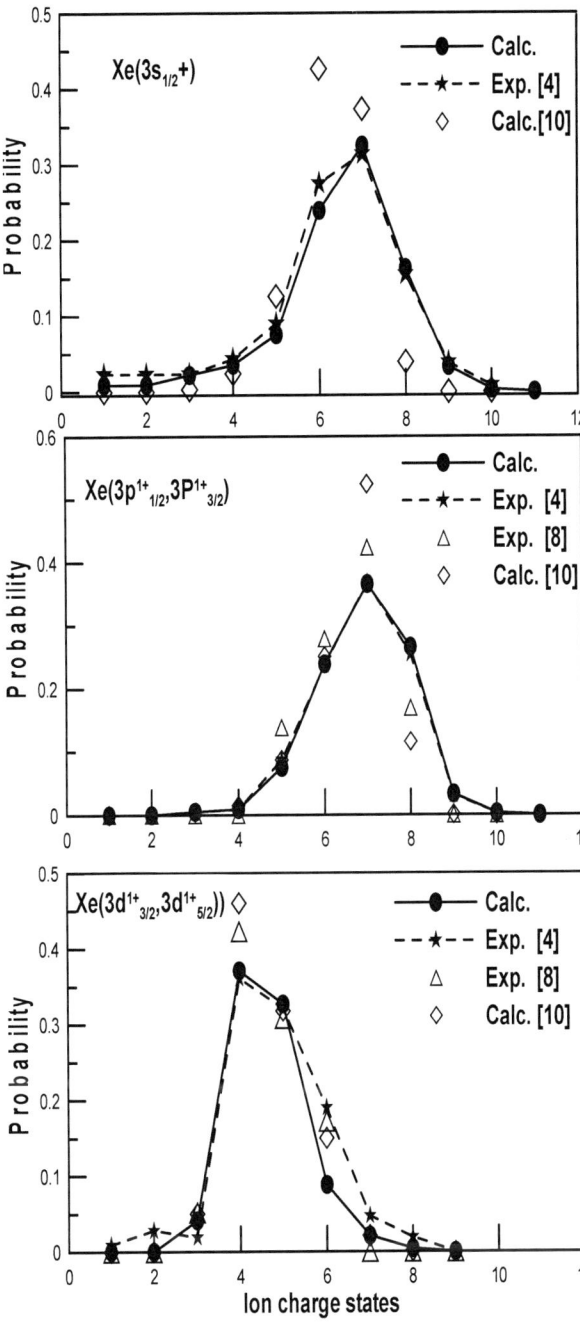

Figure 4. Final charge state distributions of Xe^{i+} following after de-excitation decay of M_1, $M_{2,3}$ and $M_{4,5}$ vacancies in Xe atom.

Fig.4 shows the probabilities of multiple Xe^{i+} ions yield after de-excitation decay of initial M_1, $M_{2,3}$ and $M_{4,5}$ vacancies. The distributions of Xe^{i+} ions are compared with the available theoretical [10] and experimental data [4,8]. The decay of the M subshell holes mainly proceeds through Auger and Coster- Kronig transitions producing electrons from N and O subshells. The charged ions mainly yield from Xe in $3s_{1/2}$ and $3p_{1/2,3/2}$ hole states are found to be X^{6+}, Xe^{7+} and

Xe^{8+} ions. The 3s and 3p vacancies are presumed to decay mainly into $M_{4,5}N$ hole states (Coster-Kronig transitions) which further de-excite through sequential Auger transitions to form Xe^{i+} ($6 \leq i \leq 8$). The single and double charged Xe^{i+} ions are yield from direct N - or O- subshells. As shown in Fig. 4, Xe in $3d_{3/2,5/2}$ hole states mainly turns into Xe^{4+} and Xe^{5+} ions, and the charged Xe^{3+} formed from de-excitation decay through $N_{4,5}O$ Auger transition. After production Xe^{2+} ions the $N_{4,5}$ vacancies de-excites into OO sushells through Auger transitions and produce Xe^{4+} ion. The Xe^{5+}, Xe^{6+} ions formed from $N_1N_{4,5}$ de-excitation through Auger and /or Coster -Kronig transitions. The present calculations agree well with the experimental data as shown in Figure 4.

CONCLUSIONS

The charge state distributions of Xe ions produced after de-excitation decays of inner shell vacancies are calculated using Monte- Carlo (MC) simulation technique. At K and $L_{2,3}$ hole states, the yield of Xe^{8+} ions are predominate. The L_2 hole state mainly turns to Xe^{9+} and the M_1 and $M_{2,3}$ hole states mainly yield Xe^{7+} ions. The charged Xe^{4+} ions are found mainly after de-excitation decay of $M_{4,5}$ vacancies. The present results of charge state distributions of ions agree well with the experimental and theoretical data.

ACKNOWLEDGEMENT

The authors would like to thank Kuwait Foundation for the Advancement of Sciences (KFAS) for financial support of the present work.

REFERENCES

[1] M. O. Krause, M. V. Vestal, W. H. Johnson, T. A. Carlson; Phys. Rev. **133,** A385 (1964).

[2] T.A. Carlson and M.O. Krause; Phys. Rev. **137,** A1655 (1965).

[3] T.A. Carlson and M.O. Krause; Phys. Rev. Letters **14,** 390 (1965).

[4] T.A. Carlson, W. E. Hunt, and M.O. Krause; Phys. Rev. **151,** 41(1966).

[5] T.A. Carlson and M.O. Krause; Phys. Rev. **140,** A1054 (1965).

[6] D.M. P. Holland, K. Coding, J. B. West, and G. V. Marr; J. Phys. B: At. Mol. Phys. **12,** 2465 (1979).

[7] T. Tonuma, A. Yagishita, H. Shibata, T. Koizumi, T. Matsuo, K. Shima, T. Mukoyama and H. Tawara; J. Phys. B: At. Mol. Phys. **20,** L31 (1987).

[8] N. Saito and I. H. Suzuki; J. Phys. B: At. Mol. Opt. Phys. **25,** 1785 (1992).

[9] H. Tawara, T. Hayaishi, T. Koizumi, T. Matsuo, K. Shima and A. Yagishita; J. Phys. B: At. Mol. Opt. Phys. **25,** 1476 (1992).

[10]A. G. Kochur, A. I. Dudenko, V. L. Sukhorukov and I. D. Petrov; J. Phys. B: At. Mol. Opt. **27,** 1709 (1994).

[11] A. G. Kochur, A. I. Dudenko, V. L. Sukhorukov and I. D. Petrov; J. Phys. B: At. Mol. Opt. **28,** 387 (1995).

[12] A.G.Kochur and V.L.Sukhorukov; J.Phys.B:At. Mol. Opt. **29,** 3587 (1996).

[13] A. Kochur; J. Synchrotron Rad. **8,** 218 (2001).

[14 J. W. Cooper, S. H. Southworth, M. A. MacDonald and T. LeBrun Phys. Rev. A **50,** 405 (1994).

[15] F.von Buch, J. Doppelfeld, C. Gunther and E. Hartmann; J. Phys. B: At. Mol. Opt. Phys. **27,** 2151 (1994).

[16] G. Omar and Y. Hahn; Phys. Rev. A **43,** 4695 (1991).

[17] G. Omar and Y. Hahn; Phys. Rev. A **44,** 483 (1991).

[18]. T. Mukoyama; Bull. Inst. Chem. Res. Kyoto Univ. **63,** 373 (1985).

[19] T. Mukoyama, T. Tonuma, A. Yagishita, H. Shibata, T. Matsuo, K. Shima and H. Tawara; J. Phys. B: At. Mol. Phys. **20,** 4453 (1987).

[20] N.Mirakhmedov and E.S.Parilis;J. Phys. B:At. Mol. Opt. Phys. **21,**795 (1988).

[21] A. M. El-Shemi , A. A. Ghoneim and Y. A. Lotfy; Turk. J. Phys. **27,** 51 (2003).

[22] A. H. Abdullah, A. M. El-Shemi and A. A. Ghoneim; Rad. Phys. and Chem. **68,** 697 (2003).

[23] I. P. Grant, B. J. Mckenzie, P. Norrington, D. F. Mayers and N. C. Pyper; Comput. Phys. Commun. 21, 207 (1980).

[24] M. Lorenz, E. Hartmann; Report ZFI-109, Leipzig, 27 (1985).

[25] V. L. Jacobs, J. Davis, F. B. F. Rozsnyai and J. W. Cooper; Phys. Rev. A **21,** 1917 (1980).

[26] O.Keski-Rahkonen and M.O.Krause; At. Nucl. Data.Tables **14,**139 (1974).

[27] S. Drees; Diploma thesis University of Bonn (1993) (Bonn-IR-93-50).

SANS STUDIES INSIGHT INTO IMPROVING OF YIELD OF BLOCK COPOLYMER-STABILIZED GOLD NANOPARTICLES

Debes Ray and V. K. Aswal*

Solid State Physics Division, Bhabha Atomic Research Centre, Mumbai 400 085, India

*Corresponding author. E-mail: vkaswal@barc.gov.in

ABSTRACT

Triblock copolymer poly(ethylene oxide)-poly(propylene oxide)-poly(ethylene oxide) (PEO-PPO-PEO) are well known as dispersion stabilizers. It has also been recently found that they can act as reducing agents along with stabilizers and these two properties of block copolymers together have provided a single-step synthesis and stabilization of gold nanoparticles at ambient temperature. We have studied the synthesis of stable gold nanoparticle solutions using block copolymer P85. Gold nanoparticles are prepared from 1 wt% aqueous solution of P85 mixed with varying concentration of $HAuCl_4.3H_2O$ salt in the range 0.001 to 0.1 wt%. Surface plasmon resonance (SPR) band in UV-visible absorption spectra confirm the formation of the gold nanoparticles and the maximum yield of the nanoparticles is found to be quite low at 0.005 wt% of the salt solution. Small-angle neutron scattering (SANS) measurements in these systems suggest that a very small fraction of the block copolymers ($< 1\%$) is only associated with the gold nanoparticles and remaining form their own micelles, which probably results in the low yield. This can be explained as on an average a high block copolymer-to-gold ion ratio r_0 (22) is required for 1 wt% P85 in the reduction reaction to produce gold nanoparticles. Based on this understanding, a step-addition method is used to enhance the yield of gold nanoparticles by manifold where the gold salt is added in small steps to maintain higher value of r ($> r_0$) and therefore continuous formation of nanoparticles.

Keywords: Block copolymer; Gold nanoparticle; Small-angle Neutron Scattering

PACS: 61.05.fg, 61.25.H-, 62.23.St, 64.70.pv, 64.75.Yz

INTRODUCTION

Recently, nanoscale materials have attracted considerable attention as a continuously growing field of research (Daniel et al., 2004). Gold nanoparticles (AuNPs) have got a significant attention due to their size-dependent electronic, thermal, catalytic, magnetic and optical properties (quantum size effect). These have got promising applications in various fields like optics, catalysts, electronics, sensors, biology (e.g., biosensors, biomarkers etc.) and medical sciences (e.g., drug delivery in cancer therapy, antibacterial treatment etc.) (Jain et al., 2007; Shan et al., 2007). Hence, there is a lot of interest to synthesize AuNPs of different shapes and sizes to optimize their properties for different applications.

Various physical and chemical methods for the synthesis of AuNPs have been reported. Among all the synthesis techniques, the chemical reduction of the metal salt in aqueous, organic phase or two phases is one of the most popular routes as nanoparticles of a wide range of sizes and shapes can be prepared by controlling the reaction conditions. However, due to the high surface energy of metal nanoparticles the reactivity is very high and therefore particles without capping have a tendency to aggregate. This makes the presence of solvent, metal salt, reducing agent along with stabilizers to prevent the agglomeration of the nanoparticles must in these reactions. Recently, chemical route using block copolymer has provided a very simple method for the synthesis of AuNPs, where the block copolymer acts as a reducing agent as well as a stabilizing agent.

AuNPs using block copolymers are synthesized by mixing it with hydrogen tetrachloroaureate (III) hydrate ($HAuCl_4.3H_2O$) in aqueous solution. The formation of AuNPs from $AuCl_4^-$ has three main steps (Sakai et al., 2005a): (1) reduction of $AuCl_4^-$ ions by block copolymers in the solution and formation of gold clusters, (2) adsorption of block copolymers on gold clusters and reduction of $AuCl_4^-$ ions on the surfaces of these gold clusters, and (3) growth of gold particles in steps and finally the stabilization by block copolymers. This method has some key advantages of being fast, single-step and biocompatible as well as works at ambient conditions. The role of block copolymers in the synthesis varies with their molecular weight, PEO/PPO block length and polymer concentration. These factors significantly control the reduction of $AuCl_4^-$ ions and hence the growth of the AuNPs. It has also been found that along with these factors, the molar ratio of block copolymer molecules to the gold ions taking part in the reduction reaction is important. This ratio increases almost linearly with the block copolymer concentration. This suggests very low yield of nanoparticles at higher

CP1202, *Neutron and X-Ray Scattering in Advancing Materials Research: International Conference – 2009*
edited by A. Saat, H. A. Kassim, M. H. H. Jumali, J. M. Saleh, M. R. Othman, A. Ibrahim, F. M. Idris, and M. H. A.-R. M. Ahmad
© 2009 American Institute of Physics 978-0-7354-0739-8/09/$25.00

block copolymer concentrations required for larger production. It is therefore of interest to understand the reasons for low yield of gold nanoparticles at higher block copolymer concentrations and if the yield can be enhanced by some means.

In the present work, we have proposed a method to enhance the yield of the gold nanoparticles using block copolymer P85 ($EO_{26}PO_{39}EO_{26}$), which is one of the most commonly used (Aswal et al., 2007) and $HAuCl_4.3H_2O$ as the gold salt. The system has been characterized by UV-visible spectroscopy, small-angle neutron scattering (SANS) and dynamic light scattering (DLS). The formation of the gold nanoparticles in the system is confirmed by the UV-visible absorption spectra. The scattering techniques SANS and DLS which are suitable for the particle in the size range of ~ 10-1000 Å, can give information about the micellization of the block copolymers as well as their role in the formation of the gold nanoparticles. The results of these studies suggest that the step-addition method can be used to enhance the yield of nanoparticles by manifold.

MATERIALS

PEO-PPO-PEO triblock copolymers are well-known non-ionic surfactants (with commercial name of Pluronics) consisting of two dissimilar moieties, i.e., hydrophilic PEO block and hydrophobic PPO block, within the same molecule. These triblock copolymers possess symmetrical structure $(EO)_x(PO)_y(EO)_x$ where x and y denote the number of ethylene oxide and propylene oxide monomers per block and are available in a range of x and y values in the form of pastes, flakes and liquids. In aqueous solution, they aggregate to form micelles, the hydrophobic blocks of the block copolymers (PPO) form the core of the aggregates whereas the hydrophilic ones (PEO), with the surrounding water molecules, form the corona. Due to their unique aggregational characteristics, Pluronics are used as dispersion stabilizers, emulsifier, pharmaceutical ingredients, biomedical materials, and templates for the synthesis of mesoporous and nanomaterials (Hamley, 2005). We have used Pluronic P85 ($EO_{26}PO_{39}EO_{26}$, Mw = 4600) as obtained from BASF Corp., New Jersey while the gold salt of hydrogen tetrachloroaureate (III) hydrate ($HAuCl_4.3H_2O$) was purchased from Sigma-Aldrich. All the solutions were prepared in D_2O (99.9 atom % D) since D_2O instead of H_2O as a solvent provides good contrast for SANS measurements.

METHODS

The gold nanoparticles were synthesized from 1 wt% (2.2×10^{-3} mol L^{-1}) P85 block copolymer solution by mixing with varying concentration 0.001 to 0.1 wt% of $HAuCl_4.3H_2O$ in aqueous solution. All the solutions were prepared at room temperature without any disturbances for about 3 hrs for the completion of the reaction. The transparent solution of P85 becomes coloured on the formation of nanoparticles.

The formation of gold nanoparticles was observed using UV-visible spectroscopy. This technique measures absorbance or transmittance to give qualitative and

quantitative information in the molecular level. Surface plasmon resonance (SPR) band determination is one of the most familiar applications of this technique, which arises due to the collective oscillations of conducting electrons of metal nanostructures. The measurements were carried out using 6505 Jenway UV-visible spectrophotometer. This instrument is suitable for measurements in the scanning wavelength range between 190 nm and 1100 nm (with an accuracy of ±1 nm). It makes use of tungsten halogen lamp as a visible light source and deuterium discharge lamp for the UV light source. The instrument was operated in spectrum mode with a wavelength interval 1 nm and the samples were held in quartz cuvettes of path length 10 mm.

The effect of varying gold salt concentration on the block copolymers in the synthesis of gold nanoparticles has been studied by SANS in which one measures the coherent differential scattering cross-section ($d\Sigma/d\Omega$) per unit volume as a function of wave transfer Q (= 4π $\sin\theta/\lambda$, here λ is the wavelength of the incident neutrons and 2θ is the scattering angle). This technique with the possibility to vary the contrast is an ideal method for studying multicomponent systems as is the present case. It provides information about the shape and size of the scattering particles in the length scale of 10 - 1000 Å. The measurements were performed on the SANS instrument at the Guide Tube Laboratory, Dhruva reactor, BARC, India (Aswal et al., 2000). This diffractometer uses a polycrystalline BeO filter as a monochromator and the mean wavelength of the incident neutron beam is 5.2 Å with a resolution ($\Delta\lambda/\lambda$) of about 15%. The angular distribution of the scattered neutron was recorded using a linear 1m long He^3 position sensitive detector. The data were recorded in the accessible Q range of 0.017-0.35 $Å^{-1}$. The sample solutions were kept in a 0.5 cm thick quartz cell with Teflon stoppers. All the measured SANS data were corrected for the background, the empty cell contributions and solvent contributions and were normalized to the cross-sectional unit using standard procedures.

DLS experiments were also performed to probe the nanostructures in P85 and gold salt solutions. DLS is a complementary technique to SANS to provide structural information of the particles in solution based on the scattering of the diffusing particles. Measurements were carried out using Autosizer 4800 (Malvern Instruments, UK) equipped with 7132 digital correlator and coherent (Innova 70C) Ar-ion laser source operated at wavelength 514.5 nm with a maximum output power of 2W. DLS provides average diffusion coefficient and hence the hydrodynamic size of the particles.

RESULTS AND DISCUSSION

It is observed that a transparent P85 solution becomes coloured on addition of gold salt. The violet colour seen in these samples is different from that (yellow) of the pure salt solutions and indicates the formation of gold nanoparticles (AuNPs) since the surface plasmon resonance of AuNPs is known to show such colour (Chen et al., 2005). We also observe that there is a variation in the colour intensity with the increase in the gold salt concentration. The maximum in the colour intensity is found at 0.005 wt% $HAuCl_4.3H_2O$ for 1 wt% of P85. The colour intensity decreases beyond this maximum and finally the colour of the samples becomes yellow at higher salt concentrations (~ 0.1 wt%). The maximum of the colour intensity occurs at the ratio of 1:22 of gold ion to block copolymer molecules. These observations thus indicate that the present method of synthesizing nanoparticles has a very limited yield.

Figure 1 show the UV-visible absorption spectra of the samples have two distinct peaks, one around 230 nm and the other at about 540 nm. The peaks centered at ~ 230 nm are due to free gold (III) chloride ions formed in the reduction of gold salt (Kuo et al., 2005) and the peaks centered at ~ 540 nm originate from the surface plasmon resonance (SPR) of the AuNPs (Shimizu et al., 2003). The build up of SPR peaks at ~ 540 nm with increase in the salt concentration confirms the formation of the nanoparticles. The absorbance at this peak increases with salt up to 0.01 wt% concentration. However, the peak is only symmetric up to 0.005 wt% salt concentration and beyond which it shifts to higher wavelengths and has a wide tail. The symmetric SPR peaks indicate the formation of spherical AuNPs. Our studies show that 0.005 wt% is stable and has the maximum yield. The TEM results in the literature have shown that the typical size of spherical particles formed using different block copolymers is 50-200 Å (Sakai et al., 2004). The departure from this behaviour with increasing salt concentration suggests formation of large non-spherical AuNPs which finally undergo phase separation (Liz-Marzan, 2006). The increase in absorbance in the salt concentration range 0.005-0.01 wt% is because of formation of these large non-spherical particles. The absorbance at SPR peaks broadens and falls for higher salt concentrations beyond 0.01 wt%. It is also interesting to note that for this region of salt concentration, the absorbance peak at ~ 230 nm which is an indication of free gold ions, starts rising and its value increases with the salt concentration.

SANS has been used to examine the role of block copolymer in the formation of gold nanoparticles. Figure 2 shows the SANS data on pure 1 wt% P85 and with the addition of $HAuCl_4.3H_2O$ around its concentration of the maximum yield. The data without and with the addition of salt have very similar features. Block copolymers in salt solutions can either participate in the formation of gold nanoparticles or form their own micelles. In the case of nanoparticles, they get coated on the particles, and follow very different distribution on the nanoparticles than that in the micelles. Therefore, the scattering from block copolymer coated gold nanoparticles is expected to be significantly different than the micelles. The fact that the scattering curve does not change with varying salt concentration even when the yield of the nanoparticles is maximum, suggests only a very small fraction of gold nanoparticles is formed in these systems. In other words, most of the block copolymers are not part of the coating on the nanoparticles and form their own micelles. It is thus not possible to separate the scattering of block copolymer coating (which is negligible) from that of the micelles. The block copolymer micelles are fitted as core-shell particles with different scattering length densities for core and shell. The structure of these micelles is described using a model consisting of non-interacting Gaussian PEO chains attached to the surface of the PPO core (Pedersen, 2004). The form factor of the micelles comprises four terms: the self-correlation of the core, the self-correlation of the chains, the cross term between core and chains, and the cross term between different chains. The structure factor is taken to be unity as valid in the case of dilute system. The analysis shows that P85 micelles have a PPO core radius of $R_c = 37 \pm 0.5$ Å and radius of gyration of PEO chain as $Rg = 12 \pm 1$ Å.

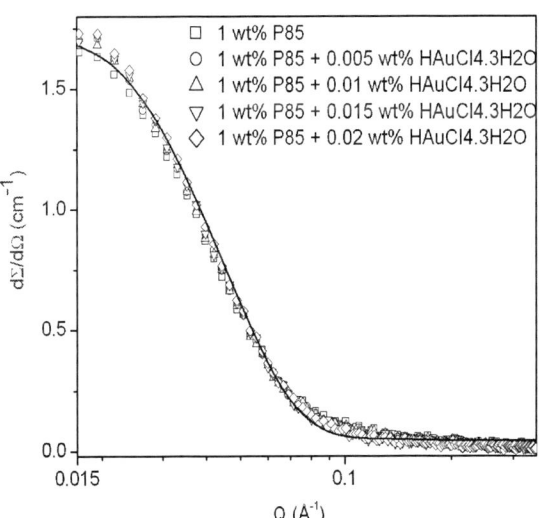

Figure 2. *SANS data of 1 wt% P85 with varying concentration of $HAuCl_4.3H_2O$ in aqueous solution. The data do not show any significant change on addition of gold salt. The solid curve is a theoretical fit to the experimental data.*

SANS has shown that for a fixed block copolymer concentration and varying the salt concentration, there is overall a very small fraction of block copolymers that are part of the gold nanoparticles. These results have

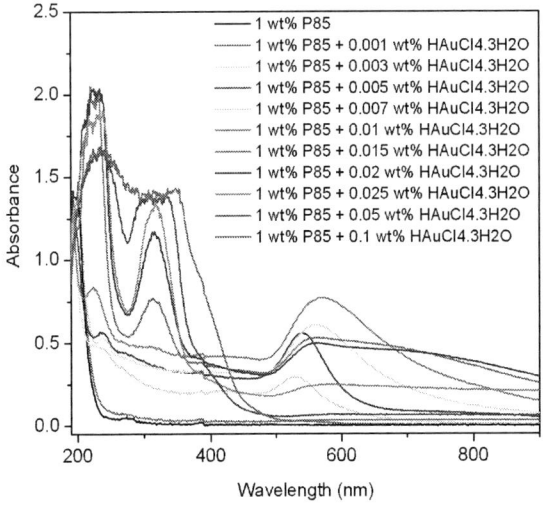

Figure 1. *UV-visible absorption spectra of the samples of 1 wt% P85 with varying concentration of $HAuCl_4.3H_2O$ in aqueous solution. The different data sets correspond to increase in absorbance peak at ~ 230 nm with the increase in the $HAuCl_4.3H_2O$ concentration.*

been further confirmed by the DLS. Figure 3 shows the DLS data of the intensity autocorrelation function [$g^2(\tau)$] of the same samples in the figure 2. The data do not show any change without and with the addition of salt. The functionality of the autocorrelation function depends on the diffusion coefficient of the particles and the data suggest that it is dominated only by one type of the particles (i.e. micelles). The analysis shows that the micelles have a diffusion coefficient of 30×10^{-6} cm^2/sec and this corresponds to the hydrodynamic size of 90 Å. This hydrodynamic size of the micelles comprises PPO core plus PEO shell along with hydration attached to the shell. Thus both SANS and DLS suggest that most of the block copolymers form their own micelles and only a very small fraction of block copolymers is associated with the coating on the AuNPs.

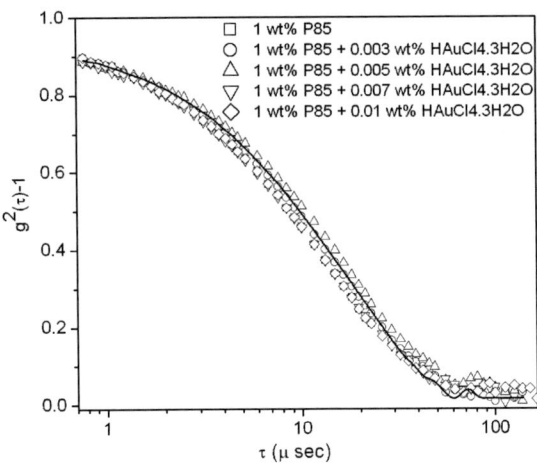

Figure 3. *Plots of the intensity autocorrelation functions of 1 wt% P85 with varying concentration of HAuCl₄.3H₂O in aqueous solution. The data do not show any significant change on addition of gold salt. The solid curve is a theoretical fit to the experimental data.*

The exact role of the block copolymers in controlling the synthesis of AuNPs can be understood in terms of the following three steps of the mechanism proposed in the literature (Sakai et al., 2005b):

(i) Reduction of $AuCl_4^-$ ions by block copolymers and formation of gold clusters:

$AuCl_4^- + n(\text{PEO-PPO-PEO}) \rightarrow (AuCl_4^-) — (\text{PEO-PPO-PEO})_n$

$m[(AuCl_4^-) — (\text{PEO-PPO-PEO})_n] \rightarrow Au_m + 4mCl^-$ +oxidation products (1)

(ii) Adsorption of block copolymers on gold clusters and reduction of $AuCl_4^-$ ions on the surfaces of these gold clusters:

$Au_m + l[(AuCl_4^-) — (\text{PEO-PPO-PEO})_n] \rightarrow Au_p—$ $(\text{PEO-PPO-PEO})_q$ + $4lCl^-$ + Oxidation products (2)

(iii) Continued growth of gold nanoparticles using the step (ii) and finally the stabilization by block copolymers.

We have observed the maximum yield of AuNPs for a fixed concentration (1 wt%) of P85 occurs at the gold salt concentration of 0.005 wt%. This corresponds to the block copolymer to gold salt molar ratio of 22 and is related to the minimum number of block copolymer molecules (n) required for the reduction [eqn. (1)] of a gold ion to occur. At lower salt concentrations when the molar ratio of block copolymer to salt ions (r) is much larger than $n = 22$, the yield increases with salt concentrations up to 0.005 wt%. Beyond this concentration the ratio r decreases and this decreases the probability of the reduction reaction in the sample to take place. The presence of large number of ions and block copolymers makes it possible that a significant fraction of ions can still find n number of block copolymers and get reduced even though average ratio r is less than n. Based on this explanation the yield is expected to be suppressed with increase in salt concentration as has been experimentally observed. UV-visible spectroscopy suggests the increase in the size of the nanoparticles with the increase in the salt concentration beyond 0.005 wt% could be understood in terms of nanoparticle aggregation occurring due to insufficient stabilization by the block copolymers on the surface of the nanoparticles as the block copolymer to gold ion ratio decreases with the increase in the gold salt concentration. As a result we have observed that nanoparticles in these systems phase separate after some time (~ 1 day). At higher salt concentrations (> 0.02 wt%) the ratio r is so low to provide any significant probability within the system to able to reduce the gold ions. Hence, UV-visible spectra only show the strong peak of unreduced gold ions and no SPR peak of AuNPs is seen (figure 1). It has been found that the gold salt (HAuCl₄.3H₂O) concentration that leads to the maximum yield for 1 wt% P85 is 0.005 wt% and corresponds to 22 times smaller in molar concentration to that from block copolymer. For the typical size of AuNPs of 100 Å, the fraction of block copolymers (R_g ~ 18 Å) coated on the particles is calculated to be only about 0.25%. It is therefore clear that only a very small fraction of block copolymers is part of the nanoparticles and most of the block copolymers form their own micelles. This explains why SANS and DLS data do not show any significant change on the formation of the nanoparticles in the present systems.

The low yield of synthesizing AuNPs is limited due to large value of n required for the reduction reaction of gold ions to occur. The fact that most of the block copolymers (~ 99.75%) remain free to form their own micelles, this can be further used for synthesis of nanoparticles if the salt concentration is added in small steps (< 0.005 wt%) to maintain the ratio r greater than n. We have used this step-addition method to show the enhancement in the yield of the AuNPs. The step-addition method is used on 1 wt% P85 with the step of addition of 0.004 wt%. The time between two steps (t_s) is decided by the time-dependent UV-visible spectroscopy as shown in figure 4. The time dependence of spectra is recorded from the freshly prepared sample by mixing the two components and from these data, integrated absorbance of SPR (proportional to the concentration of the nanoparticles) is plotted as a function of time (inset of figure 4). The value of t_s (~ 3 hrs) is decided at which the absorbance curve shows saturation.

Figure 5 shows the comparison of UV-visible spectroscopy of 1 wt% P85 with 0.005 wt% salt and prepared with the addition of 5 steps of 0.004 wt%. It is seen that the increase in absorbance as the AuNPs concentration increases with increasing salt by step addition. The overall gain of about 4 times is observed in step-addition method as expected. Figure 5 also shows comparison of the data if the salt concentration of 0.02 wt% (5 × 0.004 wt%) is added directly instead of steps, which suggests no AuNPs are formed. Thus step-addition method can be used to enhance the yield of AuNPs using block copolymers by manifold if the step addition concentration is sufficiently small in each step to maintain the reduction of gold ions.

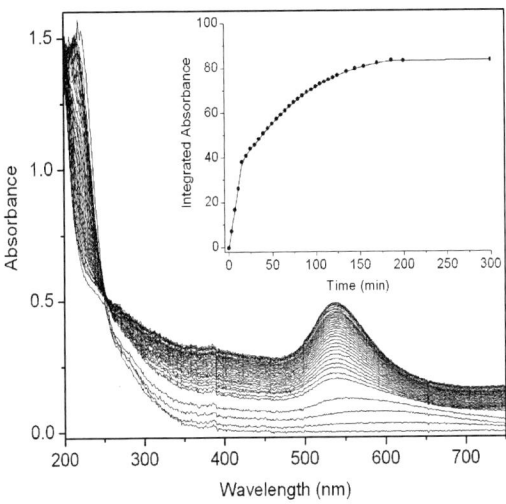

Figure 4. *Time-dependent UV-visible absorption spectra of 1 wt% P85 with 0.005 wt% $HAuCl_4.3H_2O$. The different data sets correspond to increase in absorbance with time. Inset shows the variation of the integrated absorbance (430 nm – 700 nm) of SPR as a function of time.*

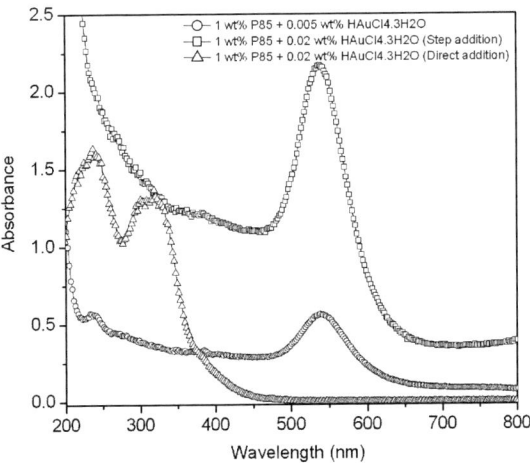

Figure 5. *UV-Visible spectra of 1% P85 with 0.005% $HAuCl_4.3H_2O$ compared with 1 wt% P85 with 0.02 wt% $HAuCl_4.3H_2O$ obtained by with and without step addition of 0.004 wt%.*

ACKNOWLEDGEMENTS

Debes Ray would like to thank ICNX 2009 organizers and IUCr for awarding him Young Scientist Grant to present the paper in the conference.

REFERENCES

Aswal, V. K. and Goyal, P. S., (2000), Small-angle neutron scattering diffractometer at Dhruva reactor, *Curr. Sci.* 79:947-953.

Aswal, V. K., Wagh, A. G. and Kammel, M., (2007), Formation of rodlike block copolymer micelles in aqueous salt solutions, *J. Phys.: Condens. Matter* 19:116101.

Chen, S., Liu, Y. and Wu, G., (2005), Stabilized and size-tunable gold nanoparticles formed in a quaternary ammonium-based room-temperature ionic liquid under γ-irradiation, *Nanotechnology* 16:2360-2364.

Daniel, M. C. and Astruc, D., (2004), Gold Nanoparticles: Assembly, Supramolecular Chemistry, Quantum-Size-Related Properties, and Applications toward Biology, Catalysis, and Nanotechnology, *Chem. Rev.* 104:293-346.

Hamley, I. W., (2005), *Block Copolymers in Solution: Fundamentals and Applications.* Wiley, New York.

Jain, P. K., El-Sayed, I. H. and El-Sayed, M. A., (2007), Au nanoparticles target cancer, *Nanotoday* 2:18-29.

Kuo, P. L., Chen, C. C. and Jao, M. W., (2005), Effects of Polymer Micelles of Alkylated Polyethylenimines on Generation of Gold Nanoparticles, *J. Phys. Chem. B* 109:9445-9450.

Liz-Marzan, L. M., (2006), Tailoring Surface Plasmons through the Morphology and Assembly of Metal Nanoparticles, *Langmuir* 22:32-41.

Pedersen, J. S., (2000), Form factors of block copolymer micelles with spherical, ellipsoidal and cylindrical cores, *J. Appl. Crystallogr.* 33:637-640.

Sakai, T. and Alexandridis, P., (2005a), Mechanism of Gold Metal Ion Reduction, Nanoparticle Growth and Size Control in Aqueous Amphiphilic Block Copolymer Solutions at Ambient Conditions, *J. Phys. Chem. B* 109:7766-7777.

Sakai, T. and Alexandridis, P., (2004), Single-Step Synthesis and Stabilization of Metal Nanoparticles in Aqueous Pluronic Block Copolymer Solutions at Ambient Temperature, *Langmuir* 20:8426-8430.

Sakai, T. and Alexandridis, P., (2005b), Size- and shape-controlled synthesis of colloidal gold through autoreduction of the auric cation by poly(ethylene oxide)–poly(propylene oxide) block copolymers in aqueous solutions at ambient conditions, *Nanotechnology* 16:S344-S353.

Shan, J. and Tenhu, H., (2007), Recent advances in polymer protected gold nanoparticles: synthesis, properties and applications, *Chem. Commun.* 4580-4598.

Shimizu, T., Teranishi, T., Hasegawa, S. and Miyake, M., (2003), Size Evolution of Alkanethiol-Protected Gold Nanoparticles by Heat Treatment in the Solid State, *J. Phys. Chem. B* 107:2719-2724.

Behaviour of Copper In Annealed Cu/Sio₂/Si Systems For On-Chip Interconnections

Thant Zin Htwe, Khin Maung Latt

Department of Physics, Yangon Technological University, Myanmar,

(Corresponding author,: e-mail: thantzin99999@gmail.com).

Department of Physics, Mandalay Technological University, Mandalay, Myanmar. (e-mail: sayarkyee9@gmail.com)

ABSTRACT

The electrical and structural properties of thin copper films attract increasing attention nowadays because of the use for on-chip interconnections. The main advantages of copper are the excellent conductivity and the relatively high stability against electro migration damaging. Interdiffusion at the copper/silicon interface can be a remarkable drawback of the interconnection quality even at room temperature which leads to the use of barrier layers between copper and silicon in technical applications. Often, thermal annealing of the as-deposited copper films is required to ensure proper process integration. In the present paper, Copper thin films of thickness 100 nm are deposited on SiO₂/Si by ionized metal plasma deposition method. Then samples are annealed at different temperatures under high vacuum condition. The behavior of copper and the mechanism of compound formation studied at different temperatures, using scanning electron microscopy SEM, X-ray diffraction XRD and four point probe method. Diffusion of Cu into SiO₂/Si layer start at 550°C and form Cu_xSi_y. Oxidation of Cu is also take place at high temperature annealing.

Keywords— Copper interconnect; Silicon dioxide; Vacuum annealing; Copper silicides,

INTRODUCTION

The research of new materials for metallization in microelectronics applications turned out to be necessary for interconnections requiring line sizes below or equal to 0.1micron technology. From this point of view, Copper (Cu) has long been considered as a promising candidate to replace aluminum (Al) as the interconnection material in ultra-large-scale integrated ULSI technology due to the main advantages of its high conductivity, Copper electrical resistivity ~1.63 μΩcm is about half that of Aluminum~2.66 μ Ωcm, while its high melting point~1085°C vs. ~550°C for aluminum endows it with a larger resistance for corrosion and electro- migration and mechanical elasticity stress.

The electrical and structural properties of copper attract increasing attention nowadays because of the use for on-chip interconnections as well as high-reliability interconnect formation is one of the key issues in the fabrication of advanced semiconductor devices.

Although initial success in Cu metalization using Dual Damascene patterning, electroplating and chemical–mechanical polishing (CMP) was announced, copper metalization is still at its early stage and much work needs to be done on the way to its maturity. For example, the high rate of Cu oxidation at relatively low temperatures and its diffusion into SiO remain the two major obstacles

for the widespread application of Cu-based interconnections. Unlike Al that forms a thin and dense layer of oxide to stop further oxidation, copper oxides are believed to be less protective and oxidation goes on at a significant rate, leading to serious degradation of the interconnection and is known as the first major obstacle.

Thus, understanding of the oxidation and passivation of thin copper films in the fabrication and applications of integrated circuits ICs at low and medium temperature is important. Therefore, the study of the copper oxidation and the development of means to prevent copper oxidation are crucial to the application of copper as an interconnection material. However, the studies of copper oxidation have been focused in the medium and high temperature range for bulk copper, and the results are conflicting. Although it is generally believed that for bulk copper the two major oxide products are cupric oxide (CuO) and cuprous oxide (Cu_2O), their formation and relative abundance depend largely on the situations in which the oxidation takes place due to the non-equilibrium conditions.

Nevertheless, apart from its renowned desired low resistivity value that gives low resistance-capacitance impedence to achieve higher device speeds, Cu is also infamous for its fast diffusion behavior mainly Cu ions in Si and the various detrimental effects on minority carrier lifetime.

CP1202, *Neutron and X-Ray Scattering in Advancing Materials Research: International Conference – 2009*
edited by A. Saat, H. A. Kassim, M. H. H. Jumali, J. M. Saleh, M. R. Othman, A. Ibrahim, F. M. Idris, and M. H. A.-R. M. Ahmad
© 2009 American Institute of Physics 978-0-7354-0739-8/09/$25.00

On the other hand, the high reactivity of copper with silicon and its easy diffusion in SiO-based dielectrics under thermal annealing and electrical field lead to device deterioration. In the case of Si substrates, interdiffusion at the copper-silicon interface can be a remarkable drawback of the interconnection quality even at room temperature which leads to the use of barrier layers between copper and silicon in technical applications and is known as the second major obstacle.

Introduction of copper interconnects in silicon integrated circuit technology has drastically increased the danger of unintentional contamination of silicon substrates with Cu. Hence, it is a common wisdom of silicon science and technology that transition metals, particularly iron, copper, and nickel, are extremely detrimental to the performance of integrated circuits and solar cells. Copper Contamination within silicon, in particular, is of great concern to us as the metal is to replace aluminum for next generation metallization. Accompanying the dimensional miniaturization of silicon-based semiconductor devices, it is mandatory to control the amount of impurities in silicon at a level lower than 10^{10} atoms/cm^3, so as to avoid any degradation in device performance and characteristics. This is due to their high electrical activity resulting from electrical levels in the band gap, the formation of precipitates in p–n junctions or at silicon:silicon dioxide interface, and the increase in the micro-roughness of the silicon surface during wet chemical cleans in contaminated solutions.

Interfacial interactions of metals with oxide and polymer surfaces play a key role in improving adhesion, inhibiting corrosion, smoothing surface, and achieving selective deposition. Among them, Cu–SiO2 system has attracted great interest recently because of its application to catalysis for alcohol oxidation and to interconnect metallization in Giga scale integration (GSI). However, Cu however, has a low sticking probability, poor wettability, and poor adhesion on clean SiO$_2$ substrates during thermal or sputter deposition, due to the lack of reactivity between Cu and SiO$_2$. By enhancing interfacial interactions, methods such as ion bombardment, heat treatment, and introduction of an intermediate layer are effective for improving such thin-film properties as surface morphology, wettability and adhesion, etc.

EXPERIMENTAL DETAILS

For all sample preparation and experiments described in this study we used 6-inch Si (100) wafers. Si wafers were cleaned in 10:1 diluted HF solution and rinsed in deionized water before SiO$_2$ deposition. First, a 500-nm thick plasma enhanced chemical vapor deposited (PECVD) SiO$_2$ dielectric was deposited by using a gas mixture of SiH$_4$, Ar and O$_2$ at 400°C on 6-inch Si wafers. Cu film of 100nm thick was then deposited onto PECVD-SiO$_2$/Si substrates by using ionized metal plasma (IMP) sputtering in a gas mixture of Ar and N$_2$. A detailed IMP deposition process has been described elsewhere [13].

Samples of 20×20 mm^2 in size were cut from wafers for further experiment. The samples were first annealed in vacuum furnace (10^{-2} torr) up to 750°C from 300°C with 50°C intervals for 60 min. The resistivities of as-deposited and annealed samples were measured by four-point probe to survey the overall reactions involving Cu. The X-ray diffraction (XRD) measurements were performed with a RIGAKU model RINT2000 diffractometer using a γ =2.5 grazing incident angle geometry. The Kα Cu X-ray (λ=1.542 °A) detection was done from 2θ =20 to 2θ =85 with scan speed of 4°/min and scan step 0.05 for the analysis of reaction product phases and the interdiffusion of the elements across the interface, respectively. The surface morphologies of the structure at various annealing temperature were observed by JEOL 5410 Scanning Electron Microscope (SEM) employing a 20 keV primary electron beam.

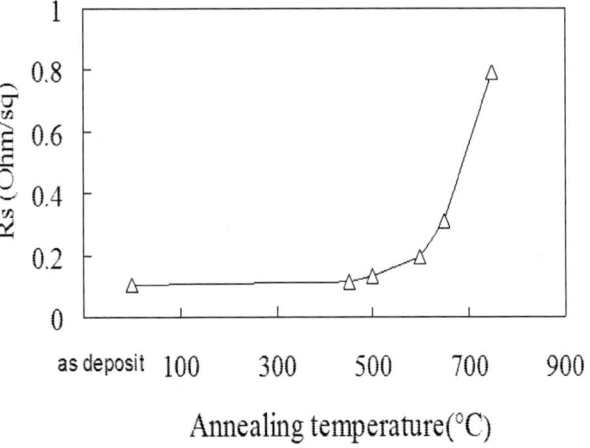

Fig. 1 Variation of sheet resistance in Cu/SiO2/Si structure as a function of annealing temperatures.

The graphs presented in Fig. 1 indicate the change in sheet resistance measured on the Cu/SiO$_2$/Si structure as a function of annealing temperature in high vacuum condition. The measured sheet resistance of the 100 nm copper film is 1.72 Ω cm^{-1} since the top Cu layer of 100 nm carries nearly all the sensor current, the sheet resistance measurements monitors the condition and the quality of the Cu overlayer. Hence, theses curves can be used to estimate the degree of intermixing, reaction, and changes of integrity across the metallization layers as well. According to this figure, all samples, annealed up to 450°C can maintain the same level of sheet resistance as the as-deposited samples. However, the sheet resistance increases slightly at temperatures just exceeding 450°C, implying that a relatively resistive substance has been produced from the highly conductive Cu layers.

As the temperature reaches 550°C and beyond, the sheet resistance of the samples rises abruptly, indicating that a severe intermixing or interfacial reactions occurred across all the Cu films.

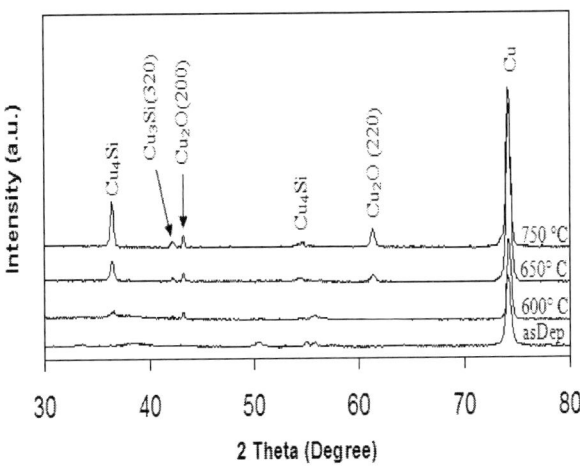

Fig. 2. The XRD spectra from EPCu/IMPCu/IMPTaN/SiO2/Si structure before (a) and after annealing at (b) 600°C; (c) 650°C; (d) 750°C; for 60 min in high vacuum ambient.

Figure 2 shows XRD patterns of the $Cu/SiO_2/Si$ samples as-deposited and annealed at various temperatures. It displays the X-ray diffraction patterns for the samples exhibiting sheet resistivity curve presented in Fig.1. The Cu film sputter-deposited on the SiO_2 has a strong (220) preferred orientation at 74.13°. But low intensity peaks of (111) and (200) are observed at 43.30°, and 50.43° respectively. As shown in Fig. 2, there is a distinction in XRD spectra between samples annealed below and above 600°C. All annealing temperatures below 600°C, Si peak and Cu peaks were observed.

Distinctly, at 650°C, several new peaks were found at 36.4, 42.40, 54.9 and 61.30°. They were identified as Cu_4Si, Cu_3Si(320), Cu_2O(200), and (220), respectively. Probably, the formation of Cu_xO_y and Cu_xSi_y are the main cause for the jump in sheet resistance value after 600°C annealing as shown in Fig. 1. XRD confirms that the reaction take place during annealing not only produce Copper Silicide (Cu_xSi_y) but also Cu_xO_y while the copper is not completely consumed. After annealing at 750 °C, the Cu_3Si peaks intensity increased, indicating the occurrence of a severe intermixing of Cu and Si. Cu and Si. The XRD results also revealed that the increase in sheet resistance discussed in the previous section was likely caused by the loss of copper due to a Cu_xSi_y formation.

Surface morphology examined by SEM micrograph of the samples at annealing temperature lower than 450°C revealed no obvious changes and pinholes and white spots on the copper surface were observed thereafter. Increasing the annealing temperature, the number and size of pinholes increased and the surface of the sample became much. rougher Fig. 2c–d . The cracks were also observed in the white spot at the annealing temperature of 750°C, as shown in Fig. 2d. In the case of, the change of copper surface morphology started to occur at 500 °C. Increasing the annealing temperature to 750°C results the surfaces shown in Fig. 3(d). Increasing the annealing temperature to 650 °C resulted in a highly damaged surface indicating a severe reaction of the Cu/Si. The agglomeration of Cu films were clearly seen and exposing part of SiO_2/Si.

Cu has been identified as the moving species in the reaction between Cu and Si, and has a very high affinity towards Si, it is likely that the diffusion of Cu through the SiO_2 layer is the mechanism which brings the copper silicide (Cu_3Si) into equilibrium. On the other hand, copper has been observed to diffuse interstitially in Si, it is probable that when Cu atoms penetrate the Si lattice their presence in interstitial sites decreases the Si–Si bond strength nearby, thus making it easier to release Si atoms from the lattice. Cu to react directly with the Si. It is generally assumed that the diffusion of Cu would lead to the formation of Cu_3Si precipitates and their further growth would result the permanent failure of the structure. There would be enough time for the formation of Cu_3Si during sample cooling, since the sample stayed above room temperature at least for 20–30min. It is known that the formation of Cu_3Si takes place even at temperatures around 150 °C.

In the case of Cu_xO_y formation, the sputtering deposited copper is first oxidized to Cu_2O below 550°C under high vacuum annealing and then transformed to CuO at higher temperatures ($\geq$550 °C) near the surface of the copper.

SEM results show that the as-deposited Cu film contains grains of similar sizes. After heat treatment at 450 °C, small grains are observed due to the nucleation of the Cu_2O phase. Further annealing even in vacuum leads to the aggregation and growth of the grains whereas the grain boundary of the grains before coalescence is still evident. When the annealing temperature is 600 ° C, lamellar CuO grains are observed due to the separation of the CuO phase from the Cu_2O phase. Here, three sources of oxygen atoms forming Cu_xO_y will be considered. Firstly, oxygen atoms from the SiO_2, which would diffuse into and/or react with the Cu layer, as the annealing temperature increased. The second source is oxygen atoms incorporated with Cu metal film from the deposition ambient during the Cu deposition and decorating the grain boundaries of each film [15]. Lastly, oxygen atoms incorporated from annealing ambient, but were excluded because all the samples were annealed in an vacuum condition. However, the formation of Cu_2O and/or cannot be fully prevented. This reveals the fact that formation of Cu_2O and Ta_2O_5 in the structure is mainly due to the oxygen incorporated during the deposition process and from SiO_2.

CONCLUSION

1. The oxidation of Cu starts at the temperature which is lower than 550°Cby forming Cu_2O first and transformed into CuO at elevated temperature. This is higher than stated in literatures since this experiment in done in high vacuum condition.

2. It also can not avoid the formation of Cu_xSi_y due to the high mobility of Cu atoms.

3. However the structure can maintain its integrity upto 400°C without showing an increase in sheet resistance. This temperature is the optimum temperature that the industry requires during the fabrication process.

REFERENCES

[1]   C.A., Schuler and W.L. Mc Namee, "Modern Industrial Electronics" New York: McGraw Hill, 1993.

[2] Microchip. 2001. PIC16F873A Data Sheet. Microchip

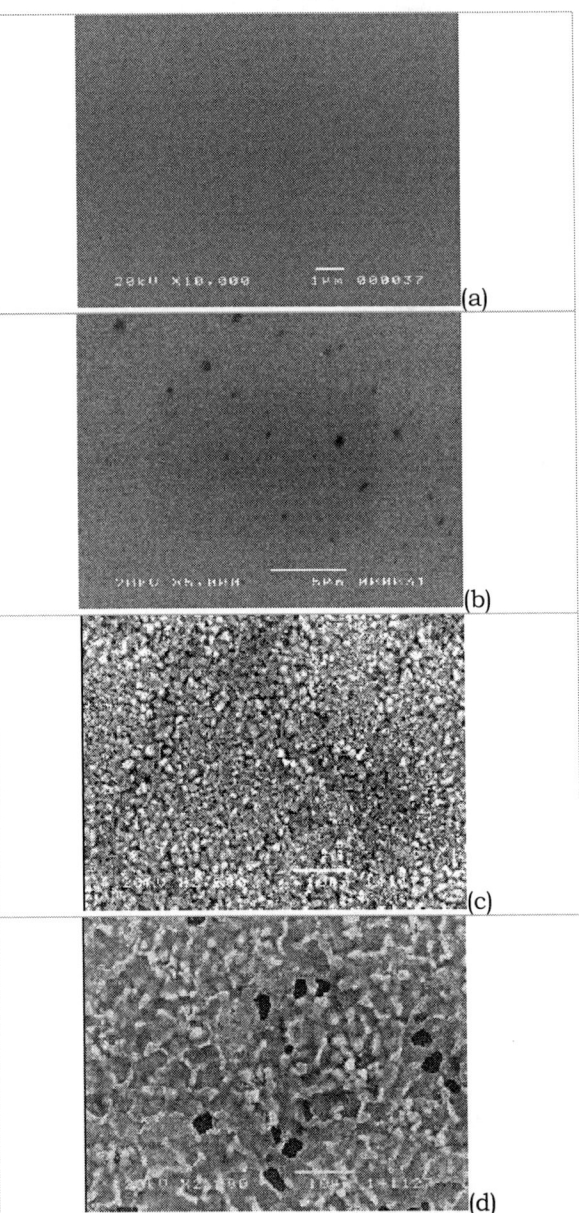

Figure 3 SEM images of Cu/SiOx/ Si(111) samples annealed for 60 min at vacuum condition: (a) as deposit (b) 450 °C (c) 650 C (d) 750 C.

Technology Inc. January 2006.

http://www.microchip.com

[3] H. Rongen, "Introduction to Microprocessors and Microcontrollers" Forschungszentrum Julich, Germany. 2000.

[4] W. Stallings, "Data and Computer Communications". Sixth Edition, Prentics-Hall, Inc. 2000

[5] Bob Harbour, "Dual Motor Bidirectional Electronic Speed Control", April 30, 1999.

http://www.circuits.Lab.com

[6] Thomas Scarborough, "Everyday Practical Electronics", September, 2000, UK.

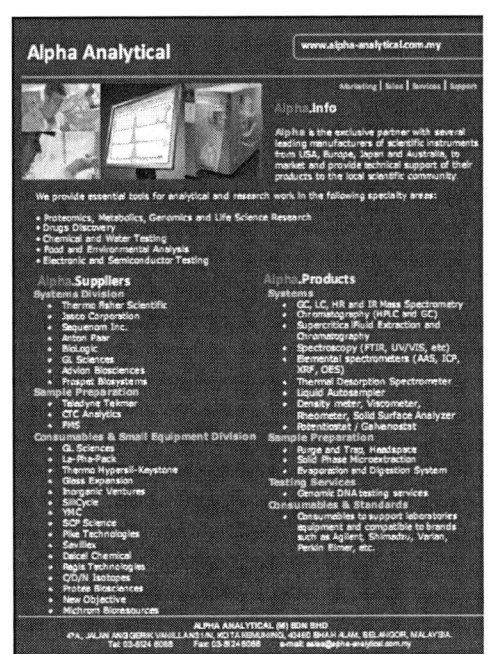

TERAS INSTRUMENTS SDN BHD (673741-A)

"Commitment To Excellent"

No. 22-2-2A, Jalan Medan PB2A, Seksyen 9, 43650 Bandar Baru Bangi, Selangor Darul Ehsan. MALAYSIA

Tel: 03-8926 9511   Fax : 03-8926 9522   Email: iteras@streamyx.com

Bumiputera Status & Registered With Ministry of Finance

- General Laboratory Equipments
- General Laboratory Consumables
- General Laboratory Chemicals
- Test & Measurement Equipments
- Optics, Opto-electronics & Laser Instruments
- Machines & Tools

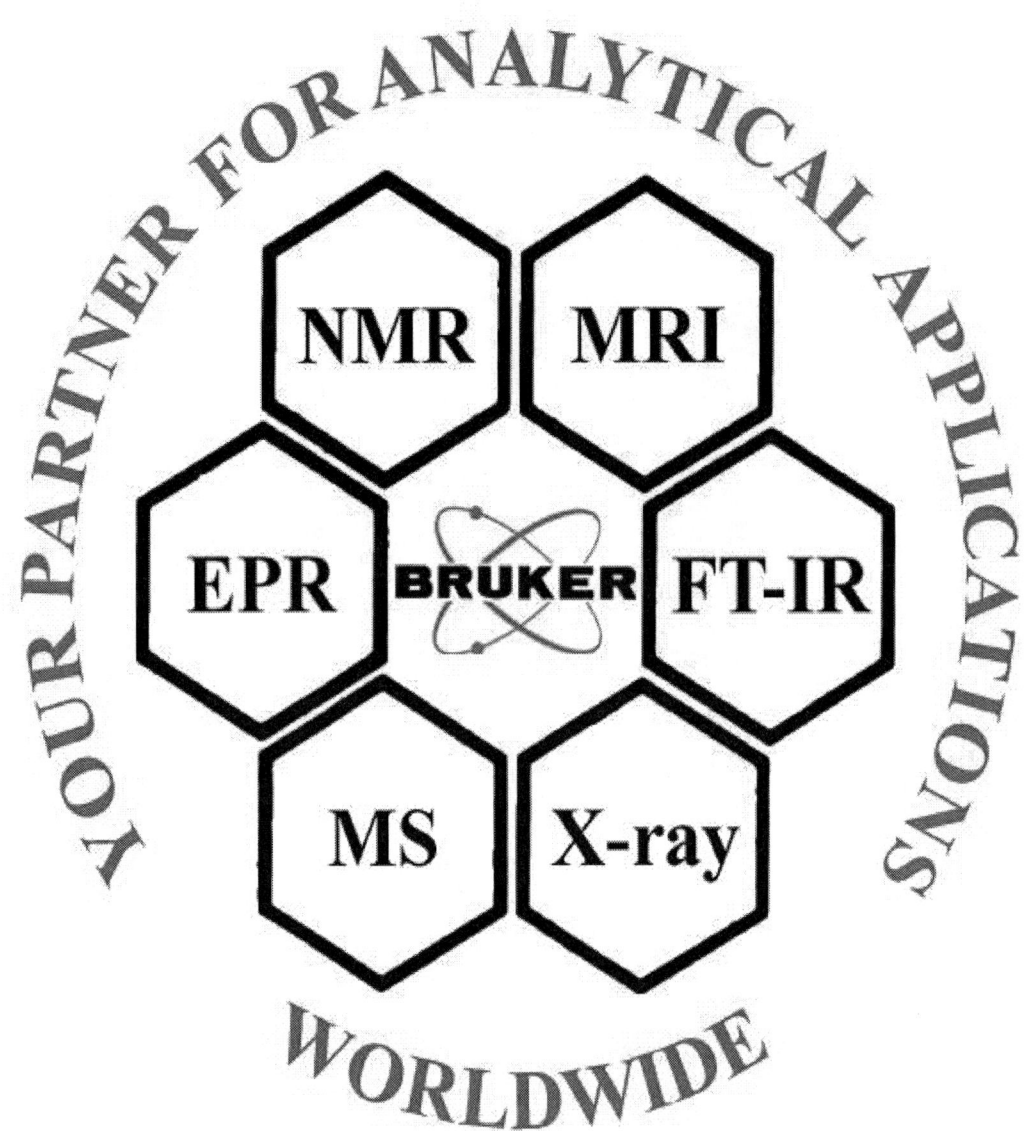

Bruker (Malaysia) Sdn Bhd (537365-D)
303, Block A,
Mentari Business Park,
No. 2, Jalan PJS 8/5,
Dataran Mentari,
46150 Petaling Jaya,
Selangor, Malaysia.
Tel: 03-5621 8303   Fax : 03-5621 9303   E-mail : bruker1@tm.net.my
www.bruker.com

INTERNATIONAL TABLES ONLINE

INTERNATIONAL TABLES for CRYSTALLOGRAPHY

- 6000 pages
- 300 chapters
- 680 tables of fundamental data
- 1100 tables of symmetry information
- interactive features and resources

it.iucr.org

AUTHORS CONTACTS

Bil	Authors	Affiliation	Email
1	A Lounis	U.S.T.H.B	zlounis@yahoo.com
2	A.R.West, Prof.	University of Sheffield, UK	A.R.West@sheffield.ac.uk
3	Abdul Aziz Mohamed, Dr.	Nuclear Malaysia	aziz_mohd@nuclearmalaysia.gov.my
4	Abdelmadjed Bouhemadou, Mr.	University of Setif, Algeria	a_bouhemadou@yahoo.fr
5	Abhijit Chatterjee, Dr.	Accelrys, Singapore	AChatterjee@accelrys.com
6	A D Fortes, Dr.	University College London	andrew.fortes@ucl.ac.uk
7	Adel Aly Ahmed Ghoneim, Dr.	College of Technological Studies, Kuwait	aa.ghoniem@paaet.edu.kw
8	Adel M. M. Mohammedein, Dr.	College of Technological Studies, Kuwait	admohamed@yahoo.com
9	Adolf Asih Supriyanto, Mr.	UKM, Malaysia	adolf@legendagroup.edu.my
10	Alexander Loffe, Dr.	JCNS Garching, Germany	a.ioffe@fz-juelich.de
11	Ali Pazirandeh	Azad University, Tehran Iran	pzrud193y@srbiau.ac.ir
12	A.M. Shaikh	BARC, India	shaikham@barc.gov.in
13	A. M. VORA, Dr.	Humanities and Social Science Department, South Gujarat, India	voraam@yahoo.com
14	Anandaraj V, Mr.	Indira Gandhi Center for Atomic Research, India	anandv@igcar.gov.in
15	Andi Idhil Ismail, Mr.	Brawijaya University	andi.idhil@gmail.com
16	André L. C. Conceição, Mr.	Universidade de São Paulo, Brazil	andre_conceicao@yahoo.com.br
17	Arum Patriati	BATAN, Indonesia	arum@batan.go.id
18	Azraf Azman	Nuclear Malaysia	azraf@nuclearmalaysia.gov.my
19	Azreena Binti Mastor, Ms.	USM, Malaysia	azreena.mastor@gmail.com
20	Binoy Kumar Saikia, Dr.	Tezpur University, Tezpur-784028, India	
21	Chih-Hao Lee, Dr.	National Tsing Hua University, Taiwan	chlee@mx.nthu.edu.tw
22	Chin Wei Wang, Mr.	National Central University, Taiwan	952402004@cc.ncu.edu.tw
23	David V. Baxter, Prof.	Indiana University, USA	baxterd@indiana.edu
24	DESERT Sylvain Dr	CEA, France	sylvain.desert@cea.fr
25	Djamel Maouche, Dr.	University of Setif Algeria	djmaouche@yahoo.fr
26	Dobrin P. Bossev, Dr. Prof.	Indiana University, USA	dbossev@indiana.edu
27	Edy Giri Rachman PUTRA, Dr.	BATAN, Indonesia	giri@batan.go.id
28	Eleftheria Mavredaki, Mr.	University of Leeds, UK	menema@leeds.ac.uk
29	Ervina Efzan Mhd Noor, Ms.	USM, Malaysia	ervina_mhdnoor@yahoo.com
30	Faridah Mohd Idris	Nuclear Malaysia	faridah_idris@nuclearmalaysia.gov.my
31	Francoise Mulhauser	IAEA, Vienna, Austria	F.Muelhauser@iaea.org
32	Hafizal B. Yazid, Mr.	Nuclear Malaysia	hafizal@nuclearmalaysia.gov.my
33	Hafiz Zin, Mr.	Institute of Cancer Research, London, UK	hafiz.zin@icr.ac.uk
34	Hazizan Md Akil Dr	USM, Malaysia	hazizan@eng.usm.my
35	Hermann Stelzer, Dr.	Forschungszentrum Jülich, ZAT, Germany	h.stelzer@fz-juelich.de
36	Hirokazu Hasegawa, Dr.	Kyoto University, Japan	hasegawa@alloy.polym.kyoto-u.ac.jp
37	Hishamuddin bin Husain, Mr.	Nuclear Malaysia	hishamuddin@nuclearmalaysia.gov.my
38	Intikhab Ulfatt, Mr.	Chalmers University of Technology, Lund Sweden	intikhab@chalmers.se

39	It Meng LOW, Dr.	Curtin University of Technology, Australia	j.low@curtin.edu.au
40	Ivan Bobrikov, Dr.	JINR, Dubna Russia	bobrikov@nf.jinr.ru
41	J. Rolando Granada, Dr.	Centro Atómico Bariloche, Comisión Nacional de Ene, Bariloche, Argentina	granada@cnea.gov.ar
42	Jason S. Gardner, Dr.	NIST Center for Neutron Research and Indiana University, USA	jsg@nist.gov
43	Jim Low, Dr.	Curtin University of Technology, Australia	J.Low@curtin.edu.au
44	Jin-Won Shin, Mr.	KAERI, Korea	jwshin@kaeri.re.kr
45	Kamel TAIBI	U.S.T.H.B, Algeria	zlounis@yahoo.com
46	Karami Mohsen, Mr.	Azad University of Iran, Tehran, Iran	mkarami_3@yahoo.com
47	KAOUA Sid Ali	Laboratoire des Sciences et de Génie des Matériaux, Algeria	sakaoua@gmail.com
48	Khaled Ali, Mr.	Al-Marghib University, Libya	khaledali222@yahoo.com
49	Khairiah Yazid, Ms.	Nuclear Malaysia	khairiah@nuclearmalaysia.gov.my
50	Khine Nyunt, Dr.	Multimedia University, Malaysia	khine.nyunt@mmu.edu.my
51	Kim Young Jin	KAERI, Korea	youkim@kaeri.re.kr
52	Lai San Kiong, Dr.	National Central University,Taiwan	sklai@coll.phy.ncu.edu.tw
53	Leonid Skatkov, Dr.	PCB, Israel	sf_lskatkov@bezeqint.net
54	Livia Eleonora Bove, Dr.	CNRS-IMPMC Universite` Pierre et Marie Curie, France	v-bove@ill.fr
55	M. Azzaz	USTHB, Algeria	azzazusthb@gmail.com
56	M.M.Sinha, Dr.	Sant Longowal Institute of Engineering and Tech. Longowal, India	mm_sinha@rediffmail.com
57	Martin RusÅˆÃ¡k, Mr.	Charles University in Prague, Czech Republic	mr.hawk777@gmail.com
58	Masashi Kitamura (Dr Nishida)	Kobe City College of Technology, Kobe, Japan	nishida@kobe-kosen.ac.jp
59	Masatoshi Arai, Dr.	J-PARC Center, JAEA, Japan	masaarai@hotmail.co.jp
60	Masayuki Nishida, Dr.	Kobe City College of Technology, Kobe, Japan	nishida@kobe-kosen.ac.jp
61	Meor Yusoff M. S.,Dr.	Nuclear Malaysia	meor@nuclearmalaysia.gov.my
62	Michel Kenzelmann, Dr.	Paul Scherrer Institut (PSI), Switzerland	michel.kenzelmann@psi.ch
63	Michihiro Furusaka, Dr.	Hokkaido University, Japan	furusaka@eng.hokudai.ac.jp
64	Mikihito Takenaka, Dr.	Kyoto University, Japan	takenaka@alloy.polym.kyoto-u.ac.jp
65	Mitsuharu Yonemura Dr.	Sumitomo Metal Industries, Ltd. Japan	yonemura-mth@sumitomometals.co.jp
66	Mohammad Abdul Bareque,Dr.	National Institute of Advanced Industrial Science Tokyo, Japan	barique184@yahoo.com
67	Mohammed Noori Ridha, Mr.	USM, Malaysia	mohammednoori71@yahoo.com
68	Mousavi Shirazi	Islamic Azad University, Tehran, Iran.	alireza_moosavi@yahoo.com
69	Mukesh Sharma, Dr.	State Forensic Science Laboratory Physics Division, India	mksphy@gmail.com
70	N. Raghu, Dr.	Indira Gandhi Center for Atomic Research Kalpakkam, India	nraghu@igcar.gov.in
71	Ned Blagojevic, Mr.	ANSTO, Australia	nbx@ansto.gov.au
72	Norio Ogata Dr.	Taiko Pharmaceutical Co., Ltd.Osaka, Japan	nogata7@yahoo.co.jp
73	Norriza Mat Isa	UTM, Malaysia	norriza@nuclearmalaysia.gov.my
74	Norariza binti Ahmad, Ms.	USM, Malaysia	nriza@hotmail.com
75	Norasiah Binti Ab. Kasim	Nuclear Malaysia	norasiah@nuclearmalaysia.gov.my
76	Nor Hayati Alias, Ms.	Nuclear Malaysia	norhayati@nuclearmalaysia.gov.my

77	Norlida Kamarulzaman, Dr	UiTM, Malaysia	norlyk3@yahoo.co.uk
78	Nur Shafiza A. Sharif	USM, Malaysia	mrsabar@eng.usm.my
79	Nurhidayatullaili Muhd Julkapli, Ms	USM, Malaysia	hid_laili@yahoo.com
80	Osman Murat Ozkendir, Dr.	Tarsus Technical Faculty of Education, Mersin University, Turkey	ozkendir@gmail.com
81	Pavol Mikula Dr	Nuclear Physics Institute AS CR, Czech Republic	mikula@ujf.cas.cz
82	Peter Holden, Prof.	ANSTO, Australia	phx@ansto.gov.au
83	Peter Timmins Prof.	Institut Laue-Langevin, Grenoble, France	timmins@ill.fr
84	Peter Willendrup, Mr.	Risø DTU, Roskilde Denmark	pkwi@risoe.dtu.dk
85	R. Brajpuriya	UGC-DAE Consortium for Scientific Research - University Campus, India	ranjeetbjp1@yahoo.com
86	Rajewska Aldona, Dr.	Institute of Atomic Energy, Poland	aldonar@cyf.gov.pl
87	Refaat Mahmoud Ali Maayouf, Dr.	Nuclear Research Center ,Atomic Energy Authority, Egypt	maayouf@hotmail.com
88	Robert Knott, Dr.	ANSTO, Australia	rbk@ansto.gov.au
89	Rosli Darmawan	Nuclear Malaysia	roslid@nuclearmalaysia.gov.my
90	Ryukhtin Vasyl, Dr.	Helmholtz-Zentrum Berlin for Materials and Energy, Germany	vasyl.ryukhtin@helmholtz-berlin.de
91	S. Nair	Université d'Oran Es-senia, Algeria	sam3100132@yahoo.fr
92	Sabar D. Hutagalung, Dr.	USM, Malaysia	mrsabar@eng.usm.my
93	Saeed S. Jahromi, Mr.	K.N. Toosi University of Technology, Tehran, Iran	saeed_s_jahromi@sina.kntu.ac.ir
94	Saibal Basu, Dr.	BARC, India	sbasu@barc.gov.in
95	Sangappa, Dr.	Mangalore University, India	sangappa@mangaloreuniversity.ac.in
96	Sergey Kulikov, Dr.	JINR, Dubna, Russia	ksa@nf.jinr.ru
97	Sergey Kuznetsov, Dr.	Lebedev Physical Institute, Russian Academy of Sci, Moscow, Russia	ckuz@sci.lebedev.ru
98	Sergey Danilkin, Dr.	ANSTO, Australia	
99	Shigeru OKAMOTO, Dr.	Nagoya Institute of Technology, Japan	okamoto.shigeru@nitech.ac.jp
100	Shraddha S. Desai	BARC, India	shraddha
101	Sistin Asri Ani	Sepuluh Nopember Institute of Technology Kampus ITS Sukolilo, Surabaya, Indonesia	sistin.aa_16@yahoo.com
102	Siti Radiah Bt Mohd Kamarudin	Nuclear Malaysia	radiah@nuclearmalaysia.gov.my
103	Sohrab Abbas	BARC, India	abbas@barc.gov.in
104	Suminar Pratapa, Dr.	Institute of Technology Sepuluh November (ITS), Surabaya, Indonesia	suminar_pratapa@physics.its.ac.id
105	Sung-Min Choi, Dr.	KAIST, Korea	sungmin@kaist.ac.kr
106	Sunil K. Karna, Mr.	Natiaonal Central University, Taiwan	karna_sk2000@yahoo.com
107	Thant Zin Htwe, Mr.	Yangon Technology University Insein, Yangon, Myanmar	hantzin99999@gmail.com
108	Tommy Nylander Dr.	Lund University, Sweden	Tommy.Nylander@fkem1.lu.se
109	Ulyanenkov, Dr.	Bruker AXS GmbH, Germany	alex.ulyanenkov@bruker-axs.de
110	Valerio Scagnoli, Dr.	European Synchrotron Radiation Facility, France	scagnoli@esrf.fr
111	Wan Saffiey Wan Abdullah Dr	Nuclear Malaysia	wansaffiey@nuclearmalaysia.gov.my
112	Weifeng Shang, Dr.	European Molecular Biology Laboratory Hamburg Outs, Germany	shang@embl-hamburg.de

113	Wei Kong Pang	Curtin University of Technology, Australia	weikong.pang@postgrad.curtin.edu.au
114	Weng On YAH	Kyushu University, Japan	alstonyah@yahoo.com
115	Widya Sari, Ms.	Andalas University , Indonesia	widyasari44@yahoo.com
116	Yahia Ahmed Lotfy Abdel-Hady, Mr.	Minia University,Egypt	Yahialotfy59@yahoo.com
117	Yee Mon Thu	Yangon Technology University, Myanmar	yeemonthu.2008@gmail.com
118	Yohanes Edi Gunanto, Mr.	University of Indonesia, Jakarta Indonesia	ye_gunanto@yahoo.com
119	Yusof Abdullah, Mr.	Nuclear Malaysia	yusofabd@nuclearmalaysia.gov.my
120	Zalina Laili, Ms.	Nuclear Malaysia	Liena@nuclearmalaysia.gov.my
121	Zulhadjri	Institut Teknologi Bandung, Indonesia	s306zulhadjri@mail.chem.itb.ac.id

AUTHOR INDEX